Security, Conflict and Cooperation in the Contemporary World

Series Editors
Effie G. H. Pedaliu
LSE Ideas
London, United Kingdom

John W. Young
University of Nottingham
Nottingham, United Kingdom

The Palgrave Macmillan series, Security, Conflict and Cooperation in the Contemporary World aims to make a significant contribution to academic and policy debates on cooperation, conflict and security since 1900. It evolved from the series Global Conflict and Security edited by Professor Saki Ruth Dockrill. The current series welcomes proposals that offer innovative historical perspectives, based on archival evidence and promoting an empirical understanding of economic and political cooperation, conflict and security, peace-making, diplomacy, humanitarian intervention, nation-building, intelligence, terrorism, the influence of ideology and religion on international relations, as well as the work of international organisations and non-governmental organisations.

More information about this series at
http://www.springer.com/series/14489

Malcolm M. Craig

America, Britain and Pakistan's Nuclear Weapons Programme, 1974–1980

A Dream of Nightmare Proportions

palgrave
macmillan

Malcolm M. Craig
Liverpool, United Kingdom

Security, Conflict and Cooperation in the Contemporary World
ISBN 978-3-319-51879-4   ISBN 978-3-319-51880-0 (eBook)
DOI 10.1007/978-3-319-51880-0

Library of Congress Control Number: 2017935968

Cover illustration © Trinity Mirror / Mirrorpix / Alamy Stock Photo and © stefano carniccio / Alamy Stock Photo.
Designed by Fatima Jamadar

Printed on acid-free paper

This Palgrave Macmillan imprint is published by Springer Nature
The registered company is Springer International Publishing AG
The registered company address is: Gewerbestrasse 11, 6330 Cham, Switzerland

Acknowledgements

First and foremost, this book would not exist without the support of Fabian Hilfrich and Robert Mason. Both of them shepherded me through my doctorate and have been a constant source of wisdom and encouragement ever since. I am also indebted to the fellow scholars who have given help, criticism, and friendship. I would therefore like to express my sincere thanks to Pertti Ahonen, Patrick Andelic, James Campbell, Frank Cogliano, Michael Devine, Owen Dudley-Edwards, Alastair Duthie, James Ellison, Andrew Futter, Mark Heise, Frances Houghton, Alex Imrie, Dolores Janiewski, Rhodri Jeffreys-Jones, Jane Judge, David Kaufman, Iain Lauchlan, Fraser McCallum, Stephen McDowell, Malcolm McKinnon, Mark McLay, Owain Mason, Robbie Maxwell, Adrienne Miller-McLaughlin, Chris Moran, Philip Morgan, David Motadel, Benoit Pelopidas, Andrew Preston, Paul Quigley, Nick Radburn, Nick Ritchie, Louise Settle, David Silkenat, Jon Singerton, Hande Watt, and Patrick Watt.

No academic endeavour would be possible without considerable administrative and technical support, and it would be remiss of me not to thank Anne Brockington, Margaret Forrest, Karen Howie, Richard Kane, Sophie Lockwood, Michael O'Reilly, Niko Ovenden, Marie-Therese Talensby, and Lindsay Scott. Likewise, without skilled archivists around the world, our task as historians would be that much harder. The staff of the BBC Written Archives Centre (Caversham, Berkshire), the Gerald R. Ford Presidential Library (Ann Arbor, MI), the Jimmy Carter Presidential Library and Museum (Atlanta, GA), the US National Archives and Records Administration (College Park, MD), the National

Security Archive (George Washington University, Washington D.C.), and The National Archives of the United Kingdom (Kew, London) have all given invaluable assistance during the research for this book. In particular, I would like to thank Jim Clancy and Keith Schuyler (Jimmy Carter Presidential Library), Bill McNitt (Gerald Ford Presidential Library), Steven Twigge (The National Archives of the United Kingdom), and Jessica Hogg and Louise North (BBC WAC). The completion of this book would not have been possible without generous funding from the President Gerald R. Ford Foundation, the University of Edinburgh's Professor James F. McMillan Scholarship, and the Institute for Advanced Studies in the Humanities (IASH).

There are also numerous groups representing the wider academic community who must be mentioned. This includes all the undergraduate students whom I have had the pleasure of teaching, including—but not limited to—Dan Austin, Joe Baldwin, Lauren Botha, Kelly-Leigh Cooper, Fergus Deery, Charlotte de Vitry de Avaucourt, Caroline Dodds, Robin Drummond, Max Furse, Jo Hagan, Alistair Hamilton, Katrin Heilmann, Megan James, Bethan John, Spike Lister, Tom Lusuardi, Saoirse McKeon, Sarah Manavis, Jodie Mitchell, Livvy Moore, Ashley Dee Paton, Lucy Robb, Charlie Thompson, Dan Troman, and Connie Wakefield. I'd also like to thank the attendees of the University of Edinburgh American History Workshop and Centre for the Study of Modern Conflict Seminar Series, all the participants in the annual Historians of the Twentieth Century United States (HOTCUS) conferences, and the attendees at the annual Scottish Association for the Study of America (SASA) conferences for their valuable feedback on various elements of this project.

No endeavour such as this would be possible without the support of family and friends. With that in mind, I must thank my father John Craig, my mother Irene Craig, my sister Lesley McLean, and my friends from all around the world, including—but most certainly not limited to—Paul Bourne, Morgan Davie, Scott Dorward, Cam Eeles, Leanne Fulton, Jonny Hodgson, Gregor Hutton, Andrew Kenrick, Debbie Murphy, Joe Murphy, Robert Robertson, Doug Whitely, Lucy Whitely, and others too numerous to mention.

Finally, there is one person without whom this book would have been impossible to complete: my wife, Jaqueline Booth. Above all others, she has been the source of strength, support, and love.

CONTENTS

LIST OF ABBREVIATIONS

ACDA	Arms Control and Disarmament Agency (USA)
ACDD	Arms Control and Disarmament Department (UK)
BBC	British Broadcasting Corporation
BDOHP	British Diplomatic Oral History Programme, Churchill College, Cambridge
BNFL	British Nuclear Fuels Limited
CENTO	Central Treaty Organisation
CIA	Central Intelligence Agency (USA)
DoD	Department of Defense (USA)
DoE	Department of Energy (UK)
DoI	Department of Industry (UK)
DoT	Department of Trade (UK)
EEIC	Emerson Electrical Industrial Controls Ltd.
ERDA	Energy Research and Development Administration (USA)
FCO	Foreign and Commonwealth Office (UK)
FRG	Federal Republic of Germany (West Germany)
GRFPL	Gerald R. Ford Presidential Library, Ann Arbor, Michigan
IAEA	International Atomic Energy Agency
INFCE	International Fuel Cycle Evaluation project
JCPL	Jimmy Carter Presidential Library, Atlanta, Georgia
JIC	Joint Intelligence Committee (UK)
JNU	Joint Nuclear Unit (UK)
MLF	Multi-lateral Force
MoD	Ministry of Defence (UK)
NARA	United States National Archives and Records Administration, College Park, Maryland
NATO	North Atlantic Treaty Organisation

NNPA	Nuclear Non-proliferation Act of 1978 (USA)
NNWS	Non-nuclear Weapon State
NPT	Treaty on the Non-proliferation of Nuclear Weapons
NRC	Nuclear Regulatory Commission (USA)
NSAEBB	National Security Archive Electronic Briefing Book
NSC	National Security Council (USA)
NSG	Nuclear Suppliers' Group
NSSM	National Security Study Memorandum
NWS	Nuclear Weapon State
OIC	Organization of the Islamic Conference
OPD	Defence and Oversea Policy Committee (UK)
PAEC	Pakistan Atomic Energy Commission
PNE	Peaceful Nuclear Explosion
PRC	People's Republic of China
PUS	Permanent Under Secretary (UK)
RST	Regional Security Treaty
SAD	South Asia Department (UK)
SANWFZ	South Asian Nuclear Weapon Free Zone
SEPECAT	*Société Européenne de Production de l'Avion d'École de Combat et d'Appui Tactique* (the European company for the production of a combat trainer and tactical support aircraft)
SGN	Saint Gobain Nucléaire
SNIE	Special National Intelligence Estimate (USA)
TNA	The National Archives of the United Kingdom, Kew
TOW	BGM-71 tube launched, optically tracked, wire guided missile
UKE	United Kingdom Embassy
UKHC	United Kingdom High Commission
UN	United Nations
UNSC	United Nations Security Council
USE	United States Embassy

CHAPTER 1

Introduction

In 2004 Pakistani authorities placed scientist Abdul Qadeer ('A.Q.') Khan under house arrest in Rawalpindi. The self-styled 'Father of the Pakistani Bomb' had finally gone too far for the liking of American and British intelligence services, who exerted influence on their South Asian partners in the War on Terror. The 'A.Q. Khan Network' had supplied sensitive nuclear technology to Iran, Libya, and North Korea, via a web that went over, around, and under national borders.[1] This was not the first time that Washington and London had collaborated to undermine a nuclear project centred in Pakistan.

Khan's network had its roots in Islamabad's national nuclear programme, a programme that emerged and developed in the 1970s. Chastened by their defeat in the 1971 Indo-Pakistani War and alarmed

---

[1] See Chaim Braun and Christopher F. Chyba, 'Proliferation Rings: New Challenges to the Nuclear Nonproliferation Regime', *International Security*, 29:2 (2004), 5–49; Justin V. Hastings, 'The Geography of Nuclear Proliferation Networks', *Nonproliferation Review*, 19:3 (2012), 429–450; Gaurav Kampani, 'Second Tier Proliferation: The Case of Pakistan and North Korea', *Nonproliferation Review*, 9:3 (2002), 107–116; Alexander H. Montgomery, 'Ringing in Proliferation: How to Dismantle an Atomic Bomb Network', *International Security*, 30:2 (2005), 153–187.

© The Author(s) 2017
M.M. Craig, *America, Britain and Pakistan's Nuclear Weapons Programme, 1974–1980*, Security, Conflict and Cooperation in the Contemporary World, DOI 10.1007/978-3-319-51880-0_1

by New Delhi's acquisition of nuclear capability in 1974, Pakistani leaders embarked on a skilful clandestine programme aimed at achieving nuclear status. Reacting to this, politicians, civil servants, diplomats, and intelligence officers from both sides of the Atlantic worked together to avert the interlocking nightmares of a sub-continental nuclear showdown and the collapse of the global non-proliferation regime.

After the Partition of the British Raj in 1947 US and UK alliances with the now-divided former colony had been plagued with instability.[2] In the years after Partition Harry Truman's administration paid little heed to events on the sub-continent. However, from the mid-1950s onwards there emerged a regional battle for influence between America, Britain, the People's Republic of China (PRC), and the USSR.[3] The Cold War arrived in South Asia as a result of Dwight D. Eisenhower's 1954 military treaty with Islamabad and the inevitable Soviet response. The Indo-Pakistani wars of 1948, 1965, and particularly 1971 were complicated by Cold War superpower rivalries. Pakistan's defeat and dismemberment by India in 1971 was a critical contributing factor to driving Islamabad towards the bomb.[4]

This book narrates and analyses the story of US and UK efforts to shut down Pakistan's nuclear programme in its early stages, between the catalytic Indian nuclear test of May 1974 and the decline of sustained non-proliferation activity in the years 1979 and 1980. It is not only a tale of cooperation between Washington and London, but also a story of differences and disputes. The brutal economic realities of the

---

[2] On US-UK relations and South Asia, see Paul M. McGarr, *The Cold War in South Asia: Britain, the United States, and the Indian Subcontinent, 1945–1965* (Cambridge: Cambridge University Press, 2013). On the US perspective, see Robert J. McMahon, *The Cold War on the Periphery: The United States, India and Pakistan* (New York: Columbia University Press, 1994), and Dennis Kux, *India and the United States: Estranged Democracies, 1941–1991* (Washington, DC: National Defense University Press, 1993), and *The United States and Pakistan, 1947–2000: Disenchanted Allies* (Baltimore: Johns Hopkins University Press, 2001). As McGarr points out, British policy towards the sub-continent remains remarkably under-studied.

[3] McGarr, *Cold War in South Asia*, 4.

[4] Gary J. Bass, *The Blood Telegram: Nixon, Kissinger, and a Forgotten Genocide* (New York: Alfred A. Knopf, 2013), 333–32.

decade, 'oil shocks', and wider geopolitical challenges all served to complicate this 'non-proliferation special relationship'.[5] Policy and action were also complicated by changes elsewhere in the world. The 1979 Iranian Revolution brought a new form of political Islamic radicalism to prominence. The fears engendered by the Ayatollah and his followers, coupled to the blustering rhetoric of Pakistani leaders, gave rise to the meme of the 'Islamic bomb', a nuclear weapon created by Pakistan to be shared amongst the Muslim *ummah*.

Since the Trinity atomic test on July 16, 1945, controlling the spread of nuclear capability has been a constant—if not always well articulated or effected—component of US foreign policy. Early attempts to deal with the international issues posed by 'the bomb' through global cooperation or outright prohibition ended in failure. The 1946 'Baruch Plan' foundered on the rocks of duplicity, lack of commitment, and the emerging Cold War.[6]

---

[5] The so-called special relationship is the subject of a vast literature. For representative studies on nuclear issues, see Jan Melissen, *The Struggle for Nuclear Partnership: Britain, the United States, and the Making of an Ambiguous Alliance, 1952–1959* (Groningen: Styx Publications, 1993); Len Scott and Stephen Twigge, *Planning Armageddon: Britain, the United States, and Command of Western Nuclear Forces, 1945–1964* (Amsterdam: Harwood Academic, 2000); and John Simpson, *The Independent Nuclear State: The United States, Britain, and the Military Atom*, 2nd edition (London: Macmillan, 1986). One volume that does analyse the US-UK arms control relationship is J. P. G. Freeman, *Britain's Nuclear Arms Control Policy in the Context of Anglo-American Relations, 1957–68* (Basingstoke: MacMillan, 1986). On the US-UK 'special intelligence relationship' see Richard J. Aldrich, *The Hidden Hand: Britain, America and Cold War Secret Intelligence* (London: John Murray, 2001) and Rhodri Jeffreys-Jones, *In Spies We Trust: The Story of Western Intelligence* (Oxford: Oxford University Press, 2013). On the 1970s see, for example, R. Gerald Hughes and Thomas Robb, 'Kissinger and the Diplomacy of Coercive Linkage in the "Special Relationship" between the United States and Great Britain, 1969–1977', *Diplomatic History*, 37:4 (2013), 861–905; Niklas H. Rossbach, *Heath, Nixon and the Rebirth of the Special Relationship: Britain, the US and the EC, 1969–74* (Basingstoke: Palgrave Macmillan, 2009); Andrew Scott, *Allies Apart: Heath, Nixon, and the Anglo-American Relationship* (Basingstoke: Palgrave Macmillan, 2011).

[6] On international control, see Campbell Craig and Sergei Radchenko, *The Atomic Bomb and the Origins of the Cold War* (New Haven: Yale University Press, 2008), 111–161; Shane J. Maddock, *Nuclear Apartheid: The Quest for American Atomic*

The US Atomic Energy Act of 1946 failed to put the nuclear genie back in the bottle, as spies and scientists spread the, so-called, secret around the world.

Into the 1950s non-proliferation policy slammed up against an ever-increasing number of nuclear weapon states. The USSR and the UK had both gained capability, in 1949 and 1952 respectively. US policy was focused on Western Europe, where fear of a resurgent West Germany, armed with nuclear weapons, was a preoccupation of American and British non-proliferation policymakers.[7] Furthermore, Dwight D. Eisenhower's 'Atoms for Peace' project represented an ultimately ill-judged policy of spreading civilian nuclear power around the world in the hope of winning the 'hearts and minds' of developing world nations.[8]

The twin crises of Berlin and Cuba—as well as the ascent to nuclear status of France in 1960 and the PRC in 1964—persuaded the USA, UK, USSR, and other states that action was required to arrest the bomb's spread. The NATO allies were mired in the seemingly endless debates over the Multilateral Force (MLF), a nuclear-sharing proposal that had at

Supremacy from World War II to the Present (Chapel Hill: University of North Carolina Press, 2010), 47–79; and Ronald E. Powaski, *March to Armageddon: The United States and the Nuclear Arms Race, 1939 to the Present* (New York: Oxford University Press, 1987), 29–45. On the 'atom spies' see Jonathan Haslam, *Near and Distant Neighbours: A New History of Soviet Intelligence* (Oxford: Oxford University Press, 2015), 134–142 for a Soviet perspective; John Earl Haynes, Harvey Klehr, and Alexander Vassiliev, *Spies: The Rise and Fall of the KGB in America* (New Haven: Yale University Press, 2009), 33–144; Amy Knight, *How the Cold War Began: The Igor Gouzenko Affair and the Hunt for Soviet Spies* (New York: Carroll and Graf, 2005); Kathryn S. Olmsted, *Real Enemies: Conspiracy Theories and American Democracy, World War 1 to 9/11* (Oxford: Oxford University Press, 2009), 83–110.

[7] See Marc Trachtenberg, *A Constructed Peace: The Making of European Settlement* (Princeton: Princeton University Press, 1999) for a classic study of this issue.

[8] Since the turn of the century, 'Atoms for Peace' as been the subject of considerable reassessment, for example by Ira Chernus, 'Operation Candor: Fear, Faith, and Flexibility', *Diplomatic History*, 29:5 (2005), 779–809; Mara Drogan, 'The Nuclear Imperative: Atoms for Peace and the Development of U.S. Policy on Exporting Nuclear Power, 1953–1955', *Diplomatic History*, 40:5 (2016), 948–974; and John Krige, 'Atoms for Peace, Scientific Internationalism, and Scientific Intelligence', *Osiris*, 21:1 (2006), 161–181.

its heart fears about West Germany.[9] In terms of international agreements, the post-Cuban Missile Crisis Limited Test Ban Treaty was a start, but it was moves in the mid-1960s towards what became the 1968 Treaty on the Non-proliferation of Nuclear Weapons (NPT) that founded the modern non-proliferation regime. The NPT set the groundwork for détente, as both sides found common ground and common interests as the USA abandoned the MLF, much to the Kremlin's relief.[10]

As co-authors and depository powers of the NPT (along with the USSR), the USA and UK had confidently led the way on non-proliferation. As the 1970s dawned, Washington's enthusiasm for preventing the spread of nuclear weapons waned. For Richard Nixon and Henry Kissinger, political forces and national interests were more powerful and important than a small matter like the spread of nuclear capability.[11] Nixon tacitly approved Israel's nuclear weapons programme, despite the vocal warnings of key advisers.[12] Kissinger was at least flexible in his approach to non-proliferation policy; the Indian test of May 18, 1974 forced the then

---

[9] See Hal Brands, 'Non-Proliferation and the Dynamics of the Middle Cold War: The Superpowers, the MLF, and the NPT', *Cold War History*, 7:3 (2007), 389–423, and 'Rethinking Nonproliferation: LBJ, the Gilpatric Committee, and U.S. National Security Policy', *Journal of Cold War Studies*, 8:2 (2006), 83–113; Andrew Priest, 'The President, the "Theologians" and the Europeans: The Johnson Administration and NATO Nuclear Sharing', *International History Review*, 33:2 (2011), 257–275; J.J. Widén and Jonathan Colman, 'Lyndon B. Johnson, Alec Douglas-Home, Europe and the NATO Multilateral Force, 1963–64', *Journal of Transatlantic Studies*, 5:2 (2007), 179–198; and John Young, 'Killing the MLF? The Wilson Government and Nuclear Sharing in Europe, 1964–66', *Diplomacy & Statecraft*, 14:2 (2003), 295–324.

[10] See Hal Brands, 'Progress Unseen: U.S. Arms Control Policy and the Origins of Detente, 1963–1968', *Diplomatic History*, 30:2 (2006), 253–285; Roland Popp, 'Introduction: Global Order, Cooperation Between the Superpowers, and Alliance Politics in the Making of the Nuclear Non-Proliferation Regime', *International History Review*, 36:2 (2014), 195–209; and Dane Swango, 'The United States and the Role of Nuclear Co-operation and Assistance in the Design of the Non-Proliferation Treaty', *International History Review*, 36:2 (2014), 210–229.

[11] Francis J. Gavin, *Nuclear Statecraft: History and Strategy in America's Atomic Age* (Ithaca: Cornell University Press, 2012), 104–105.

[12] National Security Archive Nuclear Vault Electronic Briefing Book 'Israel Crosses the Threshold II: The Nixon Administration Debates the Emergence of

Secretary of State to reassess his position and embark on new international initiatives in support of non-proliferation.

When Harold Wilson returned to 10 Downing Street in February 1974 it was at the head of a Labour Party that—despite frequent differences between the grass roots and the parliamentary party—maintained a commitment to non-proliferation.[13] This commitment was, however, bound up in domestic British energy and industrial needs.[14] Likewise, successive governments sought to maintain Britain's credibility as a nuclear weapon state while encouraging wider NPT membership. These factors would come to have a critical influence on US-UK interactions with Pakistan's nuclear programme. Although US-UK nuclear relations were not always stable, by the time of the Indian test the two nations had worked together to build the foundations of an international nuclear non-proliferation regime. That regime—and the cooperation that led to its existence—would be sorely tested by the South Asian events to come.

In the context of US-UK non-proliferation relations, it is important not to overstate the 'special' nature of the transatlantic non-proliferation relationship that emerged in the 1970s. Although US-UK non-proliferation cooperation was stronger than that between Washington and Paris or Bonn, it was also beset by problems.[15] There were significant clashes between American non-proliferation policy and British commercial interests during the 1970s, as the Pakistani case brought the allies into conflict over domestic nuclear policy and international arms sales. This study thus highlights

the Israeli Nuclear Programme', http://nsarchive.gwu.edu/nukevault/ebb485/ (accessed June 13, 2016).

[13] Rhiannon Vickers, *The Labour Party and the World, Volume 2: Labour's Foreign Policy 1951–2009* (Manchester: Manchester University Press, 2011), 3–5, 105.

[14] Robert Boardman and Malcolm Grieve, 'The Politics of Fading Dreams: Britain and the Nuclear Export Business', in Robert Boardman and James F. Keeley, *Nuclear Exports and World Politics: Policy and Regime* (London: MacMillan, 1983), 107–108.

[15] During the 1970s, animosity characterised US-FRG nuclear relations. See William Glenn Gray, 'Commercial Liberties and Nuclear Anxieties: The US-German Feud over Brazil, 1975–7', *International History Review*, 34:3 (2012), 449–474 and Fabian Hilfrich, 'Roots of Animosity: Bonn's Reaction to US Pressures in Nuclear Proliferation', *International History Review*, 36:2 (2014), 277–301.

how the UK's domestic economic and energy imperatives affected global attempts to prevent proliferation.

British economic difficulties not only had a deleterious effect on the relationship with the United States, but they also contributed to Islamabad's sense of insecurity concerning Indian military power, feeding the desire for nuclear capability. The UK's negotiations and eventual deal with India in 1978 to sell Jaguar strike aircraft—worth billions of pounds and thousands of jobs—complicated non-proliferation activity on the sub-continent. In comparison, the US government—particularly Jimmy Carter's administration—chose to abandon commercial advantage in favour of non-proliferation. The sale of potentially nuclear-capable aircraft to Pakistan's regional arch-rival demonstrated that, in times of economic crisis, London was unwilling to prioritise non-proliferation over economic self-interest. The UK also pursued an expansive domestic nuclear reprocessing industry while working against Pakistan's quest for similar facilities. This created quadrilateral tension between Washington, London, Islamabad, and New Delhi. The British pursuit of reprocessing facilities also conflicted with Carter's stated desire to see such capabilities eliminated, despite his public assurances that Western European states would not find their reprocessing ambitions inhibited by his policies. In the UK-Pakistan case, British actions left successive governments open to accusations of hypocrisy.

In addition to examining these conflicting facets of the, so-called, special relationship, this book analyses in detail how and why America and Britain attempted to influence the nuclear affairs of a sovereign third party, and how such efforts related to other foreign policy priorities. It critically adds to the literature on the foreign policy of the presidencies of Gerald Ford and Jimmy Carter, and the premierships of Harold Wilson, James Callaghan, and Margaret Thatcher.[16] This is particularly necessary in the case of the foreign policy of British governments in the mid to late

---

[16] For representative studies, see Betty Glad, *An Outsider in the White House: Jimmy Carter, His Advisors, and the Making of American Foreign Policy* (Ithaca, NY: Cornell University Press, 2009); Scott Kaufman, *Plans Unraveled: The Foreign Policy of the Carter Administration* (DeKalb: Northern Illinois University Press, 2008); Daniel J. Sargent, *A Superpower Transformed: The Remaking of American Foreign Relations in the 1970s* (Oxford: Oxford University Press, 2015); Robert A. Strong, *Working in the World: Jimmy Carter and the Making of American Foreign Policy* (Baton Rouge: Louisiana State University Press, 2000); Odd Arne Westad,

1970s, which have not been the subject of such vast scholarship as that of their American counterparts.[17]

In the face of a state determined to acquire nuclear capability, were there alternatives short of direct intervention that would move Pakistan away from the nuclear path? Two broad options were open to the US-UK alliance: placing diplomatic pressure on Pakistan, or putting pressure on Islamabad's nuclear technology suppliers. Attempting to influence Pakistani leaders had little effect. In the face of a determined drive for the bomb that remained consistent across the governments headed by Zulfikar Ali Bhutto and Muhammad Zia ul-Haq, failure was the norm.

Provoked by concerns over the—actual or imagined—nuclear programmes of Iran, Iraq, and North Korea, and by concerns about non-state actor access to nuclear materials, the role of globalisation in nuclear proliferation is a matter of considerable twenty-first century debate and discussion.[18] The intersections between globalisation—in the specific sense of the expansion and interconnectedness of global trade and communication—and proliferation go back many decades. Détente in the

---

The Global Cold War: Third World Interventions and the Making of Our Times (Cambridge: Cambridge University Press, 2005).

[17] See John Callaghan, *The Labour Party and Foreign Policy* (London: Routledge, 2007); Simon Tate, *A Special Relationship?: British Foreign Policy in the Era of American Hegemony* (Manchester: Manchester University Press, 2012); Vickers, *The Labour Party*, Anthony Seldon and Kevin Hickson (eds.), *New Labour, Old Labour: The Wilson and Callaghan Governments, 1974–79* (London: Routledge, 2004).

[18] Chaim Braun, 'The Nuclear Energy Market and the Nonproliferation Regime', *Nonproliferation Review*, 13:3 (2006), 627–644; Thomas W. Graham, 'Weapons of Mass Destruction: Does globalisation mean proliferation?', Brookings Institution, Fall 2001, http://www.brookings.edu/research/articles/2001/09/fall-weapons-graham (accessed June 10, 2016); Kenneth N. Luongo and Isabelle Williams, 'The Nexus of Globalization and Next-generation Nonproliferation', *Nonproliferation Review*, 14:3 (2007), 459–473; Chris Schneidmiller, 'Globalization outpaces nuclear, biological non-proliferation regimes, expert says', Nuclear Threat Initiative, December 18, 2007, http://www.nti.org/gsn/article/globalization-outpaces-nuclear-biological-nonproliferation-regimes-expert-says/ (accessed June 10, 2016); Ban Ki-moon, 'Globalization Expands Access to Technology, Materials Needed for Mass Destruction Weapons, Creating New Disarmament Challenges', January 18, 2013, United Nations Meetings Coverage and Press Releases, http://www.un.org/press/en/2013/sgsm14770.doc.htm (accessed June 10, 2016).

1970s took the lid off the Cold War pressure cooker, allowing previously suppressed national, international, and transnational forces to emerge. Globalisation also began to hit its stride. International trade tripled from 1964 to 1980, computerisation began to revolutionise communication and the markets, and the oil shocks focused attention on alternative energy sources.[19] Thus, a burgeoning international trade emerged in nuclear reactors, reprocessing plants, and the other parts of the nuclear fuel and energy cycle.

The expansion of international trade, widening communications networks, and the ability to tap into the intensely competitive globalising marketplace allowed Pakistan's clandestine nuclear programme to flourish as the first truly modern, second-generation nuclear proliferation effort. Instead of requiring the multibillion dollar industrial mega-systems that had created the American, Soviet, British, French, and Chinese bombs, Islamabad used blueprints stolen from the British-Dutch-German URENCO uranium enrichment plant, off-the-shelf parts, and nominally non-nuclear industrial equipment to build the uranium centrifuges that would give it the raw material for atomic bombs.

In order to understand how Pakistan became a nuclear weapon state, it is vital to examine the interactions between Washington, London, Bonn, Bern, and other Western capitals in the 1970s. Efforts to combat the covert purchasing programme were complicated by diffuse and differing understandings of when something was or was not a 'nuclear thing'.[20] Washington and London eventually resolved this between themselves, but it also caused problems when dealing with Western European supplier states. Whether something was or was not nuclear, domestic economic priorities, the often toxic nature of alliance relationships, and the grey areas of international non-proliferation agreements all slowed down the efforts to prevent Pakistan from acquiring the means to create a bomb.

---

[19] Sargent, *A Superpower Transformed*, 5–6. For analyses of the multifaceted impacts of globalisation in the 1970s, see Niall Ferguson, Charles S. Maier, Erez Manela, and Daniel J. Sargent (eds.), *The Shock of the Global: The 1970s in Perspective* (Cambridge: Belknap Press of the Harvard University Press, 2011). On the importance of this period of seeming decline to later US resurgence in the post-Cold War world, see Hal Brands, *Making the Unipolar Moment: US Foreign Policy and the Rise of the Post-Cold War Order* (Ithaca: Cornell University Press, 2016).

[20] Gabrielle Hecht, 'The Power of Nuclear Things', *Technology and Culture*, 51:1 (2010), 1–30.

For the USA and UK, placing pressure on Pakistan's mostly Western European nuclear technology suppliers was potentially the most promising non-proliferation tactic. US diplomatic pressure, at least in part, persuaded France not to sell a nuclear reprocessing plant to Pakistan, although French economic considerations played a role in this decision. Despite America's and Britain's efforts to prevent Pakistani acquisition of uranium enrichment capacity, the skilful purchasing programme, the recalcitrance of key nuclear supplier states operating in a global marketplace, and the difficulty of enforcing vague international standards for nuclear trade hampered those efforts. Thus, this book offers an opportunity to study relative success and failure, and demonstrates that, short of military intervention against Pakistan, there was no chance of diminishing Islamabad's nuclear aspirations.

This study also argues that during 1979 a key change in non-proliferation policy towards Pakistan occurred. The Soviet intervention in Afghanistan in December 1979 is frequently cited as the turning point for proliferation policy towards Pakistan. Various scholars have observed that, after the intervention, the Carter administration abandoned attempts to prevent Pakistani acquisition of nuclear capability in favour of bolstering their South Asian ally and fighting a proxy war against the USSR.[21] This book contends, however, that the major change from a policy of *prevention* of Pakistani acquisition to one of *mitigation* of that eventual acquisition took place in mid-1979, several months *before* the Afghanistan crisis. Key to this mitigation approach were requests by the USA and UK that Pakistan should not test a nuclear device. This would be the most public expression of nuclear capability, and could cause considerable loss of credibility for the USA and the UK, who had invested so much in preventing Pakistani nuclear capability. Thus, considerations of national standing and 'face' played a significant role in shaping non-proliferation strategy and policy.

A theme that unites much of the specific literature on non-proliferation is the positioning of the United States as the critical—and often solo—player in global non-proliferation activity. While it is obvious that the USA was a pre-eminent force in Cold War nuclear affairs, the history of

---

[21] For example, Kaufman, *Plans Unraveled*, 217–218; and Kux, *Disenchanted Allies*, 245–255.

non-proliferation is an international one. Other actors, even those considered insignificant in the past, must be given due attention.

While there is an extensive literature on non-proliferation as it relates to the United States and Pakistan in the 1970s, Britain has, so far, received limited attention. More specifically, Britain's contribution to anti-proliferation action in relation to Pakistan in the 1970s has frequently been ignored or placed in the shadow of the United States.[22] This book demonstrates that Britain played a significant—at times critical—role in combating Pakistani nuclear ambitions. It redresses the balance by moving Britain towards the centre of non-proliferation discussions over Pakistan during the 1970s, and makes plain that the governments of Harold Wilson (1974–76) and James Callaghan (1976–79), and the initial eighteen months of Margaret Thatcher's first administration (1979–83), played an important role in creating, shaping, and at times adversely affecting, non-proliferation policy on the sub-continent. In so doing, this book illuminates the contours of US-UK relations regarding the spread of nuclear weapons to the developing world.

In 2006–07 a slew of books was released with provocative titles such as *The Nuclear Jihadist* and *Allah's Bomb*. It was, however, in the late 1970s that revelations about an Islamic bomb—a nuclear weapon originating in Pakistan but which would allegedly be proliferated to other Muslim states because of the bonds of faith—emerged. The news media tied this to stories about A.Q. Khan's role in appropriating nuclear technology from Europe, thus ensuring that the religious and clandestine elements of Pakistani nuclear ambitions became dominant, intertwined with public narratives. These real and imagined revelations about Islamabad's atomic

---

[22] For studies examining British non-proliferation efforts, see John Krige, 'US Technological Superiority and the Special Nuclear Relationship: Contrasting British and US Policies for Controlling the Proliferation of Gas Centrifuge Enrichment', *International History Review*, 36:2 (2014), 230–251; Matthew Jones and John W. Young, 'Polaris, East of Suez: British Plans for a Nuclear Force in the Indo-Pacific, 1964–1968', *Journal of Strategic Studies*, 33:6 (2010), 847–870; Susanna Schrafstetter, 'Preventing the "Smiling Buddha"; British-Indian Nuclear Relations and the Commonwealth Nuclear Force, 1964–68', *Journal of Strategic Studies*, 25:3 (2002), 87–108: and Susanna Schrafstetter and Stephen Twigge, *Avoiding Armageddon: Europe, the United States, and the Struggle for Nuclear Nonproliferation, 1945–1970* (Westport: Praeger, 2004).

aspirations energised the media in a way that the more pedestrian details of non-proliferation diplomacy had not.

Despite this, there has been little historiographical investigation into governmental responses.[23] This book argues that while the concept generated considerable public heat, policymakers saw it as lacking in merit and much more of a propaganda issue that an imminent reality. In our post-9/11 world, infused with Islamophobia, the War on Terror's legacy, and apprehension about Islamic nuclear terrorism, asserting that fear of pan-Muslim nuclear capability was insignificant seems counter-intuitive.[24] Indeed, there is an impressive body of historical research that demonstrates the importance of cultural factors in the Cold War era and,

[23] Journalistic investigations include David Armstrong and Joseph Trento, *America and the Islamic Bomb: The Deadly Compromise* (Hanover: Steerforth Press, 2007), 89–91; Gordon Corera, *Shopping for Bombs: Nuclear Proliferation, Global Insecurity and the Rise and Fall of the A. Q. Khan Network* (London: Hurst & Co., 2006); Douglas Frantz and Catherine Collins, *The Nuclear Jihadist: The True Story of the World's Most Dangerous Nuclear Smuggler* (New York: Hachette, 2007); Adrian Levy and Catherine Scott-Clark, *Deception: Pakistan, the United States, and the Secret Trade in Nuclear Weapons* (New York: Walker, 2007); and Al Venter, *Allah's Bomb: The Islamic Quest For Nuclear Weapons* (Guilford: Lyons Press, 2007); The ur-text for all of these is Steve Weissman and Herbert Krosney's *The Islamic Bomb: The Nuclear Threat to Israel and the Middle East* (New York: Times Books, 1981). Academic studies include those by cultural anthropologist Hugh Gusterson in 'Nuclear Weapons and the Other in the Western Imagination', *Cultural Anthropology*, 14:1 (1999), 111–143, 125–126; physicist Pervez Hoodbhoy in 'Myth Building: The "Islamic Bomb,"' *Bulletin of the Atomic Scientists* (June, 1993), 42–49; policy analyst Rodney W. Jones in *Nuclear Proliferation: Islam, the Bomb, and South Asia* (Beverly Hills: Sage, 1981); and political scientist Samina Yasmeen in 'Is Pakistan's Nuclear Bomb an Islamic Bomb?', *Asian Studies Review*, 25:2 (2001), 201–215. The mainstay of analysis is over whether or not Pakistan's nuclear programme was Islamic, and does not delve into Western responses.

[24] See Graham Allison, *Nuclear Terrorism: The Ultimate Preventable Catastrophe* (New York: St Martin's Press, 2004); Charles Ferguson and William Potter, *The Four Faces of Nuclear Terrorism* (Monterey: Center for Nonproliferation Studies, 2004); and Rolf Mowatt-Larssen, *Islam and the Bomb: Religious Justification for and Against Nuclear Weapons* (Cambridge: Belfer Center, 2011).

more specifically, nuclear policies.[25] Nevertheless, an analysis of official archives from 1977–80 indicates that scares about the Islamic bomb failed to meaningfully influence policy.

The following chapters interweave an analysis of the Islamic bomb scare with US-UK non-proliferation policy towards Pakistan, and investigate how policymakers in Washington and London responded to it. This demonstrates that although the Islamic bomb *was* present in official discussions of Pakistan's nuclear programme, understandings were more nuanced than in public discourses. There were those in officialdom who emphasised the possibility of an Islamic bomb, but on the whole Washington and London concluded that there was little—if any—evidence to back up these accusations. In the media, the Islamic bomb threat became the reality, even if that reality was largely a manufactured one.

The meme wove together religion, geography, and ideology at a time when Islam was becoming the dominant signifier of the vague territory that is the Middle East.[26] What originated as a nuclear programme catalysed by a regional conflict grew and metamorphosed into a construct encompassing a vast swathe of territory, including Pakistan, Iran, Iraq, the Arabian Peninsula, central Africa, and Libya. The common

---

[25] Matthew Jones, *After Hiroshima: The United Sates, Race, and Nuclear Weapons in Asia, 1945–1965* (Cambridge: Cambridge University Press 2010); Maddock, *Nuclear Apartheid*; George Perkovich, *India's Nuclear Bomb: The Impact on Global Proliferation* (Berkeley: University of California Press, 1999). On religion, see Philip Muehlenbeck (ed.), *Religion and the Cold War: A Global Perspective* (Nashville: Vanderbilt University Press, 2012); Andrew Preston, *Sword of the Spirit, Shield of Faith: Religion in American War and Diplomacy* (New York: Anchor Books, 2012), 411–600.

[26] Melani McAlister, *Epic Encounters: Culture, Media, and U.S. Interests in the Middle East Since 1945* (Berkeley: University of California Press, 2001), 200. On the US encounter with the 'Middle East', also see Douglas Little, *American Orientalism: The United States and the Middle East Since 1945* (Chapel Hill: University of North Carolina Press, 2008) and on the 1970s in particular, Salim Yaqub, *Imperfect Strangers: Americans, Arabs, and U.S.-Middle East Relations in the 1970s* (Ithaca: Cornell University Press, 2016). On the evolution of the 'Middle East' as a concept for American and British policymakers, see Osamah F. Khalil, 'The Crossroads of the World: U.S. and British Foreign Policy Doctrines and the Construct of the Middle East, 1902-2007', *Diplomatic History*, 38:2 (2014), 299–344.

factor amongst this diverse set of territories was their perceived belonging to the imagined community of the Islamic world. The Islamic bomb therefore took Pakistan's nuclear programme and splashed it across a geographical arc from the mountains of South Asia to the deserts of North Africa, putting Islamabad's atomic ambitions into a transcontinental context. The meme was also a vessel containing not just an alleged Muslim nuclear unity, but also Libya's anti-Western 'fanaticism', Iraq's socialist Ba'athism, Iran's revolutionary ideology, Pakistan's military-Islamic thinking, and Middle Eastern terrorism. The Islamic bomb implied a common *identity*—not just a unifying faith—that challenged Western interests, eliding nuanced understandings of the manifest differences between states, the ideological positions of their leaders, and their national desires.

During the Cold War there were pro-American and pro-Soviet states, states representing various Islamic sects, and states—such as Libya and Pakistan—vying for leadership positions within the Muslim world. Pakistan used its nuclear programme to make itself unique, with nuclear weapons forming part of a quest for regional leadership based on religious community.[27] For Bhutto and Zia the Islamic bomb legitimised the nationalist reasoning behind Pakistan's nuclear programme.[28] National security and a political desire for nuclear power were the main driving forces, but the rhetoric allowed both leaders to present Western nonproliferation efforts as anti-Muslim, not just anti-Pakistani. This is not, however, to assert a hard division between the sacred and the secular. As Husain Haqqani notes, since its founding in 1947 Pakistan has found itself in tension between "mosque and military".[29] Islam remains a fundamental component of Pakistani identity, often a means of differentiating itself from India.[30] Bhutto in particular aimed to remodel Pakistan's image in foreign eyes, emphasising his socialist ideals and ties with the Muslim world.[31] Zia

[27] Yasmeen, 'Pakistan's Nuclear Bomb', 202.

[28] Ibid., 202–203.

[29] Husain Haqqani, *Pakistan: Between Mosque and Military* (Washington D.C.: Carnegie Endowment for International Peace, 2005).

[30] Aparna Pande, 'Foreign Policy of an Ideological State: Islam in Pakistan's International Relations' (Ph.D. dissertation, Boston, 2010), xi.

[31] Feroz Hassan Khan, *Eating Grass: The Making of the Pakistani Bomb* (Stanford: Stanford University Press, 2012), 97–98.

emphasised Islam's role in political and civil society, striving to mould Pakistan to his vision of an Islamic state and appealing to the military, harder-line Islamic radicals, and widespread anti-Indian sentiment.[32] As Andrew Rotter observes, US policymakers viewed Pakistan's Muslims as fellow monotheists who were "manly, energetic, and tough-minded in the face of the Communist threat".[33] Although the relationship was frequently rocky, Washington saw Islamabad as a more reliable partner into the 1970s.

Since this study investigates American and British attitudes and policies towards Pakistan, it is not primarily concerned with the evolution of the Pakistani nuclear weapons programme itself. There are several reasons for this. Primarily, this is a study of US and UK policy. Secondly, there is very little—if any—extant official primary source documentation available from Pakistan. The government of Pakistan is rigorous in preventing foreign scholars from examining materials related to the nuclear programme. Thus, researchers must rely on secondary sources, often written by former Pakistani 'insiders' and frequently unhelpfully speculative. Useful studies of the insider variety include those by Feroz Hassan Khan, Kamal Matinuddin, and Saeem Nalik.[34] As former senior Pakistani military officers, these scholars bring substantial personal experience to their respective studies, and there is considerable reliance on interviews that could only be gained by insiders. There are, however, problems inherent in such works in that there is a reticence on the part of the authors to delve into certain areas and, significantly, they rely largely on popular works on the Pakistani nuclear programme. However, by making careful use of such accounts, Pakistani voices are granted a presence as far as possible.

[32] Ibid., 99; Giles Kepel, *Jihad: The Trail of Political Islam* (London: I.B. Tauris, 2004), 98–105.

[33] Andrew J. Rotter, 'Christians, Muslims, and Hindus: Religion and U.S.-South Asian Relations,, 1947–1954', *Diplomatic History*, 24:4 (2000), 593–613; Andrew J Rotter, 'Saidism Without Said: Orientalism and U.S. Diplomatic History', *The American Historical Review*, 105:4 (2000), 1214.

[34] Khan, *Eating Grass*; Kamal Matinuddin, *The Nuclearization of South Asia* (Oxford: Oxford University Press, 2002); Naeem Salik, *The Genesis of South Asian Nuclear Deterrence: Pakistan's Perspective* (Oxford: Oxford University Press, 2009).

CHAPTER 2

"No hope of preventing proliferation" From the Indian Nuclear Test to the Politics of Limited Choice, May 1974 to December 1975

The eighteen months after India's May 1974 nuclear test witnessed a series of fruitless American and British attempts to persuade Pakistan not to head down the nuclear path. This contributed to downbeat assessments about the spread of nuclear weapons. By the end of 1975 CIA analysts were arguing that there was "no hope of preventing proliferation".[1]

During 1974 and 1975 notable differences existed between the United States and the United Kingdom on non-proliferation matters. Richard Nixon's and Gerald Ford's administrations concentrated on regional factors, in particular the sub-continental rivalry between India and Pakistan. The US government's regional focus was rooted in a desire not to highlight Nixon's and Henry Kissinger's *laissez-faire* attitude towards non-proliferation. British politicians and officials concentrated on the Indian test's wider effects on global non-proliferation efforts,

---

[1] CIA Office of Political Research, 'Managing Nuclear Proliferation: the Politics of Limited Choice', December, 1975 (exact date unknown), National Security Archive Electronic Briefing Book (hereafter NSAEBB) 'National Intelligence Estimates of the Nuclear Proliferation Problem: the First Ten Years, 1957–1967' (hereafter 'NIENP'), www2.gwu.edu/~nsarchiv/NSAEBB/NSAEBB155/index.htm (accessed March 5, 2013), Doc.15, 40.

© The Author(s) 2017
M.M. Craig, *America, Britain and Pakistan's Nuclear Weapons Programme, 1974–1980*, Security, Conflict and Cooperation in the Contemporary World, DOI 10.1007/978-3-319-51880-0_2

unwilling to damage relations with its two ex-colonies, and aware of the potential hypocrisy of a nuclear weapon state (NWS) being openly critical of a new entrant to the 'nuclear club'. Arms sales to the sub-continent also played their part during this period. Washington used weapons sales as a non-proliferation tool in an effort to either bribe or coerce Pakistan into foregoing nuclear capability. At the same time, London's determination to land a lucrative Indian contract for strike aircraft actively inhibited non-proliferation by alarming Islamabad, as British commercial considerations took precedence over anti-proliferation imperatives.

This eighteen-month period highlights many issues that remained constant during the 1970s. These included the deleterious effects of British economic self-interest on non-proliferation goals and American political decisions in favour of Pakistan. Provoked by Indian actions, Pakistan would become the most significant US and UK proliferation concern as the decade progressed.

INDIA'S TEST

The Indian detonation of May 18, 1974 was a nuclear Rubicon as the test obliterated the false but durable idea of a hard line between civilian and military nuclear technology.[2] It was also a turning point that set Pakistan irrevocably along the nuclear weapons path, Islamabad having initiated moves towards the atomic option after the trauma of 1971's Indo-Pakistani War.[3] The USA and the UK reacted to the explosion in a modest fashion, but for quite different reasons. Nixon and Kissinger, despite an outwardly exemplary arms control record, lacked enthusiasm for preventing the further proliferation of nuclear weapons.[4] While from 1969 onwards the administration would publicly support the NPT, little would be done to persuade other nations to sign up.[5] The Nixon administration saw the

---

[2] Michael J. Brenner, *Nuclear Power and Non-proliferation: The Remaking of U.S. Policy* (Cambridge: Cambridge University Press, 1981), 64.

[3] Samina Ahmed, 'Pakistan's Nuclear Weapons Program: Turning Points and Nuclear Choices', *International Security*, 23:4 (1999), 183–184.

[4] Francis J. Gavin, *Nuclear Statecraft: History and Strategy in America's Atomic Age* (Ithaca: Cornell University Press, 2012), 104–105.

[5] Francis J. Gavin, 'Blasts from the Past: Proliferation lessons from the 1960s', *International Security*, 29/3 (2004/5), 132.

Indian test as a predominantly regional issue, and treated it as such, seeking to de-emphasise administration laxity on non-proliferation and avoid setting a precedent for future action. Yet, the Indian test eventually drove Kissinger to reassess his position on non-proliferation, a re-evaluation that birthed the Nuclear Suppliers' Group (NSG).[6] Harold Wilson's government emphasised the case's global consequences in an effort to avoid antagonising either India or Pakistan by singling them out for criticism. Furthermore, Foreign Secretary James Callaghan wanted to avoid the suggestion of hypocrisy inherent in a NWS criticising another nation for gaining atomic capability.[7]

The Indian government drew upon the language of the Eisenhower-era Atoms for Peace programme and the Operation Plowshare nuclear civil engineering tests as it presented 'Smiling Buddha' as a peaceful nuclear explosion (PNE) designed to gain scientific and civil engineering data. The distinction between peaceful and non-peaceful explosions built into the NPT, and the legacies of the Eisenhower and Kennedy eras non-proliferation efforts allowed this characterisation.[8] Indian Prime Minister Indira Ghandi insisted that there was "nothing to get excited about".[9] Behind the scenes, New Delhi was driven towards testing by a constellation of factors, including Gandhi's domestic troubles, India's declining status in the non-aligned world, and the emerging Sino-Pakistani alliance that increased tensions with nuclear-armed China.[10] A majority of Indians welcomed the test and it gave Gandhi a much-needed political boost.[11] Journalists reported rejoicing at all levels of Indian society, despite the

---

[6] William Burr, 'A Scheme of "Control": The United States and the Origins of the Nuclear Suppliers' Group, 1974–1976', *International History Review*, 36:2 (2014), 2.

[7] Foreign and Commonwealth Office (hereafter FCO) to United Kingdom Embassy (hereafter UKE) Washington, 'Nuclear Test Explosion in India', May 20, 1974, The National Archives of the United Kingdom (hereafter TNA) Records of the Foreign and Commonwealth Office (hereafter FCO) 66/653, 2–3.

[8] William C. Potter, *Nuclear Power and Nonproliferation: An Interdisciplinary Perspective* (Cambridge: Oelgeschlager, Gunn & Hain, 1982), 44.

[9] 'India Joins Nuclear Club', The *Guardian* (hereafter *TG*), May 19, 1974, 1.

[10] Jayita Sarkar, 'The Making of a Non-Aligned Nuclear Power: India's Proliferation Drift, 1964–8', *International History Review*, 37:5 (2015), 933.

[11] George Perkovich, *India's Nuclear Bomb: The Impact On Global Proliferation* (Berkeley: University of California Press, 1999), 179.

shadows of economic deprivation and labour unrest.[12] US analysts noted that alongside the explosion's effect on morale, New Delhi hoped that nuclear capability would have a deterrent effect on China. However, this was at the cost of provoking alarm in Pakistan.[13]

The State Department drafted a response excoriating Gandhi's government, and informed Kissinger during the twenty-first day of his exhausting Middle Eastern shuttle diplomacy. The State Department's response was supported by Daniel Patrick Moynihan, the US Ambassador to India, then enjoying a holiday in England.[14] Kissinger's reaction—displaying his usual focus on the international order's stability—was to order his subordinates to remain silent but, if pressed, to offer a low-key response based on the formula that, "the United States has always been against nuclear proliferation for the adverse impact it will have on world stability".[15] Bilateral relations were one reason for this, with Kissinger concluding, "public scolding would not undo the event, but only add to US-Indian bilateral problems and reduce the influence Washington might have on India's future nuclear policy".[16] Relations with India had been problematic at least since the 1971 Indo-Pakistani War, when the United States had 'tilted' towards Islamabad.[17]

A follow-up State Department analysis emphasised the regional preference by stressing that containing an adverse Pakistani reaction was the most immediately pressing issue, an interpretation confirmed when Prime

---

[12] 'Atomic Jubilation Sweeps India', *TG*, May 20, 1974, 3.

[13] US Mission to NATO, 'Assessment of Indian Nuclear Test', June 5, 1974, NSAEBB 'India and Pakistan–On the Nuclear Threshold' (hereafter IPNT), www2.gwu.edu/~nsarchiv/NSAEBB/NSAEBB6 (accessed on March 5, 2013), Doc.18, 1–2.

[14] Daniel Patrick Moynihan, *Daniel Patrick Moynihan: A Portrait in Letters of an American Visionary* (New York: Public Affairs, 2010), 336–337.

[15] United States Embassy (hereafter USE) Damascus to State Department (hereafter 'State'), 'Indian Nuclear Test', May 18, 1974, US National Archives and Records Administration (hereafter NARA) Access to Archival Databases system (hereafter AAD), 1974DAMAS00764.

[16] Dennis, Kux, *Estranged Democracies: India and the United States, 1941–1991* (Washington D.C.: National Defense University Press, 1993), 315.

[17] Robert Dallek, *Nixon and Kissinger: Partners in Power* (London: Allen Lane, 2007), 349–352.

Minister Zulfikar Ali Bhutto angrily stated that Islamabad would never be the victim of Indian "nuclear blackmail".[18] Nixon and Kissinger's lack of belief in the significance of nuclear weapons and the subdued response to the test provoked concern within the arms control community over the test's long-term impact on proliferation. Arms Control and Disarmament Agency (ACDA) Director Fred Iklé warned of the dire consequences for the NPT's further ratification, of the potential impetus that India's test could give to other nations' nuclear programmes, and the impact on regional stability.[19] This was counsel that Iklé had been offering for years. As far back as 1965 he had argued that if proliferation cascaded beyond the "middle powers" then the result would be "owners of nuclear weapons who cannot be deterred because they feel they have nothing to lose".[20] Led by Iklé, the ACDA had been the only organisation within the US government to maintain a persistent institutional and intellectual commitment to non-proliferation.[21] Kissinger approved Iklé's assessment as a response to individual enquiries, with the proviso that communications did not suggest US attempts to marshal opinion against India or comment on the wider proliferation implications.[22] The State Department recognised their dichotomous position: the need to avoid being seen as orchestrating a worldwide campaign against India, while evading perceptions of acquiescence to Indian nuclear status. Foggy Bottom instructed US diplomatic posts around the world to just avoid discussing the issue.[23]

---

[18] State to Mission to the International Atomic Energy Agency (hereafter M-IAEA), 'Indian Nuclear Tests', May 18, 1974, *Foreign Relations of the United States 1969–1976* (hereafter *FRUS 69–76*), Volume E8 'Documents on South Asia, 1973–1976', http://history.state.gov/historicaldocuments/frus1969-76ve08 (accessed November 13, 2013), Doc.162, 2; UKE Islamabad to FCO, 'Indian Nuclear Explosion', May 20, 1974, TNA FCO 66/653.

[19] State to USE Jerusalem, 'U.S. Position on Indian Nuclear Test', May 18, 1974, NARA AAD, 1974STATE104621, 1–2; US Mission to the United Nations to USE Damascus, 'Indian Nuclear Test', May 20, 1974, NARA AAD, 1974USUNN01875.

[20] Iklé, quoted in Gavin, *Nuclear Statecraft*, 7.

[21] Brenner, *Nuclear Power*, 83.

[22] USE Jerusalem to State, 'US Position on Indian Nuclear Test', May 19, 1974, NARA AAD, 1974JERUSA00966.

[23] State to USE Jerusalem, 'US Position on Indian Nuclear Test', May 19, 1974, NARA AAD, 1974STATE104655.

Back in New Delhi, Ambassador Moynihan thought this wise, arguing that a more critical response of the kind he had initially supported would undermine gradual moves towards better Indo-US relations.[24]

Prior to the test Kissinger had a fatalistic view of proliferation's inevitability that de-emphasised non-proliferation as a foreign policy priority.[25] Coupled to this, Kissinger *and* Nixon believed that nuclear weapons did not actually transform the way nations behaved. Atomic bombs, according to their worldview, were important, but they did not transcend more powerful political forces. Arms control was thus a useful tool, but not an end in itself.[26] Furthermore, the global south was significant for them only if it served to support wider goals (as Pakistan had done with the opening to China), threatened to reduce US credibility, invited communist expansion, or if events in the 'developing world' limited Washington's freedom of action by arousing public opinion.[27] Framing the Indian test as a regional issue and downplaying its significance allowed Kissinger to achieve three broader political goals. Treating a nuclear India as just another state mitigated the 'prestige' effect of atomic acquisition. Secondly, a strong stance against New Delhi would have required a strong stance against all other proliferators. Nixon and Kissinger were well aware that they—and previous administrations—had allowed Israel to clandestinely develop its nuclear arsenal.[28] Criticising India could lead to uncomfortable questions about Tel Aviv's nuclear status. Such broader non-proliferation action was

---

[24] Moynihan, *A Portrait in Letters*, 337.

[25] Peter A. Clausen, *Nonproliferation and the National Interest: America's Response to the Spread of Nuclear Weapons* (New York: HarperCollins, 1993), 115; Burr, 'A Scheme of "Control"', 3.

[26] Gavin, *Nuclear Statecraft*, 105–106.

[27] Mark Atwood Lawrence, 'Containing Globalism: The United States and the Developing World in the 1970s', in Niall Ferguson et al. (eds.), *The Shock of the Global: The 1970s in Perspective* (Cambridge: Belknap Press of the Harvard University Press, 2011), 208.

[28] Avner Cohen, *Israel and the Bomb* (New York: Columbia University Press, 1998); Matteo Gerlini, 'Waiting for Dimona: The United States and Israel's Development of Nuclear Capability', *Cold War History*, 10:2 (2010), 143–161; Seymour Hersh, *The Sampson Option: Israel's Nuclear Arsenal and American Foreign Policy* (New York: Random House, 1991); Or Rabinowitz, *Bargaining on Nuclear Tests: Washington and Its Cold War Deals* (Oxford: Oxford University Press, 2014), 70–98.

not a desired part of Kissinger's finely tuned spider web of policies and alliances. Finally, plans were in train to export nuclear technology to Israel and Egypt, plans that were enacted a mere month after the Indian detonation.[29] Hence, the Indian test was presented as a relatively minor problem, rather than placed in its correct, global context.

London's response to the test had a similar, but subtly different, tone. Wilson's government initially reacted along analogous lines to the USA, stating: "We would regret the explosion of any nuclear device, even for peaceful purposes, outside the context of the NPT."[30] Foreign Secretary Callaghan stated that this event demonstrated the need for the NPT.[31] Thus, from the outset Britain placed the situation within a global context. Bilateral relations also coloured the British response. Financial assistance to India amounted to nearly twenty per cent of the UK foreign aid budget and Wilson's government wished to avoid an embarrassing perceptual entanglement of aid and nuclear weapons, an entanglement that Islamabad was actively promoting.[32] London was far less concerned about the test's influence on Pakistan when compared to its impact on the global non-proliferation regime, reflecting Callaghan's and the parliamentary Labour party's pragmatic internationalism that emphasised global cooperation, the strengthening of international institutions, and multilateral disarmament.[33] Pakistani Foreign Secretary

[29] J. Samuel Walker, 'Nuclear Power and Nonproliferation: The Controversy Over Nuclear Exports, 1974–1980', *Diplomatic History*, 25:2 (2001), 222.

[30] UK Delegation Geneva to FCO, 'Indian Nuclear Explosion', May 18, 1974, TNA FCO66/653.

[31] Ann Lane, 'Foreign and Defence Policy', in Anthony Seldon, and Kevin Hickson (eds.), *New Labour, Old Labour: The Wilson and Callaghan Governments, 1974–79* (London: Routledge, 2004), 155; 'Pakistan Enters Fray on Indian A-Test', *TG*, May 22, 1974, 4.

[32] 'UK aid: where does it go and how has it changed since 1960?' *TG*, April 14, 2011, data spreadsheet https://spreadsheets.google.com/ccc?key=0AonYZs4MzlZbdGY4VXR1RmRkY0dOUG9jejNEM0JNaEE&hl=en (accessed November 13, 2013); UK Delegation to UNICEF to FCO, 'Indian Nuclear Test', May 23, 1974, TNA FCO66/653.

[33] Lane,'Foreign and Defence Policy', 156; Rhiannon Vickers, *The Labour Party and the World, Volume 2: Labour's Foreign Policy 1951–2009* (Manchester: Manchester University Press, 2011), 105.

Agha Shahi's ominous statement that, "The barrier to nuclear proliferation interposed by the NPT has been demolished. A precedent has been set" gave London's concerns increased salience.[34] While the Cabinet had to deal with global nuclear issues, there were vocal anti-nuclear elements in the wider party, typified by MPs Frank Allaun and Tam Dalyell, who would both come to prominence as critics of Pakistan's nuclear weapons programme in the late 1970s.

The Foreign and Commonwealth Office (FCO) advised Callaghan—in Washington for a Central Treaty Organisation (CENTO) foreign ministers meeting—that the NPT's status was paramount and that the Indian explosion represented a tipping point that threatened to lead to a proliferation cascade. Like Kissinger, the FCO cautioned restraint, but for quite different reasons, arguing that:

> We do not wish to take a strong line against the Indian government.
> A) because as a nuclear weapon state such a posture might appear hypocritical and
> B) because any appearance of taking sides between India and Pakistan would certainly adversely affect our own position in the sub-continent.
> Nor should we wish to commit ourselves to any kind of open-ended guarantee of the kind which Mr Bhutto appears to envisage.[35]

This articulated much more ephemeral anxieties than those voiced by Kissinger and Iklé. The essential hypocrisy of the nuclear weapon states when it came to nuclear development in the third world, and lingering British loyalty to its former colonies, came to the fore. Furthermore, London was reluctant to acquiesce to Bhutto's demands for nuclear security assurances. The main tool for reassuring India's neighbour, the FCO argued, should be the NPT and the security assurances *it* guaranteed.[36] Observers at the British Embassy in Islamabad saw a great intensity of feeling in Pakistan about the nuclear security question and argued that

---

[34] Shahi, in Sumit Ganguly, 'The Indian and Pakistani Nuclear Programmes: A Race to Oblivion?', in Raju G.C. Thomas (ed.), *The Nuclear Non-proliferation Regime: Prospects for the 21st Century* (Basingstoke: Macmillan, 1998), 278.

[35] FCO to UKE Washington, 'Nuclear Test Explosion in India', May 20, 1974, TNA FCO66/653, 2–3.

[36] Ibid.

if London did not take a strong stance, British interests in the country would suffer.[37] Despite this warning, the Wilson government delayed, choosing to sit and watch the unfolding superpower reactions.[38]

As the test's impact was absorbed, officials in Washington and London sought to evaluate each other's positions. The assessment from the US Embassy in London characterised Wilson's government as torn between its commitment to international disarmament and the need to maintain ties with India. US officials saw Wilson and his cabinet as likely to base their actions around NPT concerns, rather than making direct points to the Indians about the impact of their nuclear actions.[39] Additionally, nuclear questions began to tie into questions of conventional arms supplies to the sub-continent. Britain was liable to come under increasing pressure from Pakistan (then—like India—labouring under the arms embargo put in place by the Johnson administration during the 1965 Indo-Pakistani war) to sell advanced aircraft.[40] US diplomats in London were unsure which way the UK would jump, but felt that the American stance on arms sales to the sub-continent would be a major contributing factor in any British decision.[41] Sir Peter Ramsbotham (the highly respected British Ambassador to the United States) noted three days after the test that the Nixon administration had not yet formulated a definitive view on the issue.[42] In conversation with Ramsbotham, Dennis Kux (the State Department's

[37] UKE Islamabad to UKE Washington, 'Nuclear Test Explosion in India', May 21, 1974, TNA FCO66/653.

[38] FCO to UK Delegation Geneva, 'Indian Nuclear Explosion', May 20, 1974, TNA Records of the Prime Minister's Office (hereafter PREM) 16/1182.

[39] USE London to State, 'British Reaction to Indian Nuclear Testing', May 20, 1974, NARA AAD, 1974STATE104647, 2; USE London to State, 'Further British Concern Over Indian Nuclear Actions', May 28, 1974, NARA AAD, 1974LONDON07588.

[40] On the embargo, see Dennis Kux, *Disenchanted Allies: The United States and Pakistan, 1947–2000* (Washington D.C.: Woodrow Wilson Center Press, 2001), 161–165.

[41] USE London to State, May 20, 1974, 3.

[42] On Ramsbotham's career in Washington, see Raj Roy, 'Peter Ramsbotham', in Michael F. Hopkins, Saul Kelly, and John W. Young (eds.), *The Washington Embassy: British Ambassadors to the United States, 1939–77* (Basingstoke: Palgrave Macmillan, 2009), 209–228.

Indian Section Chief) had placed Pakistan at the top of the priority list, suggesting that the explosion would send a "shiver down their spines". It would take time, Ramsbotham suggested, for the various US government departments to determine the full impact on Pakistan and on the wider non-proliferation environment.[43]

In Washington on May 22, US, UK, and Pakistani representatives came together at the CENTO foreign ministers meeting.[44] CENTO had been a source of disappointment for Pakistan, with the other treaty countries—UK, Iran, Iraq, and Turkey—refusing to involve themselves in the Indo-Pakistani wars of 1965 and 1971. This upset stemmed from Pakistan's misunderstanding of CENTO's nature; Islamabad viewed CENTO as a bulwark against the USSR *and* India, while the other members saw it as an anti-Soviet measure. Nixon argued that because of India's nuclear test, Pakistani security was of prime importance and suggested the USA and the UK had a "restraining influence on those who might be tempted—to go under or over a border to destroy a country's independence". Nixon contended that if his administration went so far as to re-establish dialogue with India, it would be to encourage Indian restraint.[45] Aziz Ahmed (Pakistani Minister of State for Defense and Foreign Affairs) attempted to pressure the nuclear weapon states into adopting a tone more critical of India, to no avail. Ahmed's pleas that the USSR was abusing détente to infiltrate Afghanistan, India, and Iraq likewise fell on deaf ears.[46]

Despite the dismissal of Ahmed's entreaties, reassuring Pakistan was key to a private meeting between Nixon and his guest. Although US policy broadly supported Pakistan, prior to the meeting the State Department suggested that at this stage little could be done by the administration other than offer a bland commitment to do something once the test's full

---

[43] UKE Washington to FCO, 'Indian Nuclear Explosion', May 21, 1974, TNA FCO66/653, 1–2.

[44] The United States was CENTO's architect and sat on the alliance's Military Council, but was not a full member.

[45] Memcon, Nixon and CENTO Foreign Ministers, May 22, 1974, *FRUS* 69–76, Vol.E8, Doc.74, 1–2.

[46] State to USE Kabul, 'The Pakistanis at the CENTO Ministerial', May 25, 1974, *FRUS* 69–76, Vol.E8, Doc.75, 2.

implications had been assessed.[47] The Pakistani diplomat—reflecting on the Indo-Soviet treaty of friendship and what he now realised was CENTO's true nature—again portrayed the test as part of a Soviet scheme to gain influence in South Asia. Ahmed sought two things from Nixon: conventional arms to defend Pakistan and nuclear security assurances to "reduce the incentive [for Pakistan] to get nuclear weapons".[48] Although Ahmed spoke in general terms, this was a clue about Islamabad's intentions, presaging much more explicit future hints that insufficient US help would force Pakistan's nuclearisation. Nixon agreed in principle to Ahmed's requests for arms and assurances, but commented that nothing could be done immediately. The President noted that he would make the US commitment to Pakistan clear to the Kremlin and, if possible, encourage restraint in Moscow. Nuclear security guarantees were a complex area that, although agreeable to Nixon, was difficult to implement. The President thought it difficult to put such guarantees into a treaty, harking back to his and Kissinger's lack of faith in the credibility of nuclear-based commitments.[49]

Follow-up letters from Bhutto to Nixon and Wilson gave the strongest indication yet of his nuclear ambitions. Similar missives were sent to scores of world leaders in a major diplomatic campaign.[50] Bhutto disparaged Gandhi's PNE claims, arguing that the detonation would provoke many more national nuclear weapons programmes. He portrayed Pakistan as the victim of continual Indian aggression and emphasised his attempts at cross-border compromise. Concluding, Bhutto stated, "Pakistan will do

---

[47] Memorandum for Scowcroft, 'President's Meeting with Pakistan's Minister of State', May 22, 1974, The Gerald R. Ford Presidential Library (hereafter GRFPL), National Security Adviser Files (hereafter NSAF), Presidential Country Files for Middle East and South Asia, Box 26, Pakistan, 1–3.

[48] Memcon, Nixon, Scowcroft, Ahmed, May 23, 1974, *FRUS 69–76*, Vol.E8, Doc.164, 2–3.

[49] Ibid.; Pakistan also raised nuclear security guarantees at every possible international venue. See Bhumitra Chakma, *Strategic Dynamics and Nuclear Weapons Proliferation in South Asia: A Historical Analysis* (Bern: Peter Lang, 2004), 146fn48.

[50] CIA, 'Bhutto Seeks Nuclear Policy Assurances', *National Intelligence Daily*, May 30, 1974, CIA Freedom of Information Electronic Reading Room (hereafter CIA-FOIA), Doc.0000845825.

its utmost to resist pressures toward a nuclear option. But these pressures will increase if there is no credible political insurance against nuclear blackmail."[51] Here the leader of the country most affected by the Indian explosion made a barely veiled threat to pursue nuclear capability if not offered adequate protection. This should not have come as any great surprise. Bhutto had signposted his intentions as far back as 1965 when he stated that Pakistanis would "eat grass" in order to match any future Indian nuclear capability.[52] Bhutto's 1965 comments—well publicised in the test's aftermath—led Agha Shahi to contend that although Islamabad could not bear the costs of a bomb programme, the domestic pressure for nuclearisation might be too much to resist.[53]

Wilson, drawing on FCO Arms Control and Disarmament Department (ACDD) advice on the test's implications, attempted to mollify Bhutto by agreeing with many of his remarks and sharing his concerns about international peace and security.[54] Wilson was less alarmist than his Pakistani counterpart about the impact on the NPT, trying to cajole Bhutto into signing the treaty by emphasising its security guarantees and the greater need for signatories in the test's wake. Wilson remarked that he was heartened to hear that Bhutto would "resist pressure" for the nuclear option and continued to sell the NPT's benefits.[55] Still, for British policymakers, the most pressing concern resulting from the test was ensuring the non-proliferation regime's survival, rather than "sub-continental rivalry or the security concerns of any particular country".[56]

[51] Bhutto to Nixon, 'Text of Message from Mr. Zulfikar Ali Bhutto to Richard M. Nixon', Letter, May 24, 1974, GRFPL, National Security Council Institutional Files (Hereafter NSCIF), Box 105, Chronological File–Fazio, James (1), 3; Bhutto to Wilson, Letter, May 24, 1974, TNA PREM16/1182, Doc.11, 3. The Bhutto letter arrived at 10 Downing Street on May 28.

[52] 'The Brown Bomb', *TG*, March 11, 1965, 10.

[53] UKE Islamabad to FCO, 'Indian Nuclear Explosion', May 28, 1974, TNA FCO66/653, 1–2.

[54] Summerhayes to Seaward et al., 'Indian Nuclear Test', June 3, 1974, TNA FCO66/654. The ACDD still positioned the wider global non-proliferation implications of the Indian test above the regional consequences.

[55] Wilson to Bhutto, 'Indian Nuclear Test', Letter, June 28, 1974, TNA PREM16/1182, 1–2.

[56] Dales to Bridges, Letter, June 24, 1974, TNA PREM16/1182, 1.

With Watergate to consider, the Nixon White House replied far more succinctly. The President offered to carefully consider nuclear security assurances via the United Nations Security Council (UNSC) and expressed his continued sympathy for Pakistan's situation. Nixon assured Bhutto that his country's independence and security was "a cornerstone of American policy".[57]

The test also gave impetus to a study that was under way to prepare the United States for the 1975 NPT Review Conference. National Security Study Memorandum-202 (NSSM-202) had been "limping along" before May, mostly because of Atomic Energy Commission Chief Dixy Lee Ray's lack of cooperation.[58] The study was expansive, but also specifically considered whether the USA should emphasise renewed support for the NPT by existing signatories, and for accession to the treaty by states—such as Pakistan—that had not yet signed up.[59] By the end of 1974 the NSSM-202 study's results had been presented to Ford. The report struck a demoralised tone, bemoaning the decline of US primacy in the field of international nuclear sales. This loss of dominance could have dire effects on proliferation, as would-be nuclear states turned for supplies to other nations with much less rigorous controls on sensitive materials and technology.[60] NSSM-202, however, failed to answer crucial questions on technology transfer and the risks of a plutonium-based nuclear economy.[61]

In their initial reactions to the test, the American and British governments reacted in subtly different ways. Britain, by placing the test within a global NPT context, contrasted with American fatalism and emphasis on regional issues. Where the two states came together was in their assessment that the test could well be a tipping point that led to further proliferation. The prospect of a proliferation cascade was something that threaded its way through discussions of Pakistan's now invigorated

---

[57] Nixon to Bhutto, Letter, June 1, 1974, GRFPL, NSCIF, Box 105, Chronological File–Fazio, James (1), 1–2.

[58] Brenner, *Nuclear Power*, 53.

[59] Kissinger to Secretary of Defense et al., 'National Security Study Memorandum 202', May 23, 1974, Digital National Security Archive (hereafter DNSA), PD01462.

[60] 'U.S. Non-proliferation Policy', December 4, 1974, DNSA, PR01262, 4.

[61] Brenner, *Nuclear Power*, 75.

programme over the next five years. In the test's immediate aftermath, Washington and London offered Islamabad little more than vague reassurances. Through the rest of 1974 and into 1975, Bhutto would demand more concrete signifiers of Western commitment to Pakistani security. Most notably, the issue of arms sales to the sub-continent would come to the fore as a means for retarding Pakistani nuclear development *and* a factor pushing Bhutto's country towards the atomic option.

ARMS SALES

Conventional arms sales were tied to Pakistan's bomb programme from the very instant of the Indian test. Not only did successive US administrations attempt to use weapons sales as a means to discourage Islamabad's atomic ambitions, but the Wilson and Callaghan governments also found themselves caught up in the issue through their efforts to sell India Jaguar strike aircraft.[62] Stark differences were apparent in the reasons for, and effects of, conventional arms sales to the sub-continental neighbours. US offers of arms for Pakistan were an anti-proliferation measure, used in attempts to bribe or coerce Islamabad into abandoning its nuclear ambitions. British arms sales to India were vital within the context of the post-1973 oil price rises and their resultant damaging effect on Britain's balance of payments, rising inflation, gradually increasing unemployment, and—in 1974—a genuine recession. In this situation, the British aircraft industry was an essential revenue generator and employer.[63]

The 1970s were a lean time for Britain's aerospace industry, having been a successful exporter for the previous two decades. Manufacturers found that potential purchasers would not consider replacing their equipment until the 1980s, so any major sales opportunity was vital.[64] In the case of British sales to India, advanced weapons were only ever an

---

[62] The information on the Jaguar sales issue contained in this book is derived in part from Malcolm M. Craig, "'I Think We Cannot Refuse the Order": Britain, America, Nuclear Non-Proliferation, and the Indian Jaguar Deal, 1974–78', *Cold War History*, 16:1 (2016), 61–81.

[63] Harold Wilson, *Final Term: The Labour Government, 1974–1976* (London: Weidenfeld and Nicholson, 1979), 13.

[64] Mark Phythian, *The Politics of British Arms Sales Since 1964* (Manchester: Manchester University Press, 2000), 19.

impediment to non-proliferation policy, as London's reluctance to abandon the deal affected the US-UK relationship and alarmed Islamabad, contributing to the desire for nuclear capability.

After the test, senior Pakistani government officials requested that the United States ease Lyndon Johnson's 1965 arms embargo so they could defend themselves from what they saw as an aggressive, nuclear-armed India. Having fought three major wars with India since Partition in 1947, Pakistani fears were understandable. However, US officials argued that Islamabad was overplaying the situation in the hope of getting the embargo lifted. Aziz Ahmed, in his first post-test meeting with Nixon and Kissinger, had already raised the subject, persisting with this theme in early June when he again met Kissinger. The Pakistani representative also invoked the spectre of Soviet involvement in the Indian nuclear programme as a precursor to requests for "defensive" weapons: anti-aircraft and anti-tank missiles, RADAR systems, and submarines. Ahmed pressed Kissinger for a swift announcement on this, but the Secretary of State demurred, blaming Congress for his inability to immediately take action stating, "I've always believed in military supply for Pakistan. It's absurd that the Soviets can arm India while our hands are tied. It's a massive problem, but I don't believe the Congress would let us do it."[65]

In the face of Kissinger's dissembling, Ahmed sought nuclear security assurances from friendly powers. Kissinger offered to make a statement supporting "Pakistan's independence and territorial integrity" and making clear that "the use of nuclear weapons against Pakistan would be a very grave matter". This was a bold statement that Ahmed misinterpreted as including the *threat* of use, but it fell far short of the assurances he sought. Kissinger reiterated that he would probe Congress on military aid, even though his advisers suggested that any change in arms supply policy came with unbearable congressional costs.[66] Moynihan objected to lifting the embargo, contending that supplying Pakistan would cause the immediate

---

[65] Memcon, Ahmed, Yaqub-Khan, Sober, Constable, 'Military Supply for Pakistan', June 3, 1974, *FRUS 69–76*, Vol.E8, Doc.166, 3–4.

[66] Memcon, June 3, 1974, 7–8; Sober to Kissinger, 'Military Supply for Pakistan', June 4, 1974, Declassified Documents Reference System (hereafter DDRS), DDRS-270991-i1-4, 3.

failure of sub-continental non-proliferation policy.[67] Henry Byroade—Moynihan's counterpart in Islamabad—asserted that the opposite would happen, arguing that arms sales would reassure Bhutto and improve the chances of normalisation in Indo-Pakistani relations.[68]

As the Pakistanis attempted to have the arms embargo lifted, the Wilson government assessed British arms sales policy on the sub-continent. Sales to both Pakistan and India were of "immediate concern," although in the long run it was sales to the latter that were the most problematic and introduced the greatest complexity into non-proliferation efforts.[69] This situation was not without precedent for, since Partition, successive British governments had faced the challenging task of balancing sales to the two regional rivals and resolving arms sales disputes with the United States.[70] The FCO South Asia Department (SAD)—with ACDD, Far Eastern Department, and East European and Soviet Department support—argued that the potential British sale of a nuclear-capable strike aircraft such as the SEPECAT Jaguar (which had just come into service with the Royal Air Force in the tactical nuclear strike role) to India might be regarded by Pakistan with "great suspicion".[71] Sir Michael Cary, Permanent Under

[67] USE New Delhi to State, 'The Secretary's Meeting with Aziz Ahmed, June 3', June 6, 1974, DDRS, DDRS-270989-il-3, 1.

[68] Sober to Kissinger, June 4, 1974, 3.

[69] Seaward to White, 'Indian Nuclear Test Explosion', May 28, 1974, TNA FCO66/653.

[70] Phythian, *The Politics of British Arms Sales*, 130; Paul M. McGarr, *The Cold War in South Asia: Britain, the United States, and the Indian Subcontinent, 1945–1965* (Cambridge: Cambridge University Press, 2013), 351. On US-UK Cold War trade disputes more widely, see Frank Cain, *Economic Statecraft During the Cold War* (London: Routledge, 2007); Alan P. Dobson, *The Politics of the Anglo-American Economic Special Relationship 1940–1987* (Brighton: Wheatsheaf Books, 1988); Jeffery A. Engel, *Cold War at 30,000 Feet: The Anglo-American Fight for Aviation Supremacy* (Cambridge: Harvard University Press, 2007); Christopher Hull, 'Our Arms in Havana: British Military Sales to Batista and Castro, 1958–59', *Diplomacy and Statecraft*, 18:3 (2007), 593–616 and '"Going to War in Buses": The Anglo-American Clash over Leyland Sales to Cuba, 1963–1964', *Diplomatic History*, 34:5 (2010), 793–822.

[71] Seaward to Wilford, 'Implications of the Indian Nuclear Test', May 29, 1974, TNA FCO66/654.

Secretary (PUS) at the Ministry of Defence (MoD), contended that the Jaguar was *not* a new capability for India, which had a fleet of ageing but serviceable Canberra bombers that could be adapted for nuclear delivery.[72] Sir Thomas Brimelow, Cary's opposite number in the FCO, counter-argued that the sale of Jaguar aircraft to New Delhi was inconceivable in light of India's nuclear test. The Indians would require subtle notification so as not to wound their pride, but could be mollified by the sale of less controversial equipment, such as the Nimrod maritime patrol aircraft.[73] ACDD Chief David Summerhayes insisted that wider knowledge of British intent to sell nuclear delivery systems to India would have a significant, damaging impact on the UK's standing as an anti-proliferation voice.[74] Summerhayes' colleague FCO Assistant Under Secretary of State John Thomson agreed, stating the primacy of non-proliferation policy and arguing, "if ever there was a moral cause, it is this".[75] Thus the battle lines over the Jaguar sale's symbolism—if not its reality—were drawn.

On June 10 the peripatetic Aziz Ahmed arrived in London to meet with Wilson. Once again, the Pakistani minister characterised the May test as part of an Indo-Soviet attempt to dominate South Asia and requested British nuclear security guarantees. Wilson also stuck to a familiar line, sympathising with Pakistan's position, but always emphasising Britain's commitment to the NPT.[76] Wilson's continued calls—made to India and Pakistan—to sign up to the NPT were destined to fail. The refusal to join was not based solely on strategic considerations, but also on recent colonial history, and the lingering view that the NPT represented a condescending, possibly racist, worldview that amounted to "nuclear apartheid".[77] On the question

---

[72] Cary to Brimelow, 'Relations with India', June 10, 1974, TNA FCO66/655, 2. In RAF service the Jaguar would occupy the role formerly taken by the Canberra.

[73] Brimelow to Cary, 'Relations with India', June 13, 1974, TNA FCO66/655, 2.

[74] Summerhayes to Thomson et al., 'Indian Nuclear Test', June 13, 1974, TNA FCO66/655, 3.

[75] Thomson, attachment to: Summerhayes to Thomson et al., 'Indian Nuclear Test', June 13, 1974, TNA FCO66/655.

[76] 'Record of a Conversation Between the Prime Minister and the Pakistan Minister of State for Defence and Foreign Affairs', June 10, 1974, TNA PREM16/1182, 2–3.

[77] Sumit Ganguly and S. Paul Kapur, *India, Pakistan, and the Bomb: Debating Nuclear Stability in South Asia* (New York: Columbia University Press, 2010), 18.

of arms sales, Ahmed asked if there was any way that Britain could lift its arms embargo, a decision that might have a favourable influence on United States thinking. Although Wilson responded that he would see what he could do, both statesmen were mistaken in this case.[78] There was no British arms embargo, only a requirement that Pakistan must pay cash for any sales.

Following Ahmed's attempts to have the US arms embargo lifted, Henry Byroade added to the debates by noting "profound shock" in Pakistan over the Indian test, which had exacerbated "chronic feelings of insecurity".[79] Byroade argued that Islamabad's persistent requests to lift the embargo and give security guarantees were not just for show. Such demands reflected genuine Pakistani concerns about national security. However, the Ambassador stated that Washington's overriding concern needed to be the impact of events on the global non-proliferation regime.[80] Bhutto was resisting calls in parliament for an immediate Pakistani nuclear weapons programme (despite the fact that he had instigated such a programme in 1972), arguing there was no way for the country to build the bomb because of technological and economic disadvantages. Byroade noted that Bhutto had foreseen American anti-proliferation action against Pakistan, the Prime Minister commenting that any public call for a weapons project "could precipitate active efforts on the part of the nuclear powers to obstruct Pakistan's nuclear programme".[81]

When Kissinger made a high-profile trip to the sub-continent in October, the question of arms for Pakistan was central to his discussions. Prior to the trip, the dénouement of Watergate had brought Gerald R. Ford to power. Ford was an amateur in the field of foreign relations and relied heavily on his Secretary of State.[82] At the same time, the United States continued to demonstrate the limited consequences for India of its nuclear test, when it proceeded to ship a load of uranium fuel for the

---

[78] 'Record of a Conversation', June 10, 1974, 3.

[79] USE Islamabad to State, 'US Arms Policy Towards Subcontinent Following Indian Nuclear Test', June 12, 1974, *FRUS 69–76*, Vol.E8, Doc.167, 1.

[80] Ibid., 4.

[81] USE Islamabad to State, 'Further GOP [Government of Pakistan] Reaction to Indian Nuclear Test', June 12, 1974, NARA AAD, 1974ISLAMA05622, 2–3.

[82] Kux, *Disenchanted Allies*, 215.

nuclear reactor at Tarapur.[83] During this transitional period, Ahmed had a series of meetings with Ford and Kissinger to press again for lifting the embargo.[84] Within the US foreign policy apparatus, Pakistani pleas for a change to weapon sales policy were seen as less to do with actual security needs and more to do with the psychological pressure of India's test and affirmation of the strong political relationship with the USA that the "arms pipeline" implied.[85] Bhutto also tied conventional arms to the quest for nuclear weapons. A few weeks before the Kissinger mission, he told the *New York Times*: "If security interests are satisfied, if people feel secure and if they feel they will not be subject to aggression, they [will] not want to squander away limited resources in [the nuclear] direction."[86]

Kissinger utilised Bhutto's veiled threats when he met Gandhi at the end of October. Reporting to Ford, he noted: "I made the point that at most what was involved were limited cash sales to Pakistan and that she should reflect on the risk that, if frustrated on conventional arms purchases, Pakistan would be under even greater pressure to go nuclear."[87] Having been briefed by the CIA that Islamabad would proceed with a nuclear programme as rapidly as its limited capabilities allowed, Kissinger pressed his Indian hosts over the arms to Pakistan issue.[88] The Secretary of

[83] Perkovich, *India's Nuclear Bomb*, 184.

[84] Memcon, 'The Call of the Pakistani Minister of State for Defense and Foreign Affairs, Aziz Ahmed, on the Secretary', September 30, 1974, *FRUS 69–76*, Vol.E8, Doc.176.

[85] Memorandum for the President, 'Your Meeting with Pakistani Minister of State Aziz Ahmed', October 16, 1974, GRFPL, Presidential Country Files, Box 26, Pakistan (2), 4.

[86] 'Pakistani Presses U.S. For Arms', *New York Times* (hereafter *NYT*), October 14, 1974, 10.

[87] Scowcroft to Ford, Memorandum, October 28, 1974, GRFPL, NSAF, Kissinger Trip Briefing Books Box 3, October 20–November 9, 1974–Europe, South Asia, and Middle East, HAK [Henry A. Kissinger] Messages for the President (1), 2.

[88] CIA SNIE 4-1-74, 'Prospects for Further Proliferation of Nuclear Weapons', August 23, 1974, NSAEBB 'In 1974 Estimate, CIA Found that Israel Already Had a Nuclear Stockpile and that "Many Countries" Would Soon Have Nuclear Capabilities' (hereafter CIAINS), www.gwu.edu/~nsarchiv/NSAEBB/NSAEBB240/index.htm (accessed March 5, 2013), 38.

State challenged the Indian argument that any shipments to Islamabad would demonstrably harm Indo-American relations by stating that the United States had no interest in Indo-Pakistani military conflict but that the USA had to consider Pakistan's independence.[89]

After Kissinger's visit to New Delhi, his subsequent meeting with Bhutto was friendly and good-humoured. The Secretary joked darkly that after seeing India first hand he was, "thinking about supplying nuclear weapons, not only conventional arms, to Pakistan and even Bangladesh!"[90] The discussions ranged from the Middle East's problems to India's perceived hegemonic ambitions. The two statesmen briefly spoke of arms, Kissinger indicating that it was very much in the United States' interest to ensure the integrity of Pakistan.[91]

In London, Wilson's government debated the problematic nature of arms sales to India in the time between the Byroade telegrams and Kissinger's visit to the sub-continent. The UK was delaying giving a definitive answer to the Indians over Jaguar—using a wrangle over credit terms as the excuse—and British officials were dispatched to Washington in an effort to understand American thinking.[92] How the USA handled Pakistani arms requests—where weapon sales *were* a non-proliferation measure—would have a significant bearing on the British position regarding India, where arms sales were never anything but an *impediment* to non-proliferation.

Jaguar had become the subject of vigorous debate within the Wilson government, a government that had—according to John Thomson—taken a "strong and consistent stand" on non-proliferation.[93] Favouring

---

[89] CIA, 'Memorandum in Support of the Secretary of State's Trip to South Asia: Tab E, Regional Papers; Prospects for Long-term Stability in South Asia', October 15, 1974, GRFPL, Remote Archives Capture files (hereafter RAC), Box 16, NSAF, Briefing Books and Cables for Henry Kissinger, 29; Scowcroft to Ford, Memorandum, October 29, 1974, GRFPL, NSAF, Kissinger Trip Briefing Books, Box 3, October 20–November 9, 1974–Europe, South Asia, and Middle East, HAK [Henry A. Kissinger] Messages for the President (1), 2.

[90] Memcon, Kissinger and Bhutto, October 31, 1974, DNSA, KT01391, 4.

[91] Ibid., 7–8.

[92] 'Mr Thomson's Visit to Washington: Policy Towards India Following the Indian Nuclear Explosion: Speaking Notes', July 22, 1974, TNA FCO66/657, 4.

[93] Thomson to Walker, 'Non-proliferation and the Indian Nuclear Explosion', August 6, 1974, TNA FCO66/657, 1.

the sale were Secretary of State for Industry Tony Benn and Secretary of Defence Roy Mason. Benn recognised that the explosion was concerning, but argued that the sale helped the British aircraft industry.[94] On the other side were the Treasury—where Indian demands for generous credit terms provoked objections—and sub-departments such as the ACDD that emphasised non-proliferation's significance. All sides recognised the £300 million-plus deal's "immense benefit to British industry".[95] In representations to India's government, the economic sticking points were the main topics for discussion.[96] The Treasury—despite irritation that the High Commission in New Delhi had been encouraging the Indians—were happy to take a "hands off" approach to the entire affair and let the FCO and MoD handle matters, unless serious demands were made for credit.[97] Chancellor of the Exchequer Denis Healey eventually agreed to refer the issue to the high-level Defence and Oversea Policy Committee (OPD) for further consideration.[98]

ACDD—the department with the greatest institutional commitment to non-proliferation—concentrated on the sale's symbolism, arguing that selling Jaguars to India would negatively impact Britain's wider anti-proliferation efforts. Arms control officials contended that, "[A] British decision to supply Jaguar and thus improve India's nuclear delivery capability would be interpreted on all sides as showing that we are not really serious about containing nuclear proliferation and that we give narrow commercial and industrial interests a higher priority than this major issue of world security".[99] The ministerial OPD's civil service

---

[94] Benn to Callaghan, Letter, August 8, 1974, TNA FCO66/657.

[95] Callaghan to Benn, 'Jaguar Sales to India', Letter, August 23, 1974, TNA FCO66/658, 1; Mason to Benn, Letter, September 2, 1974, TNA FCO66/658.

[96] Walker to Cary, 'Corvettes and Jaguars for India', August 16, 1974, TNA FCO66/658, 2–3.

[97] Rich to Boothroyd, 'India: Defence Sales', June 5, 1974, TNA Records of the Treasury (hereafter T)362/53; Rich to Kelley, 'Sale of Jaguar Aircraft to India', August 13, 1974, TNA T362/53.

[98] Healy to Callaghan, 'Jaguar Sales to India', September 6, 1974, TNA T362/53.

[99] ACDD Brief, 'OPD (Official) Committee Meeting; Defence Sales to India', October 14, 1974, TNA FCO66/659, 1–2.

counterpart—OPD(O)—realised that this could thus lead to prejudice against Britain at the 1975 NPT Review Conference.[100]

There followed a paper analysing British-Indian relations drafted by the FCO Planning Staff, circulated by Brimelow, and approved by Callaghan. The paper made the case that, absent any real chance of moving India away from Soviet arms purchases:

> the justification for continued major British arms sales to India must there-fore rest on the value of such arms sales to our own economy either through the payments received for them or through the contribution they make to the workload of certain of our own industries.[101]

Analysts stressed the economic benefits and downplayed the symbolism of selling allegedly nuclear-capable aircraft and the impact this might have in Islamabad. Pakistan was, they argued, much less important than India. Pakistani objections to the deal were better countered with parallel sales rather than restricting New Delhi's options. Finally, the Planning Staff contended that should India request actual nuclear-capable aircraft, the government might need to consider a politically motivated embargo.[102]

This assessment wilfully ignored the fact that the Jaguar was already seen in Islamabad and other quarters as a nuclear bomber. Here was explicit, high-level recognition that a nuclear-capable aircraft was genu-inely and symbolically different from other arms, although the FCO, MoD, and other departments repeatedly emphasised that India would not be offered the nuclear version of Jaguar. This was a pattern through-out the Cold War, as Britain persistently downplayed security threats whenever the government saw those threats as imperilling sales.[103] The pattern can be observed with the Jaguar sale: the impact on Pakistan's security and the symbolism of selling allegedly nuclear-capable aircraft was minimised in favour of stressing the economic benefits for Britain.

---

[100] Defence and Oversea Policy (Official) Committee, Sub-committee on Strategic Exports (hereafter OPDO-SE), 'Credit for Sale of Defence Equipment to India', October 15, 1974, TNA Records of the Cabinet Office (hereafter CAB) 148/152, 2.

[101] Brimelow to Heads of Department et al., 'British Policy Towards India', November 4, 1974, TNA FCO 66/660, 9.

[102] Ibid.

[103] Engel, *Cold War at 30,000 Feet*, 298.

The Jaguar sale was still viewed by the wider government through the lens of economic necessity, as bringing in hundreds of millions of pounds to British industry, securing thousands of jobs, boosting the aircraft's further marketability, and counterbalancing Soviet arms sales. At this stage, Pakistan was only really considered as desiring equal treatment.[104] The British government gave relatively little consideration at the highest levels to the impact on the Pakistani drive for nuclear weapons of the sale of what was seen as a nuclear-capable strike aircraft to the nation that was the catalyst for Pakistani nuclear ambitions.

This stance ignored a nagging sense that Pakistan was heading towards the bomb. Harold Wilson and the FCO feared that with India having "gone nuclear", Islamabad might decide to develop a nuclear capability, putting the NPT under increasing strain.[105] As one of the NPT's architects, a depository power, and the first state to ratify the treaty, this placed the UK in a tricky position. Washington shared the view that Islamabad had committed to the nuclear option, as State Department officials were "convinced that the Pakistanis would make every endeavour to develop a nuclear capability in as short a time as possible".[106]

The vexed question of Britain selling Pakistan arms as a means of heading off the nuclear quest was mired in misperceptions and financial problems. Concerns about the UK's balance of payments, the state of the British and Pakistani economies, and Islamabad's continued defaulting on previous deals meant that London preferred to deal in cash, not credit. The Pakistanis incorrectly believed that there was an embargo in place similar to the American one, but the British position on arms sales was less to do with geo-strategic concerns about the sub-continent and everything to do with the era's brutal economic necessities. As long as Britain could maintain a good relationship with India, there might be commercial opportunities in relaxing the attitude towards Pakistan. As the FCO's Richard Dales argued, Pakistan would get arms somehow and by denying

---

[104] Cabinet: Defence and Oversea [sic] Policy (Official) Committee, Sub-committee on Strategic Exports, 'Credit for Export of Defence Equipment to India; Note by the Secretaries', November 8, 1974, TNA FCO66/660, 2–4.

[105] Bridges to Dales, 'Pakistan', September 18, 1974, TNA FCO66/664; Thomson to Edmonds, 'Pakistan', October 25, 1974, TNA FCO66/664.

[106] 'Talks in State Department, October 2', October 11, 1974, TNA FCO66/660, 3.

sales Britain was depriving itself of further revenue.[107] In early December, OPD(O) clarified the arguments: economically, the Indian deal was not particularly financially advantageous to Britain, but was extremely valuable in keeping production lines open and avoiding the loss of upwards of 1,500 jobs in the aerospace industry; politically, the main roadblock was that an early decision on Jaguar could prejudice Britain's position at the May 1975 NPT Review Conference.[108]

The arms supply debate continued on both sides of the Atlantic into 1975. The invitation to make a state visit to America delighted Bhutto and the discussions that took place in the new year resulted in much more concrete and satisfactory outcomes for Pakistan.[109] In Britain, the debate over selling Jaguar to India and the potential non-proliferation implications rumbled on, complicating South Asian and transatlantic relationships.

BHUTTO'S VISIT

The drama of India's nuclear test and its catalytic effect on Pakistan was felt into 1975, as the Ford administration began to consider the embryonic Pakistani nuclear weapons programme in a more systematic way, using bribery and persuasion in an effort to lead Islamabad from the nuclear path. The British government continued to concentrate on global non-proliferation issues while engaging in an internal debate over the Indian Jaguar negotiations. Furthermore, 1975 saw the first NPT Review Conference, an event that promised much but resulted in very little.

American suspicions about a Pakistani bomb programme hardened prior to Bhutto's visit to the United States in February 1975. The ACDA's Robert Gallucci argued that although Pakistan's current nuclear infrastructure was not worrying, the Pakistani government's future intentions were a cause for concern. Most significantly, Pakistan had re-energised its negotiations with France for the purchase of a nuclear fuel reprocessing plant, a vital component of a complete fuel cycle and a

---

[107] Dales to Bridges, 'Defence Sales to Pakistan', September 11, 1974, TNA FCO66/664, 1–2.

[108] OPDO-SE, 'Credit for Export of Defence Equipment to India', December 6, 1974, TNA CAB148/152, 3.

[109] Kux, *Disenchanted Allies*, 217.

source of fissile plutonium. Gallucci asserted that: "Given their treaty status, their determination to purchase critical nuclear facilities, and their near declaratory policy of acquisition following the Indian detonation, they [Pakistan] may well have already decided to produce a weapon, and they clearly decided to have the capability to build one." According to the ACDA man, the reasons for the Pakistani quest for a bomb were multifaceted. The Indian explosion was the defining factor, but Gallucci argued that Bhutto might well use a nuclear explosion as a means of creating national unity, in much the same way as Indira Gandhi had done.[110] Gallucci also alluded to the possibility of "Arab financing" being available to Pakistan (Bhutto had enthusiastically courted wealthy, oil-producing Muslim states to revive Pakistan's ailing economy). Finally, Gallucci urged a modicum of caution on the linkage between arms supply and the nuclear issue, arguing that although the two should not be specifically tied together if it became clear that Pakistan was undeniably aiming for nuclear capability. Washington's policy should be "sensitive to such an important and unfortunate turn of events".[111] A few days later Gallucci's boss, Fred Iklé, re-emphasised these assertions, pointing out to Ford that France was preparing to export reprocessing plants and that Pakistan may be able to divert experienced scientists and engineers from its civilian facilities towards a bomb programme.[112]

In advance of Bhutto's arrival in the USA, the perception of Pakistan's ties with the Arab, or Islamic, world, Islamabad's negotiations with France over the reprocessing plant, and the solidifying belief in Bhutto's military nuclear intentions all dominated high-level briefings and discussions. An American aim for the visit was to extract a public statement from Bhutto not to develop nuclear explosives. Alfred Atherton (Assistant Secretary for Near Eastern and South Asian Affairs) argued that Pakistan *was* clearly trying to develop an independent nuclear fuel cycle and gain the requisite technical expertise to make "the nuclear explosion option" feasible.[113]

[110] 'Pakistan and the Non-proliferation Issue', January 1, 1975, IPNT, Doc.20, 1–3.

[111] Ibid., 3–5.

[112] Memcon, Ford and Iklé, January 6, 1975, GRFPL, NSAF, Memoranda of Conversations, 1973–1977, www.fordlibrarymuseum.gov/library/guides/findin gaid/Memoranda_of_Conversations.asp#Ford (accessed March 6, 2013), 2.

[113] Atherton to Kissinger, 'Your Meeting with Prime Minister Bhutto of Pakistan, February 5, 2:30 pm', January 31, 1975, DDRS, DDRS-255407-il-15, 3, 6.

Also highlighted were Bhutto's exploitation of Pakistan's Islamic ties, particularly when it came to economic aid from fellow Muslim countries in the aftermath of the 1973 oil crisis. Using such ties paid off for Pakistan, as states such as Saudi Arabia and Iran promised $400 million in assistance.[114] Even though certain elements within the State Department characterised Bhutto as being ambivalent about the quest for a nuclear bomb, Kissinger—in his final briefing to Ford before the meeting—sided with Gallucci and Atherton.[115] He informed the President that there was "considerable evidence that Pakistan is embarking on a program that could in time give it the capability of duplicating India's nuclear explosion last May".[116]

At the Ford-Bhutto meeting on the morning of February 5, the American side arrived having decided to lift the embargo. Although Bhutto and Ahmed were mildly aggrieved at the initially limited value of arms sales, they recognised that they had gained the concessions they had travelled 7,000 miles for. To smooth the passage of arms sales through Congress, Kissinger sought Bhutto's assurance that he would make a public statement promising nuclear restraint. After pressing his hosts to agree to substantial food aid and development loans, Bhutto was not exactly forthcoming, stating: "I am not enchanted by the grandiose notion that we must explode something, no matter how dirty, if our security needs are met. I want to spend the money on something else. We will have a nuclear programme, but if our security is assured, we will be reasonable."[117]

In the afternoon Kissinger had a further meeting with Bhutto and his key advisers, aiming to obtain non-proliferation assurances in exchange for US arms sales. Pakistani Foreign Secretary Agha Shahi claimed that because of pre-existing safeguards agreements on nuclear facilities, Pakistan could not clandestinely aim for a nuclear weapons option. Kissinger jumped on this morsel, asking if the public statement could be that no nuclear development

[114] Kux, *Disenchanted Allies*, 218.

[115] 'Briefing Notes On Bhutto', February, 1975 (exact date unknown), GRFPL, NSAF, Presidential VIP Briefings, Box 6, Pakistan–Prime Minister Bhutto (6) (hereafter PVIP-Bhutto), 1–3.

[116] Kissinger to Ford, 'Meeting with Zulfikar Ali Bhutto', February 1975 (exact date unknown), GRFPL, NSAF, PVIP-Bhutto, 3.

[117] Memcon, Bhutto and Ford, February 5, 1975, *FRUS* 69–76, Vol.E8, Doc.188, 3–7.

would take place outside of internationally agreed safeguards?[118] Bhutto equivocated and segued into a discussion of his moral objections to the NPT. Kissinger steered the conversation back to safeguards, but Bhutto then diverted into comments about PNEs. Kissinger argued that, for developing countries, there was no difference between a PNE and a bomb. In response the Pakistanis sought further US assurances for a South Asian Nuclear Weapon Free Zone (SANWFZ), something that Kissinger found problematic because of the Cold War imperatives of nuclear defence in Europe. At the end of this back and forth the Pakistanis had what they wanted, while the Americans failed to gain assurances from their guests not to develop a nuclear weapon.[119]

On February 24 the US government announced the repeal of the ten-year-old Indo-Pakistani arms embargo. Bhutto warmly thanked Ford and reaffirmed his commitment to the 'Simla Process' negotiations with India and avoidance of a South Asian arms race.[120] Although the Indian reaction was restrained, External Affairs Minister Yashwantrao Chavan cancelled a pre-arranged trip to the United States and Gandhi sharply criticised the United States and Pakistan.[121] The repeal was subject to limitations. Sales would be reviewed on a case-by-case basis, credit for arms was unavailable, and (at least in the initial stages) all weapons sold must be purely "defensive".[122] Although logically explicable, it was a bitter twist of fate for Indira Gandhi that India's own act of a setting off a nuclear explosion led to the delivery of US arms to Pakistan.[123]

[118] Memcon, 'The Secretary's Meetings with Prime Minister Bhutto', February 5, 1975, *FRUS 69–76*, Vol.E8, Doc.189, 5.

[119] Ibid., 6–7.

[120] Bhutto to Ford, Letter, February 24, 1975, GRFPL, NSAF, Presidential Correspondence Files, Box 3 (hereafter PCF3), Pakistan–Prime Minister Bhutto (1), 1.

[121] Sober to Kissinger, 'Indian Reaction to Arms Supply Decision–Chavan Postpones Visit', February 26, 1975, *FRUS 69–76*, Vol.E8, Doc.191.

[122] National Security Decision Memorandum 289, 'US Military Supply Policy to Pakistan and India', March 24, 1975, GRFPL, NSAF, Study Memoranda and Decision Memoranda, Box 1, 1.

[123] Perkovich, *India's Nuclear Bomb*, 186.

JAGUAR, AGAIN

While the United States was lifting the sub-continental arms embargo, potential warplane sales to India, and the impact this had on Pakistani perceptions, remained a major issue for British politicians and civil servants through 1975. Harold Wilson had visited Washington just before Bhutto's arrival, but other than a brief discussion with Kissinger of plans for the NSG and Wilson's expression of concern over the issue in general, non-proliferation was not on an agenda packed with discussions of oil, US-UK nuclear cooperation, the Middle East, and Britain's ongoing defence review.[124]

In preparation for Wilson's visit to America, his advisers had sought to push non-proliferation on to the agenda, at least in part because of alarming intelligence reports on the danger of more nations—including Pakistan—acquiring nuclear weapons. Wilson's briefing noted that the "prospects for a stable environment had deteriorated" in light of the Indian test and Israeli President Ephraim Katzir's recent allusion to his country's nuclear weapons, perhaps driving Arab and Islamic nations towards nuclear capability. The mantra in London was that "only strong international controls universally accepted and applied" could solve the proliferation problem.[125] Even though Britain was pushing ahead with plans for the 1975 NPT Review Conference and the NSG, Wilson's government still saw cooperation with the United States as the bedrock of any successful anti-proliferation activity. Wilson hoped that the Americans would approve the release of a major statement on non-proliferation during his upcoming visit to Moscow, a visit that, at its core, was about expanded British-Soviet commercial links.[126] Part of

---

[124] 'Note of a Meeting Held in the Oval Office, White House, on Friday, January 31, 1975: Nuclear Non-proliferation', February 3, 1975, TNA PREM16/1182; Memorandum of Conversation, Ford and Wilson, January 31, 1975, GRFPL, NSAF, Memoranda of Conversations, 1973–1977, www.fordlibrarymuseum. gov/library/guides/findingaid/Memoranda_of_Conversations.asp#Ford (accessed March 6, 2013).

[125] Joint Intelligence Committee (hereafter JIC) to Wilson, 'Nuclear Proliferation', November 15, 1974, TNA PREM16/1182.

[126] Hunt to Wilson, Memo, January 20, 1975, TNA PREM16/1182; Brearley to Wright, 'Nuclear Weapons and Non-proliferation', January 28, 1975, TNA PREM16/1182.

the American agenda was to re-emphasise to the British delegation the UK's importance as an actor in global affairs, affairs that included non-proliferation. US officials therefore raised no objections to the proposed UK-USSR communiqué on disarmament and non-proliferation.[127] This statement—included in the joint declaration that came at the end of the mid-February British-Soviet summit and couched in the usual bland goodwill of such events—stressed the significance of moves towards disarmament and the importance of multilateral talks as a means of achieving non-proliferation goals.[128]

Notwithstanding the Wilson government's non-proliferation enthusiasm, Jaguar remained a complicating factor. As ministers and civil servants debated the deal's financing, MPs posed difficult questions.[129] Minister of State for Defence William Rodgers glossed over the non-proliferation implications when quizzed in Parliament on January 23, preferring to concentrate on the benefits to British industry.[130] The British government hoped that the Indians would—in light of the upcoming changes to US arms sales policy to Pakistan—still agree to purchase the aircraft, despite being denied credit. This decision was conveyed to the Indians as one made on purely economic grounds.[131]

Despite the insistence to New Delhi that decisions being made about the sale were economic, there was an intense debate in Whitehall over the non-proliferation implications, coupled to mounting realisation that this was a matter of great interest to Pakistan. The SAD contended that it was

---

[127] Kissinger to Ford, 'Meeting with Harold Wilson', January 29, 1975, GRFPL, NSAF, Presidential Briefing Material for VIP Visits, Box 5, 1/29-2/1/75–United Kingdom–Prime Minister Wilson (7), 2; 'Note of a Meeting', February 3, 1975, TNA PREM16/1182.

[128] 'Extracts from Wilson-Brezhnev communiqué signed in Kremlin yesterday', *The Times* (hereafter *TT*), February 18, 1975, 6.

[129] Seaward to Wilford, 'Credit for defence Sales to India', January 30, 1975, TNA FCO37/1626.

[130] William Rodgers, House of Commons Debate, 'Jaguar Aircraft', January 23, 1975, *Hansard Online* http://hansard.millbanksystems.com/commons/1975/jan/23/jaguar-aircraft (accessed March 7, 2013).

[131] United Kingdom High Commission (hereafter UKHC) New Delhi to MoD, 'Jaguar Sales', February 24, 1975, TNA FCO37/1626; FCO to UKHC New Delhi, 'Jaguar', February 17, 1975, TNA FCO37/1626.

possible that the aircraft would form the core of a small nuclear strike force, but that India was too far away from producing a workable bomb for that to be a concern. Moreover, if Britain did not sell the aircraft, other competitors, such as France or Sweden, would.[132] SAD also pointed out that, even though the version of Jaguar the UK proposed for India would not—without substantial modification—be able to deliver nuclear weapons, the Pakistanis would *think* that the aircraft was adding to Indian nuclear capabilities.[133] With varying degrees of reservation, the FCO's South Asian, Science, and Defence departments all agreed that the sale should go ahead.[134]

The sale's most vocal critics were the ACDD, who consistently argued that the non-proliferation implications were significant. ACDD's contention that the sale was incompatible with Britain's overall non-proliferation stance was summarised in a comprehensive submission to ministers. They argued that, if the sale went ahead, the government would be criticised domestically and internationally as symbolically validating India's nuclear status. Pakistan was regarded as the nation most likely to attack the UK over the sale. Moreover, the deal could be perceived by the NPT's non-signatories as legitimising their status; currently, non-nuclear states could go nuclear if they wished, and Britain would still be happy to sell them weapons.[135]

Senior ministers were now left to make the final decision. The response from Callaghan came quickly: "Arms Control are right to put their case, but I think we cannot refuse the order."[136] Wilson, despite misgivings about the non-proliferation implications, signed off on the deal one week later.[137] London informed the British Embassy in Islamabad of the decision, but

[132] O'Neill to Richards, 'Sale of Jaguar to India', May 28, 1975, TNA FCO37/1626, 2.

[133] Dean to Seaward & O'Neill, 'Indian Air Force Capabilities', June 2, 1975, TNA FCO37/1626.

[134] O'Neill to Wilford, 'Sale of Jaguar Aircraft to India', June 5, 1975, TNA FCO37/1627, 2.

[135] O'Neill to Wilford, 'Sale of Jaguar Aircraft to India', Annex B, June 5, 1975, TNA FCO37/1627.

[136] Dales, 'Sale of Jaguar Aircraft to India', June 9, 1975, TNA FCO37/1627.

[137] Wright to Dales, 'Sale of Jaguar Aircraft to India', June 16, 1975, TNA FCO37/1627.

asked them to keep quiet as the FCO was well aware of Pakistan's interest in the case.[138] By the winter of 1975, arms control proponents within the FCO were still trying to argue that arms sales to India would negatively affect non-proliferation. A key contention was the impact on domestic and foreign opinion, with Pakistan highlighted as a key element of this. The main non-proliferation objection to the sale was based around:

> The potential effect on opinion in this country, on opinion amongst the committed non-proliferation countries, e.g. the US, USSR, Japan, and on opinion in the potential exploders, e.g. Pakistan, Argentina, etc. It is this effect on opinion in these areas rather than in India itself which is the core of the non-proliferation argument.[139]

This was only the beginning of an issue that would affect Britain's non-proliferation standing in the years to come, as the UK sought to halt the Pakistani nuclear programme.

MULTILATERAL ACTION

During 1975—as the Ford administration used arms sales as a non-proliferation tool and Britain debated the implications of its potential weapons deal with India—there were two multilateral meetings that carried the potential for significant changes in the non-proliferation landscape. The NSG's opening meeting set in motion export controls that would prove problematic for the Pakistani nuclear programme. The NPT's first review conference was a major disappointment that highlighted the rift between the nuclear world's haves and the have nots. The initial impact of these two international non-proliferation gatherings was relatively insignificant. However, the issues considered at these meetings became vastly more important to future US-UK attempts to combat the Pakistani nuclear weapons programme.

During late 1974 and early 1975 Kissinger began a process of consultation with other Western nuclear supplier nations in the hope of putting together a comprehensive 'trigger list' of technology not to be supplied to

---

[138] O'Neill to Imray, 'Sale of Jaguar to India', June 20, 1975, TNA FCO37/1627.

[139] Thomson to Mason, 'Credit for Export of Jaguars to India', October 8, 1975, TNA FCO37/1628.

non-nuclear weapon states (NNWS). This expanded upon the agreement reached in August 1974 by the Zangger Committee in an effort to prevent a repetition of the Indian test.[140] The NSG—consisting of the UK, USA, USSR, Canada, France, Japan, and West Germany—was itself one concrete outcome of the Indian test. Flustered by the ease with which India had used civilian technology to produce a nuclear weapon, the group was created—largely by Kissinger—to monitor the export of nuclear materials and technology from the advanced nuclear supplier states to other nations. The NSG also resulted from divisions within the Ford administration: the ACDA felt that the USA should take a unilateral lead in restricting exports; while Kissinger and others argued that a multilateral approach must be taken to avoid harming the domestic nuclear industry.[141] During April the 'London Club' met in private, because of a combination of Kissinger's predilection for clandestine diplomacy and French insistence on secrecy.

For informed observers such as the Stockholm International Peace Research Institute, the NSG's very existence, and the need for even more restrictions beyond the NPT, was an open admission that the treaty had failed.[142] More widely, Kissinger's insistence on secrecy boosted the belief in the developing world that the NSG was a conspiracy to prevent 'less-developed' states from gaining access to nuclear technology.[143] Meetings to formulate a list of sensitive items that the nuclear supplier states would control the export of took place throughout 1975 and 1976. As a result of these meetings, the trigger list of sensitive exports was eventually published in 1977.[144] The Group was not without its tensions

---

[140] The Committee was named for its first chairman, Professor Claude Zangger, and began meeting after the NPT's implementation in 1970. The NSG drew upon the Committee's work, but was not initially associated with it. For more, see Fritz Schmidt, 'NPT Export Controls and Zangger Committee', *Non-proliferation Review*, 7:3 (2000) 136–145.

[141] Perkovich, *India's Nuclear Bomb*, 191.

[142] Potter, *Nuclear Power and Nonproliferation*, 44–45.

[143] John Simpson, Jenny Neilsen, Marion Swinerd, and Isabelle Anstey (eds.), *NPT Briefing Book, 2012 Edition* (London: Kings College London Centre for Science & Security Studies, 2012), 18.

[144] Ian Anthony, Christer Ahlstrom, and Vitaly Fedchenko, *SIPRI Research Report No.22: Reforming Nuclear Export Controls: The Future of the Nuclear Suppliers Group* (Oxford: Oxford University Press, 2007), 16–18.

and problems. The West German announcement of the nuclear "deal of the century" with Brazil just as the NSG convened caused considerable congressional and public outcry in the United States.[145] The FRG-Brazil and the Franco-Pakistani reprocessing plant deals were the two major nuclear contracts of the 1970s that put the greatest strains on non-proliferation diplomacy.

In May 1975 the ninety-one states that had acceded to the NPT met in Geneva for the first Review Conference. Demands from non-nuclear weapon states that the nuclear powers take greater steps towards disarmament, as they were obligated to do under the treaty's Article VI, dominated the event. The conference's Final Declaration acknowledged this aim in the most general and non-committal terms.[146] The reaction to the conference was gloomy, with the USA and USSR considered to blame for the lack of progress towards genuine nuclear disarmament.[147] Britain was seen as less blameworthy because of statements that the UK would not move towards deploying a new generation of nuclear missiles. The conference's positive outcomes were that it once more made non-proliferation a major political issue and served notice on the superpowers that serious arms control negotiations must start moving, and start moving quickly.[148]

[145] William Glenn Gray, 'Commercial Liberties and Nuclear Anxieties', *International History Review*, 34:3 (2012), 456–457.

[146] Review Conference of the Parties to the Treaty on the Non-proliferation of Nuclear Weapons, Final Document, Part 1, 'Final Declaration of the Review Conference of the Parties to the Treaty on the Non-proliferation of Nuclear Weapons', May 30, 1975, United Nations Office For Disarmament Affairs, NPT Review Conferences and Preparatory Committees, http://www.un.org/disarmament/WMD/Nuclear/NPT_Review_Conferences.shtml (accessed on March 6, 2013), 8.

[147] Jack N. Barkenbus, 'Whither the Treaty?', *The Bulletin of the Atomic Scientists*, 36:4 (April, 1980); 'Gloom ends conference on nuclear proliferation', *TT*, May 30, 1975, 4.

[148] Keith D. Suter, 'The 1975 Review Conference of the Nuclear Non-Proliferation Treaty', *Australian Outlook*, 30:2 (1976), 337–340. The British statement was disingenuous. Although the Labour government had not committed to new missiles, it was undertaking a complex and expensive upgrade to the Polaris missiles that were the nuclear deterrent's backbone. See Kristan Stoddart, 'The Labour Government and the Development of Chevaline, 1974–1979', *Cold War History*, 10:3 (2010), 287–314.

While the immediate impact of these two international non-proliferation meetings was small, the issues they addressed became significant in later US-UK efforts to combat Pakistan's nuclear programme. The increased emphasis on sensitive nuclear exports—typified by the NSG trigger list—narrowed down Pakistani options and drove them towards a diffuse clandestine programme, in turn creating new non-proliferation challenges for Washington and London.

NUCLEAR EXPORTS AND ARMS DEALS

By mid-1975 Pakistani negotiations with France over the nuclear fuel reprocessing plant were well advanced and discussions were taking place with West Germany for a heavy water plant. Both facilities were cited as a critical expansion of Pakistan's civilian nuclear network in light of the oil crisis. This was part of a global trend, with expanded reliance on nuclear power seen as one way to meet the post-1973 energy predicament.[149] France wanted to make profits from nuclear deals with developing countries and was happy to enter into negotiations for a reprocessing plant that—to informed observers—far exceeded possible Pakistani civilian needs.[150] The FRG also possessed a significant commercial nuclear industry that—like France—"sweetened" its offers to developing world countries by adding sensitive fuel cycle technologies, a result of the fierce commercial competition between nuclear supplier states.[151] Despite these moves—and notwithstanding the warnings of analysts like Gallucci and agencies such as the Energy Research and Development Administration (ERDA), who warned that they had "hard intelligence" that Bhutto had ordered a PNE to be ready in four years—little was done by the USA to prevent Pakistan from acquiring these facilities during 1975.[152]

Aziz Ahmed continued to press the issue of Soviet influence in India. At a meeting of CENTO ministers in May, five days after the Indian

[149] Naim Salik, *The Genesis of South Asian Nuclear Deterrence: Pakistan's Perspective* (Oxford: Oxford University Press, 2009), 79.

[150] Feroz Hassan Khan, *Eating Grass: The Making of the Pakistani Bomb* (Stanford: Stanford University Press, 2012), 130.

[151] Clausen, *Nonproliferation*, 128; Potter, *Nuclear Power and Nonproliferation*, 109.

[152] Jones to Nettles, Letter, April 1, 1975, DNSA, NP01401, 1.

explosion's first anniversary, he gave a lengthy speech on creeping Soviet involvement in South Asia and the need for CENTO as a bulwark against India becoming the Soviet proxy in the region. Kissinger reiterated American commitment to Pakistan, commenting, "we believe the integrity of Pakistan to be an interest of the United States".[153] In private Kissinger sympathised with Ahmed's characterisation of the Indo-Soviet relationship, encouraging the Pakistani minister to seek funds for arms from Iran and Saudi Arabia and making plain his dislike of India and individual Indian politicians.[154]

Gerald Ford was also coming under Pakistani diplomatic pressure. After the CENTO meeting Bhutto wrote a series of letters to the President that re-emphasised the points made by his Minister and continued to position India as a pawn of Soviet expansionism.[155] Ford and Kissinger attempted to calm the histrionic nature of Bhutto's missives by pointing out that the United States did indeed have Pakistan's best interests at heart, pointing to the lifting of the arms embargo and the considerable amount of food aid as the signifiers.[156]

The various parties discussed arms sales during a September meeting between Kissinger and Ahmed. Uppermost in Ahmed's thoughts were the purchase of advanced A-7 Corsair attack aircraft. Kissinger was reticent about the sales of such powerful offensive weapons, despite the fact that they were obviously core to Pakistani demands.[157] In his briefing for an October 9 meeting with Ahmed, Kissinger informed Ford that Islamabad's interest in the A-7 should be discouraged and the emphasis

---

[153] Memcon, 'Restricted CENTO Ministerial Session', May 23, 1975, *FRUS* 69–76, Vol.E8, Doc.88, 3–4, 6.

[154] Memcon, Kissinger and Ahmed, May 22, 1975, GRFPL, NSAF, PCF3, Pakistan–Prime Minister Bhutto (1), 3–5.

[155] Bhutto to Ford, Letter, June 13, 1975, GRFPL, NSAF, PCF3, Pakistan–Prime Minister Bhutto (1); Bhutto to Ford, Letter, August 17, 1975, GRFPL, NSA, PCF3, Pakistan–Prime Minister Bhutto (1).

[156] Kissinger to Bhutto, 'Message for Prime Minister Bhutto', Telegram, August 30, 1975, GRFPL, NSAF, PCF3, Pakistan–Prime Minister Bhutto (1); Ford to Bhutto, Letter, October 15, 1975, GRFPL, NSAF, PCF3, Pakistan–Prime Minister Bhutto (2).

[157] Memcon, Kissinger and Ahmed, September 30, 1975, GRFPL, NSAF, PCF3, Pakistan–Prime Minister Bhutto (2), 3.

placed on defensive armaments. Kissinger recognised legitimate Pakistani security concerns, but argued that the bluster about an imminent, Soviet-backed, Indian attack was a fiction created for the purpose of leveraging more and faster American arms sales.[158] In all these high-level discussions, not once was the Pakistani nuclear programme raised, despite the fact that the State Department became increasingly concerned about Islamabad's nuclear ambitions in the final months of 1975.[159]

The CIA highlighted the significance of Pakistan's nuclear development and raised doubts over the capacity for arms sales to change the Pakistani course. In a follow-up to August 1974's Special National Intelligence Estimate (SNIE) 4-1-74, the CIA suggested that lifting the embargo "may have reduced Pakistan's motivation to develop nuclear weapons, but we believe it did not remove it. On balance, we conclude that the Pakistanis still intend to try to acquire a nuclear capability."[160] Assuming there would be a quick conclusion to Franco-Pakistani wrangling regarding reprocessing plant safeguards, it was estimated that Pakistan could produce a crude nuclear device as early as 1978. Significantly for what was to come, the report gave little attention to uranium enrichment as a route to the bomb. Its focus was almost exclusively on plutonium—the product of reprocessing—as a means for Islamabad to gain nuclear capability. Pakistan also cropped up in 'Managing Nuclear Proliferation: The Politics of Limited Choice', a downbeat assessment of the proliferation picture. This analysis argued that there was "no hope of preventing proliferation" and that proliferation was very much a political phenomenon "strongly influenced by the growing atmosphere of confrontation between the developed and less-developed countries". The CIA contended that the Indian nuclear test had smashed the barrier between civilian and military nuclear power. Coupled with the desire of nuclear supplier states to make money from selling advanced technology to the developing world, this made

---

[158] Kissinger to Ford, 'Meeting with Aziz Ahmed', October 8, 1975, GRFPL, NSAF, PCF3, Pakistan–Prime Minister Bhutto (2).

[159] Memcon, 'Apprehensions Regarding Pakistan's Nuclear Intentions', September 3, 1975, DNSA, NP01433.

[160] Memorandum to Holders of SNIE 4-1-74, 'Prospects for Further Proliferation of Nuclear Weapons', December 18, 1975, NSAEBB 'The United States and Pakistan's Quest for the Bomb' (hereafter USPQB), www2.gwu.edu/~nsarchiv/nukevault/ebb333 (accessed on March 6, 2013), Doc.1, 8.

proliferation more likely *and* more dangerous. The United States' and Britain's muted reaction to the Indian test had demonstrated the limited political costs of setting off a nuclear explosion.[161]

Furthermore, a new threat had raised its head, nuclear terrorism, the "most puzzling and extreme aspect of the potential diversification of nuclear actors". The possible acquisition of nuclear weapons by non-state actors was, argued the CIA, intimately connected to developing world proliferation: "The same increasing availability of nuclear materials and technology which has made nuclear explosives accessible to developing states can also be expected sooner or later to bring them within reach of terrorist groups."[162] Such prognostications were not the CIA's sole domain. The fear of non-state actors gaining access to nuclear materials had been on the rise since the early 1970s and would—in part—provoke a flurry of congressional non-proliferation legislation in 1976 and beyond. Taken together, these reports presented a downbeat picture. The barriers against proliferation were perceived as crumbling and, in Pakistan's case, there were obvious signs that nuclear weapons capability was high on the national agenda. The reports also highlighted an institutional blind spot when it came to uranium enrichment as a means to gain nuclear weapons capability. This blind spot was not confined to the intelligence community. It would plague many functions of government, inhibiting strategies dedicated towards stopping the Pakistani nuclear weapons programme.

Such gloomy reporting was part of a trend in proliferation analysis that had existed since the atomic age's earliest days. Ever since the 1950s the CIA in particular had been pessimistic about the spread of nuclear weapons. As John Mueller points out, the pace of proliferation was much slower than even the most optimistic forecasts of the intelligence community, governments, and think tanks.[163] Moeed Yusuf argues that, as with the cases of India and Pakistan, the "shift from developed-world proliferation to developing-world proliferation was accompanied by greater alarm regarding the impact of proliferation. It was felt that

---

[161] 'Managing Nuclear Proliferation: the Politics of Limited Choice', December 1975, 5–7

[162] Ibid., 9–14.

[163] John Mueller, *Atomic Obsession: Nuclear Alarmism from Hiroshima to Al-Qaeda* (Oxford: Oxford University Press, 2010), 89.

developing countries were more dangerous and irresponsible nuclear states than developed countries."[164]

Throughout 1975 American policy had done little, if anything, to prevent Pakistan from moving towards nuclear weapons. The lack of interest in the issue shown by Ford and Kissinger allowed the Pakistani government to continue their nuclear negotiation with little interruption or criticism. This *laissez-faire* attitude led to the gloomy prognostications of December 1975.

CONCLUSION

In spite of May 1974's drama, Britain and America did little to persuade Pakistan not to follow its neighbour down the nuclear path. Lack of criticism of India after the nuclear test, Washington's continued support of New Delhi's civilian nuclear programme, and London's willingness to sell India advanced aircraft implied that the political ramifications of sudden 'nuclear club' membership were slight. Faced with economic turbulence at home and aware of—but de-prioritising—proliferation considerations, London pressed ahead with efforts to sell what were seen as nuclear-capable strike aircraft to the nation that had pushed Pakistan towards nuclearisation. In pursuing the Jaguar sale the Wilson government compartmentalised arms sales and nuclear issues, unwilling to acknowledge the serious impediment to non-proliferation created by the Jaguar deal. Ford and Kissinger's acquiescence to Pakistani demands for arms—while gaining nothing in return—demonstrated their belief in the ability of arms sales to assuage regional security concerns and arrest proliferation, despite all the signs that Pakistan was actively pursuing a weapons programme. Furthermore, there was little—if any—consideration of the deep-seated political and territorial issues that were driving South Asian proliferation. In 1976 Washington and London woke up to the threat of Pakistan's nuclear project. Ford took the issue far more seriously

---

[164] Moeed Yusuf, *Predicting Proliferation: The History of the Future of Nuclear Weapons* (Boston: Brookings Institution, 2009), http://www.brookings.edu/~/media/Research/Files/Papers/2009/1/nuclear%20proliferation%20yusuf/01_nuclear_proliferation_yusuf.PDF (accessed on November 28, 2013), 4.

when pushed towards an anti-proliferation stance by Congress, public opinion, and his presidential rival Jimmy Carter. In London, there were the first inklings that the reprocessing plant was not the only way Islamabad was trying to get the bomb; that there might be a second, more secretive strand to Bhutto's efforts.

CHAPTER 3

"An end to the first 'easy' phase" Pakistan's Nuclear Reprocessing Plant Deal and the Clandestine Programme's Discovery, January 1976 to January 1977

On April 7, 1976 Henry Byroade, the American Ambassador to Pakistan, sent a telegram to Henry Kissinger encapsulating his thoughts about US engagement with the Pakistani nuclear problem. Byroade—overcoming a stated reluctance to comment on State Department non-proliferation policy—characterised nuclear matters as central to the American-Pakistani relationship in the wake of Bhutto's very public deal to purchase a nuclear fuel reprocessing plant from the French nuclear supplier Saint Gobain Nucléaire (SGN). Most tellingly, the Ambassador saw "an end to the first 'easy' phase of the exercise to lead Pakistan away from the nuclear option path".[1] Byroade's comments were both telling and prescient. From his vantage point in Islamabad, he saw that attempts to stop Pakistan's emergent nuclear weapons programme were only going to become more difficult as the decade progressed.

In 1976 the Ford administration radically reassessed its attitude towards non-proliferation. The years 1974 and 1975 had been a time of

---

[1] Byroade to Kissinger, 'Pakistan and Non-proliferation', April 7, 1976, US National Archives and Records Administration (hereafter NARA) Access to Archival Databases (hereafter AAD), 1976ISLAMA03497, 1–2.

© The Author(s) 2017   57
M.M. Craig, *America, Britain and Pakistan's Nuclear Weapons Programme, 1974–1980*, Security, Conflict and Cooperation in the Contemporary World, DOI 10.1007/978-3-319-51880-0_3

laxity on the non-proliferation front, with a stark absence of a critical response to the Indian test, Pakistan being rewarded with the US arms embargo's repeal despite offering no assurances about nuclear intentions, and gloomy intelligence prognostications about the prospects for further proliferation. For Ford and Kissinger, arms sales to Pakistan were a means of bribing or coercing the Pakistanis into abandoning their nuclear aspirations. By 1976 the Ford administration—realising that Pakistan was determined to pursue the nuclear option, pressured by congressional action on proliferation, and influenced by Jimmy Carter's foregrounding of the issue during the presidential election campaign—engaged more directly with the proliferation problem.

On the other side of the Atlantic, Britain rejected the opportunity to profit from the Franco-Pakistani reprocessing plant deal. More significantly for the future of British and American non-proliferation efforts, evidence emerged that clandestine Pakistani purchasing networks were feeding into a secret uranium enrichment programme. Pakistan's emergent clandestine programme thrived because of widely differing standards for what qualified as 'nuclear', the troubled atmosphere of US-UK-Western European relations in the 1970s, and the complex dimensions of globalisation that served to complicate international efforts to address a major nuclear proliferation risk.

The period from March 1976, when Islamabad and Paris sealed the reprocessing plant deal, to January 1977, when Carter assumed the presidency, represented a crossroads for non-proliferation policy. It was a bridge between the laxity of the Nixon and early Ford years and Carter's non-proliferation determination and, as Byroade had stated, the end of the easy phase and a transition to the more bitterly fought battles to come.

Reprocessing

Kissinger and Bhutto met in New York on February 26, but the traumas of the last few years meant that the Secretary of State cut a diminished figure in foreign policy circles.[2] He retained power and influence though, and high on the agenda for his meeting was Islamabad's deal with SGN to purchase a nuclear fuel reprocessing plant. The Pakistani-French

[2] Jussi Hanhimäki, *The Flawed Architect: Henry Kissinger and American Foreign Policy* (Oxford: Oxford University Press, 2004), 443–444.

negotiations had been ongoing since 1973 (well before the Indian test), but there had thus far been little in the way of open US criticism. Kissinger now attempted to dissuade Bhutto from walking further down the nuclear path.[3] Influenced by intelligence reporting and pushed by Congress, the State Department and the National Security Council (NSC) concluded that Pakistan might genuinely be aiming for nuclear weapons capability, with dire consequences for South Asia. During 1975 and 1976 the United States managed to persuade its allies South Korea and Taiwan not to press ahead with their respective nuclear projects. Both nations were dependent on the USA for their security in the face of communist 'aggressors', giving America considerable leverage.[4] In the case of Pakistan—far less of a 'client state' than either South Korea or Taiwan—the US government had considerably less leverage and negotiations promised to be a lot more difficult.

Prior to the Kissinger-Bhutto meeting, the State Department outlined concerns about the reprocessing plant. "The reprocessing plant is particularly disturbing," Bureau of European Affairs staffer David H. Swartz, stated, "since there can be no present (or, probably, future) economic justification for such a capability." Swartz noted that even if Pakistan acquired many more reactors that could be fuelled by the reprocessing facility, it was still not economically viable.[5] His inference was that the only possible justification for the plant was to provide plutonium for nuclear weapons. Under Secretary of State for Political Affairs Joseph J. Sisco—Kissinger's right hand man—expressed these concerns to Sahabzada Yaqub-Khan, Pakistani Ambassador to the United States. Sisco had already made these points to Kissinger, reflecting on the need for negotiations with Islamabad and Paris, and noting that intransigence over the nuclear issue could jeopardise the entire US-Pakistani relationship.[6] Sisco

---

[3] Memcon, 'The Secretary's Meeting with Prime Minister Bhutto', February 26, 1976, Digital National Security Archive (hereafter DNSA), KT01902, 23–26.

[4] Lyong Choi, 'The First Nuclear Crisis in the Korean Peninsula, 1975–76', *Cold War History*, 14:1 (2014), 71–90; Rebecca K.C. Hersman and Robert Peters, 'Nuclear U-turns', *Nonproliferation Review*, 13:3 (2006), 539–553.

[5] Swartz to Hartman, 'Demarche to Pakistan on Nuclear Fuel Reprocessing', January 30, 1976, National Security Archive Electronic Briefing Book (hereafter NSAEBB), Indian and Pakistan: On the Nuclear Threshold (hereafter IPNT), http://nsarchive.gwu.edu/NSAEBB/NSAEBB6, Doc.21, 1–2.

[6] Sisco to Kissinger, Memorandum, February 12, 1976, DNSA, NP01450.

outlined to Yaqub-Khan mounting international concern over the spread of reprocessing facilities and their impact on proliferation.[7] The Under Secretary did not intend to cast aspersions on Bhutto's assurances, but he could see little economic reason for the purchase of reprocessing and heavy water plants. Congressional pressure over the issue, Sisco added, could very well complicate bilateral arrangements—such as arms sales—between the USA and Pakistan. Yaqub-Khan stuck to Islamabad's official line that, although there was no immediate need for the facilities, future energy and development requirements necessitated nuclear expansion.[8]

On February 26 Kissinger re-emphasised to Bhutto the points made by Sisco and Swartz about the danger of national reprocessing facilities and dangled the carrot of enhanced arms supplies in exchange for Pakistani acquiescence to non-proliferation demands. Bhutto reiterated his request for the powerful A-7 attack aircraft, but Kissinger once again demurred, unwilling to sponsor a high technology arms race on the sub-continent. The Secretary of State was sympathetic to Bhutto's appeal, but asserted that a combination of election year attacks on Ford's foreign policy from within the Republican Party, congressional disquiet over lifting the arms embargo, and concerns that the reprocessing plant might constitute part of a bomb programme militated against a more forthcoming response. Sisco interjected that the Ford administration was moving as fast as it could, "consistent with what Congressional and public opinion will bear".[9] Once again, Bhutto argued that, if given strong enough conventional defences, Pakistan would not have to go down the nuclear route.[10] Kissinger, Ford, and their subordinates found themselves caught in a situation not unlike that of Captain John Yossarian in Joseph Heller's *Catch-22*. To try to prevent Pakistan from going nuclear, conventional arms sales were required to enhance Pakistani security and demonstrate US commitment. However, Islamabad's drive for nuclear capability meant that Congress was reluctant to allow the arms sales that would

[7] State Department (hereafter State) to United States Embassy (hereafter USE) Islamabad, 'Approach to Pakistan Concerning Sensitive Nuclear Facilities', February 19, 1976, *Foreign Relations of the United States, 1969–1976* (hereafter *FRUS 69-76*), Vol.E8, Doc.224, 1–2.

[8] Ibid., 4.

[9] 'The Secretary's Meeting with Prime Minister Bhutto', February 26, 1976, 23.

[10] Ibid., 25.

allegedly ameliorate the sense of insecurity at the heart of Pakistan's quest for the bomb.

The administration also made moves to utilise the trilateral US-Pakistani-Iranian relationship to persuade Islamabad to agree to a joint Iranian-Pakistani regional reprocessing plant that would bring the plutonium output under rigorous safeguards.[11] Although this eventually came to naught, early 1976 saw the overenthusiastic Aziz Ahmed engender false hopes that such a solution might flourish.[12] Finally, Kissinger brought diplomatic pressure to bear on France and the Federal Republic of Germany (FRG), with mixed results. The West Germans—already under serious pressure from the USA over their nuclear deal with Brazil—were happy to adopt a 'wait and see' attitude towards the heavy water plant's sale. The French were far more intransigent, with Foreign Minister Jean Sauvagnargues making it clear that there was no going back.[13]

This early engagement with the reprocessing plant issue emphasised that the United States possessed little leverage where it mattered. The hope that conventional arms shipments would shift Pakistan away from the nuclear path proved ill founded as, despite Bhutto's claims to the contrary, he was unwilling to set aside his nuclear ambitions. France also resisted pressure to cancel the reprocessing plant contract, but at the same time quietly offered the Pakistanis various options that Paris hoped would ensure the facility was only used for peaceful purposes. Islamabad refused, believing that Paris was acting under intense pressure from Washington, even though from the American point of view it seemed that there was little that could be done to persuade a nation that had been happy to sell nuclear technology to apartheid South Africa not to sell equipment to Pakistan.[14]

---

[11] Ibid., 25–26.

[12] Byroade to Kissinger, 'Sensitive Facilities for Pakistan', March 13, 1976, Gerald R. Ford Presidential Library (hereafter GRFPL), National Security Adviser Files (hereafter NSAF), Presidential Country Files for Middle East and South Asia (hereafter PCFMESA), Box 27, Pakistan–State Department Telegrams to SECSTATE–NODIS (3) (hereafter Pakistan 3); Byroade to Kissinger, 'Pakistan Nuclear Reprocessing Facility', March 17, 1976, GRFPL, NSAF, PCFMESA, Box 27, Pakistan 3,1.

[13] Kissinger to US Mission IAEA Vienna, 'Pakistan Reprocessing Plant', April 11, 1976, DNSA, NP01461, 2.

[14] Feroz Hassan Khan, *Eating Grass: The Making of the Pakistani Bomb* (Stanford: Stanford University Press, 2012), 131–132.

Kissinger found Congress to be one of the most significant factors inhibiting his freedom of action and a means of diverting blame for a lack of aid to Pakistan away from the administration. As the Secretary noted to Bhutto, "It is important that we manage affairs so that Congress does not pass resolutions on arms sales which would inhibit us from doing the things we want to do." Kissinger's pungent turn of phrase and opinionated manner came to the fore: "One of our problems," he suggested, "is that we have some maniacs in Congress. They are violent anti-communists but at the same time reluctant to give arms to friends."[15]

The President was also in a difficult position because of the ongoing Republican presidential primaries. With Ronald Reagan challenging Ford for the Republican presidential candidacy, taking a firm stand against Congress—and thus potentially alienating key supporters—was impossible until Ford's nomination was secure. The primaries became increasingly bitter during 1976, with Reagan attacking Ford's perceived weakness on foreign policy and stating that if he became President, Kissinger would be "sent packing".[16] Thus, the domestic context of an anti-proliferationist Congress coupled to a bitter election campaign significantly influenced wider non-proliferation policy.

It was the events of 1974 and 1975 that had provoked Congress into a flurry of legislation on nuclear proliferation. Three key reasons were the Indian test, growing awareness of the dangers posed by the spread of reprocessing plants, and the realisation that civilian nuclear programmes could be turned to military purposes. The perceived laxity of domestic nuclear security and rising concerns about the potential for nuclear terrorism also prompted congressional concern.[17] As Michael J. Brenner notes,

Congressional discovery that proliferation might be an imminent danger, and one encouraged by a less-than-vigilant U.S. government, prompted activists in both houses to make a lunge for the tail of the horse they visualised cantering out the open stable door. The action was frenetic, heroic

[15] 'The Secretary's Meeting with Prime Minister Bhutto', February 26, 1976, 25.

[16] Yanek Mieczkowski, *Gerald Ford and the Challenges of the 1970s* (Lexington: University Press of Kentucky, 2005), 314.

[17] J. Samuel Walker, 'Nuclear Power and Nonproliferation: The Controversy Over Nuclear Exports, 1974–1980', *Diplomatic History*, 25:2 (2001), 215.

in intention, and not always well-aimed, and it often sent the rescuers tripping over one another's feet.[18]

Wrangling over non-proliferation between the legislature and the executive truly began in 1975 with the introduction of the Export Reorganization Bill. Drafted by Senators Charles Percy (R-IL), Abraham Ribicoff (D-CT), and John Glenn (D-OH), the bill had a tortuous legislative history and would—after numerous alterations—finally become law in 1978. The crucial questions that the bill posed—who had authority to make determinations over nuclear exports and what standards should govern those decisions—lay at the heart of congressional interest in the proliferation problem and challenged the role of the executive branch in making foreign policy in the aftermath of the Vietnam War.[19] Ribicoff was a vocal critic of existing proliferation policy and the decision by US allies such as France and West Germany to sell sensitive technology, characterising the French and German sales as a "direct threat to world peace".[20]

The congressional measures that ultimately affected the US-Pakistani relationship most significantly were the Symington Amendment to the International Security Assistance Act and Arms Export Control Act of 1976 and its sibling, the Glenn Amendment. The June 1976 Symington Amendment banned US economic and military assistance to countries that delivered, received, or acquired nuclear enrichment technology outside of International Atomic Energy Agency (IAEA) regulations. The Glenn Amendment—passed in 1977—was a companion piece, covering reprocessing. The full impact of these measures—exactly the type of legislation that would inhibit Kissinger "from doing the things we want to do"— were not felt until well after Ford and Kissinger had departed, but they demonstrated that American lawmakers now appreciated that there was little to separate civilian and military nuclear power.

After French and Pakistani officials finally signed off on the reprocessing plant deal at a meeting in Paris on March 17, Ford attempted to address

[18] Michael J. Brenner, *Nuclear Power and Non-proliferation: The Remaking of U.S. Policy* (Cambridge: Cambridge University Press, 1981), 88.

[19] Ibid., 92.

[20] Ribicoff, Congressional Record–Senate, April 26, 1976, GRFPL, Glenn R. Schleede Files (hereafter GRSF), Box 27, Nuclear Policy, 1976: Background Material (2), s5873.

this major proliferation issue with a letter to the Pakistani Prime Minister. Assistant National Security Adviser Brent Scowcroft had advised Ford that Bhutto had rebuffed Kissinger's overtures in February and argued that Pakistan was pursuing the nuclear weapons option.[21] Prompted by Kissinger, Ford attempted to use the South Korean example to influence Bhutto and dwelt upon perception, rather than reality. "My concern is not the reliability of the assurances of your Government," he wrote, "It is that the establishment of sensitive nuclear facilities under national control inevitably gives rise to perceptions in many quarters that, under circumstances which perhaps cannot even be foreseen today, non-peaceful uses may be contemplated."[22] Despite the President's entreaties, Bhutto refused to postpone or cancel the nuclear programme. The Prime Minister's response, received by Kissinger in mid-April, was to repeat his previous assurances that nuclear facilities were for peaceful purposes and to emphasise the stringent IAEA safeguards that applied to the facilities.[23] Turning to the necessity for the reprocessing plant, Bhutto played on Western fears about energy security in the wake of the 1973–74 oil crisis by reminding Ford that Pakistan was poor in terms of indigenous resources and could barely afford to keep up with the vast increases in oil prices. Concluding, he attempted to reassure the President by mentioning his discussions with the Shah of Iran on regional reprocessing, but could not resist pointing out the fact that India's test had received little diplomatic criticism.[24] The lack of a stern response to the Indian detonation was coming back to haunt the US government in the form of Pakistani recalcitrance and congressional determination.

One month later, Byroade submitted his "easy phase" report to Kissinger. The Ambassador described Bhutto as determined to hold his present course, and persistent US pressure, especially of a negative sort, could damage US-Pakistani relations. Byroade argued that bribery through conventional arms sales was pointless and it would take more than a "couple

---

[21] Scowcroft to Ford, 'Letter to Bhutto on Pakistani Nuclear Issues', March 19, 1976, GRFPL, NSAF, PCF3, Pakistan–Prime Minister Bhutto (3) (hereafter P-PMB3).

[22] Ford to Bhutto, Letter, March 19, 1976, GRFPL, NSAF, PCF3, P-PMB3, 2.

[23] Bhutto to Ford, Letter, March 30, 1976, GRFPL, NSAF, PCF3, P-PMB3, 2.

[24] Ibid., 3–5.

of squadrons of A-7s" to buy Bhutto off.[25] Even though Alfred Atherton described the Byroade telegram as "thoughtful", Byroade never received a response from Kissinger as Bhutto's letter to Ford and the Prime Minister's hints to the media about a Pakistani nuclear weapons programme usurped the Ambassador's missive.[26] He was reduced months later to plaintively enquiring if his "very important message" ever reached Kissinger.[27]

THE REGIONAL DIMENSION

The prospects for a regional reprocessing plant involving Iran and Pakistan, which had been mooted in early 1976, formed part of ongoing diplomatic efforts to persuade Islamabad not to develop a national repro-cessing facility. Ford administration hopes that Bhutto could be brought round steadily diminished, despite the 'carrot' of arms supplies. Within the US government, the issue led to friction between Kissinger and his subordinates, especially the Arms Control and Disarmament Agency (ACDA), who were opposed to any concessions while the reprocessing issue remained unresolved.

The suggestion that the Shah of Iran—an enthusiast for civilian nuclear power as a modernising force—could be coopted to prevent Pakistan from acquiring its own reprocessing plant was taken seriously at the top level.[28] The Iranian Atomic Energy Agency Chief, Dr. Akbar Etemad, had visited the United States in April and expressed anxiety about Pakistani nuclear intentions. Etemad confided that, while the Shah was against a joint plant located in Pakistan (because of personal suspicions about a Pakistani bomb programme and a desire not to

---

[25] Byroade to Kissinger, April 7, 1976, 4.

[26] Atherton to Kissinger, 'Response to Byroade Cable on Pakistan and Non-proliferation', April 8, 1976 DNSA, NP01459, 1; 'Bhutto Hint That Pakistan Seeks Nuclear Weapons', *TT*, April 8, 1976, 6.

[27] Byroade to Kissinger, 'Pakistan and Nuclear Proliferation', June 8, 1976, DNSA, NP01470.

[28] On the Shah's nuclear ambitions, see Jacob Darwin Hamblin, 'The Nuclearization of Iran in the Seventies', *Diplomatic History* 38:5 (2014), 1114–35; David Patrikarakos, *Nuclear Iran: The Birth of an Atomic State* (London: I.B. Tauris, 2012), 14–89.

associate Iran with such a venture), the Emperor was willing to at least discuss some form of plant based in his country.[29]

Kissinger was suspicious of, and doubtful about, the idea of a regional reprocessing centre. He repeatedly argued that the Pakistanis and Iranians could collude to divert nuclear material from a joint plant, no matter where it was located, and characterised the endeavour as a "fraud".[30] Kissinger claimed to find the entire idea baffling, stating, "We are the only country which is fanatical and unrealistic enough to do things which are contrary to our national interests. The Europeans are not so illogical. If you go around the world, where can you find a region where the multinational concept would apply?"[31] The situation was made all the more urgent because Congress had approved the Symington Amendment the day before. Kissinger thought that it was "a bit rough" to expect Pakistan to adhere to the Amendment's requirements when India was expected to do no such thing, despite having already tested a nuclear device. Atherton upped the ante by noting that recent intelligence reports suggested Libya had agreed to partially finance Pakistan's reprocessing plant in exchange for some unspecified future nuclear cooperation.[32] This in itself was a fearful prospect because of Muammar al-Gaddafi's blustering pan-Arabism and anti-Americanism.[33]

Kissinger was eventually persuaded, in order to avoid having the new Symington Amendment damage US-Pakistani relations. Despite his doubts about the wisdom of the regional option, Kissinger approved moves to explore it, instructing his ambassador in Iran to meet with the Shah and press the case for a regional reprocessing venture.[34] The US representative at the time was Richard Helms, the former Director of

---

[29] State to USDEL Secretary, 'Action Memorandum–Consultation with the Shah On Pakistan's Nuclear Program', May 5, 1976, DNSA, NP01466, 1–2.

[30] Memcon, 'Proposed Cable to Tehran on Pakistani Nuclear Reprocessing', May 12, 1976, DNSA, WM00187, 1, 4.

[31] Ibid.

[32] Ibid., 3.

[33] Shane J. Maddock, *Nuclear Apartheid: The Quest for American Atomic Supremacy from World War II to the Present* (Chapel Hill: University of North Carolina Press, 2010), 289.

[34] Memcon, May 12, 4.

Central Intelligence whose reputation was taking a battering as a result of recent revelations about three decades of CIA misadventures.[35] Kissinger opined to Helms that the Symington Amendment's passage might affect close cooperation between Pakistan and the United States in non-nuclear areas such as arms supply.[36] Helms reacted quickly, speaking with the Shah on May 15. The Iranian ruler agreed with the assessment that there was no economic justification for the Pakistani plant, even going so far as to state that the French had told him so when he recently visited Paris. The Shah disparaged US efforts with Pakistan so far,

> You give them a few TOW missiles and odds and ends of defensive weapons but no A-7 aircraft which they badly need. Pakistan has no air force. Do they really care if your Congress gets mad at them? Suppose I do put pressure on Bhutto. What will he say? It seems clear that he is determined to obtain a reprocessing facility.[37]

Regarding the regional reprocessing option, the Shah was more in favour of an Iran-based multinational plant than a Pakistani-Iranian one. Nevertheless, there was an element of self-interest. As Helms stated, "The successful outcome of the current negotiations between USG [United States Government] and GOI [Government of Iran] on arrangements to build American nuclear power plants in Iran would seem to be a *sine qua non* if we expect any useful assistance from the Shah in dissuading Pakistan from its drive to obtain reprocessing facilities."[38]

[35] Christopher Moran, *Company Confessions: Revealing CIA Secrets* (London: Biteback, 2015), 228–29.

[36] 'Action Memorandum–Consultation with the Shah On Pakistan's Nuclear Program', May 5, 1976, 8–9. Despite rigorous research, the cable sent to Helms remains unavailable. The version outlined in the May 5 telegram remains the most complete and, cross-referencing with the May 12 Memcon and comparing it with Helms' response, appears to be the version that Kissinger approved for transmission.

[37] USE Tehran to State, 'Sensitive Nuclear Technology in Pakistan', May 16, 1976, NARA AAD, 1976TEHRAN04920, 2.

[38] Ibid., 2–3.

Henry Byroade, despite his confusion over whether or not his messages were reaching Kissinger, was instructed to make Bhutto aware of discussions with the Shah.[39] In his talks with the Prime Minister he "played the role of personal friend", surmising that there was no way to force Bhutto to publicly renounce the reprocessing plant.[40] The plant—and by extension Islamabad's wider nuclear programme—had become a crucial part of political and public debate in Pakistan. For Bhutto especially, the entire programme was deeply connected to the Pakistani people's morale and confidence, factors that were vital in the run-up to the 1977 national elections.[41] Byroade stressed the increasing pressure from Congress and other domestic interests, and emphasised that the Symington Amendment was symptomatic of significant changes in congressional attitudes.[42] Responding to Bhutto's questioning, the Ambassador expressed fears that the nuclear issue may well affect the entire bilateral relationship. Bhutto was aggrieved, suggesting once more that India had suffered little from its nuclear test and that Pakistan was suffering discrimination. Byroade suggested that the increased emphasis on Pakistan was a direct result of the Indian test and that, given the region's troubled history, it was only to be expected. Why, Bhutto asked, had the USA not raised the reprocessing plant issue before 1976? To back down now would be a political disaster for him. Byroade agreed with Bhutto on this and made his feelings known to Kissinger in the strongest possible terms.[43]

Kissinger's response was to stress that the United States was not discriminating against Pakistan nor attempting to retard its development. He couched his remarks in terms of the global non-proliferation situation, a stark change from his attitude in May 1974. Kissinger argued that Pakistani acquisition of "sensitive facilities" could harm the good relationship between the United States and Pakistan, mainly because of intense

[39] Kissinger to Byroade, 'Sensitive Nuclear Technology in Pakistan', May 20, 1976, NARA AAD, 1976STATE123889.

[40] Byroade to Kissinger, 'Bhutto and Ambassador Discuss Nuclear Proliferation Issue', June 4, 1976, GRFPL, NSAF, PCFMESA, Box 27, Pakistan: State Department Telegrams to SECSTATE EXDIS, 1, 7–8.

[41] Ibid., 4.

[42] Ibid., 1–2.

[43] Ibid., 3–4.

congressional and public pressure on the administration, and that, while the reprocessing plant was still on the table, it would be impossible to consider arms exports.[44] Despite this, discussions about arms sales formed a parallel track to the 'Iranian option'.

While the Iranian option was being explored, the administration continued to use arms sales to coerce Bhutto into abandoning the French deal. Internal divisions between Kissinger and the ACDA—and Pakistani demands for advanced weapons—made this a fraught process. Nominally, the ACDA was tasked with formulating and advocating for arms control and non-proliferation policies. During the Kissinger years the agency was effectively sidelined and bypassed because of the centralisation of responsibility for foreign policy. On April 7 the ACDA demanded a halt to all decisions concerning arms transfers—from two *Gearing* class naval destroyers to anti-tank missiles—to Pakistan until the reprocessing and heavy water plant situations were resolved. "Such acquisitions [the reprocessing and heavy water plants] would clearly contribute to nuclear proliferation", stated career foreign service officer William Stearman, and were "a matter of major arms control concern".[45] Kissinger and the State Department reacted to ACDA demands with barely contained fury, arguing the agency was attempting to impose its desired outcomes. George Vest (Director of the State Department's Bureau of Politico-Military Affairs) characterised ACDA's refusal to authorise arms transfers to Pakistan as "usurping the function of the Secretary of State to determine U.S. foreign policy". Furthermore, the "ACDA's arbitrary intervention in this matter can only complicate our diplomatic efforts to bring about in an orderly manner the result which we all wish to achieve, i.e., the cancellation of the Pakistani reprocessing plant".[46] Matters reached the stage where senior NSC staffer Robert Oakley urged Scowcroft to contact

---

[44] Kissinger to Byroade, 'Sensitive Nuclear Technology in Pakistan', June 10, 1976, GRFPL, National Security Council, Institutional Files, Box 96, Logged Documents–1976–Log Numbers 760388-7603801.

[45] Stearman to Vest, 'Proposed Arms Transfers to Pakistan', April 7, 1976L, GRFPL, NSAF, Presidential Agency File, 1974–1977, Box 1, Arms Control and Disarmament Agency, 5/10/76–8/2/76 (hereafter PAF-ACDA).

[46] Vest to Maw, 'ACDA's "Hold" on Conventional Arms Transfers to Pakistan (GOP)', May 13, 1976, GRFPL, NSAF, PAF-ACDA.

Fred Iklé and suggest that the destroyer sale would be passed to Congress for authorisation over his agency's objections.[47]

By coincidence, presumptive Democratic Party presidential candidate Jimmy Carter made a tough stance on proliferation a plank of his presidential campaign on the same day that Vest lambasted the ACDA.[48] The candidate-in-waiting argued that nuclear energy posed a dual threat to the world, through hazardous waste and the spread of weapons technology, dramatically referring to reprocessing plants as "bomb factories".[49] Unlike Nixon, Ford, and Kissinger, Carter linked domestic reprocessing and global nuclear proliferation into a major threat to US national security.[50] Carter called for an immediate, worldwide moratorium on the purchase and sale of reprocessing and uranium enrichment plants, even going so far as to suggest that the US reprocessing plant under construction at Barnwell, North Carolina become the first multinational, IAEA controlled facility.[51]

The triple threat of Congress, Carter, and public opinion forced the Ford administration to commission a study that became known as the Fri Report, after the study leader Robert W. Fri, Deputy Administrator of the recently created Energy Research and Development Administration (ERDA). Carter's pronouncements encouraged the review, but the Ford administration had to react to a combative Congress where "the atmosphere on the Hill is still punitive as regards nuclear export policy and there is suspicion of the administration's dedication to nuclear non-proliferation". It also had to account for broader public perceptions that the President and his administration were weak on proliferation issues, a perception that was leading many states of the union to enact

[47] Oakley to Scowcroft, 'ACDA and Destroyers for Pakistan', June 25, 1976, GRFPL, National Security Council Institutional Files, Box 96, Logged Documents–1976–Log Numbers 7603688-7603801.

[48] Peter A. Clausen, *Nonproliferation and the National Interest: America's Response to the Spread of Nuclear Weapons* (New York: HarperCollins, 1993), 132.

[49] 'Carter's Nuclear Ideas: Different', *NYT*, May 16, 1976, 139.

[50] J. Michael Martinez, 'The Carter Administration and the Evolution of American Nuclear Nonproliferation Policy, 1977–1981', *Journal of Policy History*, 14:3 (2002), 264.

[51] 'Carter Enters Nuclear Arena', The *Washington Post* (hereafter *WP*), May 14, 1976, A12.

prohibitive nuclear legislation.[52] Numerous bills and draft acts were flying around Capitol Hill long before Carter took the stage, many of which the State Department and ERDA thought diminished, rather than strengthened, the United States' position as a non-proliferation actor. A proposed amendment to the ERDA Authorization Bill that required congressional review of nuclear exports to non-NPT states was a stark example of this. According to the State Department and ERDA, success in non-proliferation stemmed from global perceptions of the United States as a stable and reliable nuclear supply partner. Only by maintaining such a position—and thus discouraging potential purchasers from looking to other supplier states such as France and West Germany—could the USA maintain its influence.[53]

ERDA's Director Robert C. Seamans pushed Ford for a wholesale non-proliferation review, arguing, "I believe there is an opportunity and a need for the United States to take a major initiative to resolve uncertainties that now exist in the nuclear fuel cycle and to reduce the risk of international proliferation of special nuclear materials."[54] Not only was the proliferation of nuclear weapons an issue, but the spectre of terrorists gaining access to plutonium through diversion from national reprocessing plants in the developing world—a point raised by the CIA in the 'Politics of Limited Choice' paper at the end of 1975—arose. Seamans urged the President to take the initiative and support the formulation of a new non-proliferation policy.[55] Thus, Ford—encouraged by Scowcroft—authorised a major,

---

[52] Evening Report (Scientific Affairs), 'Possible Presidential Message on Non-Proliferation and U.S. Nuclear Export Policy', June 17, 1976, GRFPL, NSAF, Situation Room, Evening Reports from the NSC Staff: June 1976–January 1977, Box 1, June 17, 1976; Scowcroft and Cannon to Ford, 'Possible Presidential Statement and New U.S. Initiatives to Reduce Proliferation Due to Commercial Nuclear Power Activities', June 22, 1976, GRFPL, James M. Cannon Files, 1975–1977 (hereafter JMCF), Box 24, Nuclear Policy Statement June–July 1976.

[53] 'Proposed Amendment to the ERDA Authorization Bill Requiring Congressional Review of Nuclear Exports to States Not Party to the NPT', May 25, 1976, GRFPL, GRSF, Box 27, Nuclear Policy, 1976: Exports, January-May, 1–2.

[54] Seamans to Ford, Letter, June 9, 1976, GRFPL, Philip W. Buchen Files, 1974–1977, Box 35, Nuclear Regulatory Commission (2), 1.

[55] Ibid., 3–4.

interagency study that the administration hoped would take the initiative back from Carter and Reagan.[56]

In the midst of all of this debate over new anti-proliferation policies, Kissinger's anger at the obstacles his subordinates were erecting against arms transfers to Pakistan came to a head on July 9 when he gave a furious dressing-down to Atherton and Iklé. After expressing disillusionment with the proposed Pakistan-Iran reprocessing plant, Kissinger said that he had "some sympathy for Bhutto in this. We are doing nothing to help him on conventional arms, we are going ahead and selling nuclear fuel to India even after they exploded a bomb and then for this little project, we are coming down on him like a ton of bricks."[57] Iklé argued that there were significant differences between the Indian and Pakistani nuclear projects. India had an extensive nuclear power network for which a reprocessing plant made sense. In the case of Pakistan, he argued it made no sense to purchase a reprocessing plant at this stage of its nuclear development.[58] Kissinger responded that Bhutto knew the technological details and knew what he wanted, to which Under Secretary of State for Political Affairs Philip Habib responded, "That's right. And what he wants is to build a bomb." Once again, Kissinger's pre-Indian test fatalism about the potential for, and risks of, nuclear proliferation resurfaced. The Secretary of State sympathised with the Pakistani leader; "If you were in his place, you would do the same thing."[59]

Arms sales provoked Kissinger's most vitriolic attack on his subordinates. Reflecting on his discussion with Yaqub-Khan, he commented that the way in which arms sales to a valued ally were being held up was "indecent". Iklé tried to justify his position with reference to the Symington Amendment, but Kissinger refused to agree; while A-7s were not the totality of what should be offered, they must be included in the package, as it was possible that the Pakistanis could go to Britain or France to buy equivalent equipment. Similar to Kissinger's reaction to the Indian

---

[56] Scowcroft et al. to Ford, 'Nuclear Policy–Issues and Problems Requiring Attention and Potential Policy Statement', July 10, 1976, GRFPL, GRSF, Box 27, Nuclear Policy, 1976: Presidential Decision Memo, July 10 (1).

[57] Memcon, 'Pakistani Reprocessing Issue', July 9, 1976, *FRUS 69–76*, Vol.E8, Doc.231, 3.

[58] Ibid., 3–4.

[59] Ibid., 4–5.

test, non-proliferation was not the only objective in this case. His main concern was regional political stability. The Secretary did not want the USA to be party to creating an imbalance of power that favoured India, a nation whose leader he intensely disliked. In the end, Kissinger concluded that "We have to find a package which is conceivably acceptable to France, such as substituting a reactor sale for the reprocessing plant." he stated, "If we give Bhutto A-7's in return for giving up the reprocessing plant, there would be unshirted hell to pay in France. If we can get them to switch a reactor for reprocessing, this could be completed with an overall agreement like that between the FRG and Iran."[60]

The NSC contended that Kissinger was giving mixed messages about proliferation. NSC staffers pointed out that the Secretary had approved a deal whereby the FRG and Iran would jointly operate a reprocessing plant constructed after a certain number of reactors were in operation. Writing to Scowcroft, David Elliot and Robert Oakley wondered if there was there an opportunity to persuade France and Pakistan to accede to a similar deal? To "sweeten the deal," Elliot and Oakley suggested Kissinger could offer the sought-after A-7s to Bhutto in order to gain his agreement.[61] The NSC staffers concluded:

> It is not clear that this new approach will offer the French and the Paks enough of a carrot (i.e., USG agreement to facilitate early reactor sales, withdraw our objection to eventual reprocessing along the lines of the FRG/Iran agreement, and A-7s) or a stick (i.e., withhold approval for the A-7, though continuing with other military equipment) to modify their present agreement on an early reprocessing facility. Pakistan clearly wants to have, in the near term, a nuclear capability comparable to that of India, including the possibility of making a nuclear explosive device.[62]

The fact that Kissinger was broadly against an Iranian-Pakistani joint reprocessing plant but in favour of a German-Iranian facility highlights the Secretary of State's known attitudes about the developing world.

---

[60] Ibid., 8–11.

[61] Elliot and Oakley to Scowcroft, 'Kissinger's Interim Decisions Regarding Pakistan's Nuclear Acquisition', July 12, 1976, GRFPL, NSAF, PCFMESA, Box 27, Pakistan (6), 1–2.

[62] Ibid., 3.

Developing world nations were susceptible to unstable regimes and too likely—according to that vision—to cheat, lie, and divert to make inter-state nuclear cooperation a viable option.

THE END OF THE LINE

Kissinger flew to Lahore in August to show that the administration was doing *something* to prevent Pakistan from going nuclear. He aimed to persuade Bhutto to abandon the reprocessing plant and carried with him a package of pressures and inducements.[63] Byroade briefed the Secretary before his arrival, noting that the Prime Minister was in something of a troubled mood. Any effort at further persuasion on the reprocessing plant front would, the Ambassador suggested, come up against Bhutto's personal involvement in the deal. Not only did Bhutto believe that his handshake with French President Valery Giscard d'Estaing constituted a binding agreement between two statesmen, but that there was no way he could admit to his people that he had made a mistake. The only way out was to get a package that could be positioned as an alternative, but beneficial, deal for Pakistan.[64] The strategy that Kissinger took with him to Pakistan was, as Dennis Kux points out, a combination of "carrots and sticks". Kissinger planned to offer the long-awaited A-7s as the carrot, and the stick was the suggestion that if Jimmy Carter won the upcoming election, the incoming President might well decide to make an example of Pakistan.[65]

Kissinger and Bhutto met on August 9, the Secretary informing Bhutto that he ran the risk of suffering far harsher treatment at the hands of Carter than if he agreed with the Ford administration's propo-sals.[66] Bhutto refused to back down, but in the coming months his diplomatic representatives strove to persuade Kissinger and Ford to adopt a compartmentalised approach, de-linking the A-7s from the

[63] Dennis Kux, *Disenchanted Allies: The United States and Pakistan, 1947–2000* (Washington D.C.: Woodrow Wilson Center Press, 2001), 222.

[64] Byroade to Kissinger, 'Secretary's Visit–Nuclear Matters', August 2, 1976, NARA AAD, 1976ISLAMA07858, 2–3.

[65] Kux, *Disenchanted Allies*, 222.

[66] Ibid., 223.

nuclear issue.[67] Indeed, subsequent diplomatic contacts in Washington emphasised the deep-seated commitment to the reprocessing plant in Pakistan. "Once the agreement [with France] had been signed, however," stated Ambassador Yaqub-Khan, "the GOP was locked in not only with France, but with third world public opinion."[68]

Kissinger gave a press conference at Lahore airport before his departure from Pakistan where he reflected on his discussions with Bhutto. In the manner of such events, the Secretary said as little as possible. Proliferation was not a Pakistan-focused issue, but a global issue, there was no dichotomy in the treatment of India and Pakistan, and the USA was not blackmailing Pakistan regarding arms sales.[69] The press reported that Kissinger seemed unconvinced by Pakistani claims that they were buying the reprocessing plant with peaceful intent, but he later explained his visit as an attempt to inform the Pakistanis about the Symington Amendment's implications.[70]

While Kissinger was flying back to Washington, Scowcroft approved the sale of a range of military equipment, such as armoured personnel carriers, missiles, and small arms—but not the A-7s—to Pakistan.[71] When Kissinger arrived home, he expressed his regret to Ford over his demands on Bhutto but was cognisant of the good it would do for Ford's public standing on the non-proliferation issue: "I had to take Bhutto on on the non-proliferation issue. It was too bad to do that to him, but if we didn't take a stand, it would get out of control. Actually it should be good for

---

[67] Kissinger to Byroade, 'Nuclear Reprocessing–A-7 Sales: Call by Pakistani Ambassador', September 8, 1976, GRFPL, NSAF, PCFMESA, Box 27, Pakistan–State Department Telegrams, From SECSTATE NODIS (2).

[68] Ibid., 1.

[69] State to USDEL Secstate, 'Dr Kissinger's August 9 Lahore Press Conference Text', August 10, 1976, DNSA, NP01482, 2–4.

[70] 'Pakistan resists pressure from Dr Kissinger to cancel purchase of French nuclear Plant', TT, August 10, 1976, 5; 'Conventional Arms Sales and the Nuclear Push', WP, August 15, 1976, C7; 'Dr. Kissinger explains move to stop nuclear deal', TT, August 11, 1976, 4.

[71] Granger to Scowcroft, 'Foreign Military Sales to Pakistan', August 11, 1976, GRFPL, NSAF, PCFMESA, Box 27, Pakistan (6).

you for me to have been tough on it."[72] Toughness aside, the American stance created waves in France, touching on deeply held ideas of sovereignty and independence. Outgoing French Prime Minister Jacques Chirac criticised Washington's perceived "hegemonic tendencies" and the French press almost universally condemned Kissinger for trying to ride roughshod over French independence and credibility.[73]

By mid-September, things came to a head for the Ford administration. With Carter confirmed as the Democratic Party's presidential candidate and Fri having submitted his findings, Ford needed to make a decision about non-proliferation policy. In a report that echoed the problems and solutions outlined by Candidate Carter, Fri argued that the oil crisis had created a wider global interest in—and market for—nuclear power. Coupled to this was an increased range of nuclear supplier states, not all of whom was the USA in agreement with. This increased choice of nuclear technology suppliers coincided with a loss of US influence in the market because of restrictive domestic legislation. Reprocessing, Fri contended, was a central issue. It was not only substantive, in that the plutonium fuel cycle raised genuine proliferation worries, but it was also symbolic. In light of these factors there needed to be more rigorous international controls over plutonium inventories, more effective safeguards against theft and diversion, further attempts to persuade states to foreswear reprocessing, and strengthened sanctions should the rules be broken.[74] Fri offered a graduated range of policy outlines, from business as usual, through acceptance of reprocessing as a fact of life, to a conservative stance that reprocessing should not proceed until it was

---

[72] Memcon, Ford, Kissinger, Scowcroft, August 13, 1976, GRFPL, NSAF, Memoranda of Conversations, online collection, www.fordlibrarymuseum.gov/library/document/0314/1553529.pdf (accessed April 25, 2013), 3.

[73] USE Paris to State, 'Chirac Statement on Franco-Pakistani Nuclear Sale', August 11, 1976, NARA AAD, 1976PARIS23376; USE Paris to USDEL Secretary Aircraft, 'More Press On the French/Pakistan Nuclear Plant Deal', August 9, 1976, NARA AAD, 1976PARIS23116; USE Paris to State, 'Press Reports US "Blackmail" Over Pakistan's Planned Purchase of French Nuclear Plant', August 9, 1976, NARA AAD, 1976PARIS23034.

[74] Fri to Ford, 'Nuclear Policy Review', September 7, 1976, Declassified Document Reference System (hereafter DDRS), DDRS272464i1-13, 1–7.

judged to be safe, to a "very tough line" that positioned reprocessing as a critical danger to world peace.[75]

There remained serious divisions over what the eventual policy should be. ACDA wanted the administration to take the "very tough line" against domestic and foreign nuclear reprocessing. The State Department, ERDA, and the Department of Defense (DoD) all supported a less confrontational option where reprocessing could proceed as long as stringent safeguards accompanied it.[76] Kissinger sought to place his stamp on matters by submitting his own thoughts to Ford the day *before* Fri announced the review's completion. In addition to advising advanced consultation with Canada, France, West Germany, Japan, Britain, and the USSR, he cautioned against statements that might place the USA in a serious strategic dilemma. Kissinger noted, "It should be recognized that if the [nuclear] suppliers, many of whom are also our allies, do not wish to follow a U.S. initiative voluntarily, then we will either have to coerce them or jeopardize our non-proliferation policy. Clearly, we should not select a strategy which could so easily trap us in such a dilemma."[77] The corollary to this strategic dilemma was obviously that non-supplier allies could pose the same problems. Pakistan was a case in point. Either Bhutto would have to be coerced into conforming or the ongoing situation would pose a serious challenge to non-proliferation. To head off such potential problems, Kissinger advised a policy that forced "sensitive facilities" such as reprocessing plants to be located only in supplier countries, all of which were developed world states.[78]

Carter pre-empted Ford's planned nuclear policy announcement by amplifying and expanding his position on non-proliferation. Ford, Carter contended, was "failing miserably" to set an example that would encourage other nations not to seek nuclear weapons.[79] In light of Carter's

[75] Ibid., 9–12.

[76] Scowcroft et al. to Ford, 'Nuclear Policy', September 15, 1976, GRFPL, JMCF, Box 24, Nuclear Policy Statement, September 15–30, 1976, 9.

[77] Kissinger to Ford, 'Nuclear Policy Review and Non-Proliferation Initiatives', September 6, 1976, GRFPL, GRSF, Box 27, Nuclear Policy, 1976: Agency Comments on Fri Report, 3.

[78] Ibid., 4.

[79] 'Carter: Sharp Criticism of Atom Policy', *WP*, September 26, 1976, A1.

stance, Ford's advisers suggested to the press that a major announcement on nuclear policy was in the pipeline. The *Washington Post* believed that Ford intended these suggestions to reduce Carter's advantage in the upcoming candidates' debate on foreign policy.[80] That debate saw Carter offer a scathing critique of Ford's non-proliferation policy, arguing that "only with the election approaching has Ford taken an interest" and citing the cases of Pakistan and Brazil as particularly alarming. If the Ford policy was continued, the Georgian contended, twenty new nuclear weapon states would emerge by 1985 or 1990.[81]

The reprocessing plant, the upcoming election, and arms sales were all problems that collided in the last meeting between Kissinger, Aziz Ahmed, and Sahabzada Yaqub-Khan before Ford's nuclear policy statement. Kissinger had convinced the Pakistanis that if Carter won the election, the new President would make an example of them. Furthermore, Yaqub-Khan tacitly admitted that a nuclear weapons programme was, at least in part, the reason for the reprocessing plant. Kissinger had stated, "But I don't think he [Bhutto] would need reprocessing now if he had only the intention of peaceful uses", to which Yaqub-Khan responded, "Your understanding of the Prime Minister's attitude was correct. I was wrong. The Prime Minister confirmed the understanding was as you had stated it."[82] This dramatic revelation, which contradicted Bhutto's previous claims of peaceful intentions, led Kissinger to discuss the strategic dilemma he had outlined to Ford on September 6. Kissinger argued that he was against much of the congressional legislation on proliferation because it harmed allies more than hurt enemies, and the more punitive action the United States took, the *more* proliferation would take place.[83] Finally, Kissinger stressed that he was doing what he could over the A-7 issue, even though he was fighting a "reluctant bureaucracy" and Congress, both of which took a dim view of advanced weapon sales to South Asia.[84]

---

[80] 'Ford Plans Strict Curbs on Atom Fuel', *WP*, October 3, 1976, 1.

[81] 'Transcript of Foreign Affairs Debate Between Ford and Carter', *NYT*, October 7, 1976, 36–37.

[82] Memcon, Ahmed, Yaqub-Khan, Kissinger, Habib, et al., October 6, 1976, DNSA, KT02102, 6–7.

[83] Ibid., 8.

[84] Ibid., 11.

On October 28, four days before the presidential election, Ford made his nuclear policy announcement, a statement that was the most extensive and detailed review of nuclear energy and proliferation yet made by a US president.[85] It was in itself ground-breaking and reflected many of Fri's recommendations, in particular the 'conservative' option on reprocessing. For the first time the Chief Executive acknowledged the fundamental links between domestic and foreign nuclear policy, the United States needing to be seen to lead the way on non-proliferation and nuclear safety.[86] Ford stated that this was not just a US responsibility, but also a global responsibility.[87] The President argued that the spread of reprocessing plants and their associated plutonium required arresting until the international community could guarantee the security of the process *and* the product. However, an understanding of the energy needs of the developing world also ought to be core to policy, "For unless we comprehend their real needs, we cannot expect to find ways of working with them to ensure satisfaction of both our and their legitimate concerns."[88] To achieve the statement's lofty goals, the United States needed to maintain its position as a reliable supplier of nuclear technology and resources. US reprocessing efforts would be placed on hold for three years but balanced by an expansion of uranium enrichment facilities to ensure reliable global supplies.[89] The parts that potentially had the greatest impact on Pakistan came within Ford's remarks on export policies. Although adherence to the NPT was the favoured option, there were a number of get-out clauses. The United States would look favourably upon non-NPT

---

[85] Brenner, *Nuclear Power*, 113.

[86] 'Statement by the President on Nuclear Policy', October 28, 1976, GRFPL, GRSF, Box 30, Nuclear Policy, 1976: Presidential Statement–Press Release October 28, 5.

[87] Ibid., 1.

[88] Ibid., 2.

[89] Domestic reprocessing was already in a parlous state when Ford outlined his new policy. Environmental legislation, changes in government policy, and lobbying by pressure groups such as the Natural Resources Defense Council had already hobbled the nascent developments by the time Ford's statement put the final nail in the coffin of American reprocessing. See John L. Campbell, *Collapse of an Industry: Nuclear Power and the Contradictions of U.S. Policy* (Ithaca: Cornell University Press, 1988), 118–119.

states adhering to full safeguards on their nuclear fuel cycle and foregoing or postponing national reprocessing or uranium enrichment facilities.[90]

At the moment Ford was outlining his plans, British officials from the ACDD—under the aegis of Ambassador Ramsbotham—were meeting with Carter supporters to discuss non-proliferation issues.[91] Ford's brave new world of non-proliferation never eventuated under his control. Four days later he lost the presidential fight and the former nuclear engineer, peanut farmer, and Georgia Governor Jimmy Carter became President-elect.

The Ford administration was never wholly committed to a non-proliferation strategy towards Pakistan, only adopting a non-proliferation stance when congressional, electoral, and public pressure made it unwise to ignore the issue. Even then, Kissinger's response lacked clarity, frequently blaming Congress for the problematic nature of proliferation policy. His lack of commitment concerning Pakistan is demonstrated by the continued efforts to supply A-7s, even when it became apparent that the nuclear programme was military in nature and that Bhutto was unlikely to give up the reprocessing plant. The Ford administration—like many other governments the world over—viewed proliferation through a realist lens, seeing Pakistani ambitions as solely driven by national security imperatives. If Pakistan were reassured—so the thinking went—then Bhutto would cast aside his nuclear ambitions. However, as Jacques Hymans and Peter Lavoy contend, insecurity is never the *sole* driver in the quest for the bomb.[92] Domestic politics, the ephemeral notion of 'prestige', and the proclivities of individual leaders all feed into a complex

[90] 'Statement by the President on Nuclear Policy', October 28, 1976, 10.

[91] John C. Edmonds, interview by Malcolm McBain, May 21, 2009, Churchill College Cambridge British Diplomatic Oral History Programme (hereafter BDOHP), www.chu.cam.ac.uk/archives/collections/BDOHP/Edmonds.pdf (accessed May 3, 2013), 25.

[92] See Jacques E.C. Hymans, *The Psychology of Nuclear Proliferation: Identity, Emotions and Foreign Policy* (Cambridge: Cambridge University Press, 2006), 1–15; Peter Lavoy, 'Nuclear Myths and the Causes of Nuclear Proliferation', *Security Studies*, 2:3 (1993), 192–212; Stephen M. Meyer, *The Dynamics of Nuclear Proliferation* (Chicago; University of Chicago Press, 1984); Scott Sagan, 'Why Do States Build Nuclear Weapons? Three Models in Search of a Bomb', *International Security*, 21:3 (1996/97), 54–86.

network of factors.[93] The administration's—and particularly Kissinger's—attitude towards Pakistan was informed by a view of the developing world that had evolved during the Nixon years. As Michael Hunt argues, "Washington continued to see it [the developing world] as a collection of nations that were backward, plagued by turbulence, and led by touchy, stubborn men."[94] Few developing world leaders could be as touchy and stubborn as Bhutto, especially when it came to the nuclear issue.

In the interregnum, Ford's outgoing administration continued to work on the Pakistani issue. By the middle of December, the French government had committed not to sell any more reprocessing plants, but remained wedded to the Pakistani one.[95] Kissinger was determined to make a watertight arrangement to deliver the A-7s, regardless of the future Carter administration's desires, if Bhutto would even postpone the reprocessing plant. Kissinger promised to speak to incoming Secretary of State Cyrus Vance in an effort to ensure Pakistani arms supplies.[96] Kissinger later suggested that Vance would have "six heart attacks" when he saw the proposed arms package for Pakistan. Despite this, Kissinger displayed a misplaced confidence that he could persuade the new administration that the A-7s for reprocessing deal was valid and valuable.[97]

Thirteen days after Kissinger expressed confidence that he could bring the incoming administration round to his position Carter was inaugurated. Bhutto was still adamant that he needed the reprocessing plant *and* the A-7s. The issue was far from dead but, as administrations transitioned in the United States, junior officials in the British government made a startling discovery that placed Pakistan's ambitions in a whole new light.

[93] Nick Ritchie, 'Relinquishing Nuclear Weapons: Identities, Networks and the British Bomb', *International Affairs*, 86:2 (2010), 465.

[94] Michael H. Hunt, *Ideology and U.S. Foreign Policy* (New Haven: Yale University Press, 1987), 183.

[95] 'France Bans Sale of Atom Fuel Plants', *TT*, December 17, 1976, 5; 'Canada Ends Nuclear Links With Pakistan', *TT*, December 24, 1976, 4.

[96] Memcon, Kissinger, Robinson, Byroade et al., 'Pakistan Nuclear Reprocessing', December 17, 1976, DNSA, KT02151, 1–2.

[97] Memcon, Kissinger, Robinson, Sonnenfeldt, et al., 'Non-Proliferation: Next Steps on Pakistan and Brazil', January 7, 1977, DNSA, KT02157, 1–2.

BRITAIN, REPROCESSING, AND JAGUAR (AGAIN)

Direct British involvement with Pakistani nuclear affairs in 1976 was—in comparison to the United States—limited. The ramifications of the oil shocks and domestic economic turmoil continued to significantly influence non-proliferation policy towards the sub-continent. It was also a time of political tumult, as Harold Wilson announced his resignation on March 16.[98] Wilson's resignation took place at a time of gathering crisis brought on by sterling's plummeting value from March onwards.[99] James Callaghan—considered by senior civil servants in the Foreign and Commonwealth Office (FCO) to be astute on foreign policy matters— won the leadership contest and moved into 10 Downing Street.[100] His choice as Foreign Secretary was Anthony Crosland, one of Labour's heavyweight intellectuals, who remained in the job for a mere ten months before his death in February 1977.[101] The year 1976 also saw Britain suffer international economic ignominy, as Callaghan's government was forced to turn to the International Monetary Fund (IMF) for aid during the 'IMF Crisis'.[102]

British reactions to the French-Pakistani deal were of a different nature to those of the US government. Some quarters saw the contract as a potential opportunity for British industry, despite misgivings about the nuclear programme's true nature. The Americans had informed their British allies about the February 26 Kissinger-Bhutto meeting, suggesting

[98] Kenneth O. Morgan, *Callaghan: A Life* (Oxford: Oxford University Press, 1997), 475.

[99] Ben Pimlott, *Harold Wilson* (London: HarperCollins, 1993), 724.

[100] Bryan G. Cartledge, interview by Jimmy Jamieson, November 14, 2007, BDOHP, http://www.chu.cam.ac.uk/archives/collections/BDOHP/ Cartledge.pdf (accessed July 10, 2013), 34.

[101] Rhiannon Vickers, *The Labour Party and the World, Volume 2: Labour's Foreign Policy 1951–2009* (Manchester: Manchester University Press, 2011), 111; Michael Palliser, interview by John Hutson, April 28, 1999, BDOHP, www.chu.cam.ac. uk/archives/collections/BDOHP/Palliser.pdf (accessed July 26, 2013), 43.

[102] For succinct analyses of the IMF Crisis, see Kathleen Burk, 'The Americans, the Germans and the British: the 1976 IMF Crisis'; *Twentieth Century British History*, 5:3 (1994), 351–369; and Steve Ludlam 'The Gnomes of Washington: Four Myths of the 1976 IMF Crisis', *Political Studies*, 40:4 (1992), 713–727.

that there was "a 20 per cent chance" of persuading the Pakistanis to back away from the deal.[103] In the aftermath, communications from Britain's Washington Embassy re-emphasised American concerns about the potential role of a reprocessing plant in a nuclear weapons programme. This had enough of an impact on British officials to make them think twice about potentially providing reprocessing technology to allies such as Iran.[104] The Joint Intelligence Committee (JIC) accurately assessed that the Pakistanis had embarked upon a nuclear weapons programme before India's test, New Delhi's actions merely reinforcing Islamabad's decision.[105] Like January's American assessments, the JIC argued that there was no economic need for the reprocessing plant, even in the long term, and its only purpose was to extract plutonium for use in nuclear weapons, although this was an almost impossible task due to Pakistani acceptance of international safeguards on the plant and reactor facilities.[106] In the final analysis, the reprocessing plant was positioned more as a means of accessing technology and gaining expertise, than as a direct route to the bomb. In the long term, Pakistan might well repudiate safeguards agreements and employ the French plant to produce weapons-grade material.[107] Within the report were the seeds of what was to come. Although the intelligence analysts might not have known it, their comments on natural uranium reserves and the potential for a Pakistani uranium refining plant hinted at the clandestine uranium enrichment programme discovered later that year.

Arms sales also influenced affairs. Spurred on by Anglo-Indian wrangling over Jaguar and seemingly barred from purchasing the sought-after A-7 from the United States, Pakistani ministers had turned a speculative

[103] Reeve to Wilmshurst, Letter, February 26, 1976, The National Archives of the UK (hereafter TNA) Records of the Foreign and Commonwealth Office (hereafter FCO) 96/575.

[104] United Kingdom Embassy (hereafter UKE) Washington to FCO, 'Reeve's Letter of 26 February to Wilmshurst: Sale of Reprocessing Plant to Pakistan', March 2, 1976, TNA FCO96/575; Wilmshurst to Edmonds and Thomson, 'Sale of Reprocessing Plants: US Views', March 8, 1976, TNA FCO96/575.

[105] Cabinet: JIC, 'Pakistan's Nuclear Programme', March 18, 1976, TNA FCO96/575, 1.

[106] Ibid., 2–4.

[107] Ibid., 4.

eye towards their own Jaguar deal. It was an issue bound up in rumour and misunderstanding. Agha Shahi—labouring under the convenient misapprehension that India was being offered generous credit terms for its potential purchase—had asked for £100 million credit to purchase the aircraft. British representatives in Islamabad angrily rejected the embarrassing Pakistani request for credit.[108] Despite Whitehall officials having to speak "firmly" to Pakistani representatives, the FCO noted that there was no political reason why the aircraft could not be sold to Pakistan and India, but both sales must be on equal, but far from generous, terms.[109] By May the Pakistanis were still interested in Jaguar but with Indira Gandhi mired in a 'state of emergency' and her ministers repeatedly modifying their requirements and failing to make a decision on the British, Swedish, or French options, and the British government unwilling or unable to offer more generous credit terms, Foreign Secretary Crosland did not pursue the potential Pakistani deal with any enthusiasm.[110] Here was a stark example of the radically different ways America and Britain viewed conventional arms sales. For the USA, conventional arms were either a stick or a carrot to push Pakistan away from the nuclear option. In London, conventional arms sales were only ever a commercial matter. Indeed, the Indian Jaguar negotiations were an impediment to non-proliferation, rather than a means of positively influencing nuclear attitudes.

British commercial interest was piqued when worldwide concern about the reprocessing plant's seemingly unnecessary nature caused Islamabad to announce a raft of new nuclear investment. Munir Ahmad Khan of the Pakistan Atomic Energy Commission (PAEC) held a press conference on March 22, outlining plans for twenty-four new nuclear power stations in Pakistan by the twenty-first century. This led to a remarkable discussion of the potential for UK companies to profit from Pakistan's new-found nuclear largesse. The British Embassy in the Pakistani capital noted that Khan's announcement was obviously intended for international, rather

---

[108] Pumphrey to O'Neill, 'Jaguars for India', February 16, 1976, TNA FCO37/1791, 1.

[109] 'Note for the File: Call on Mr Cortazzi by the Minister, Pakistan Embassy, March 2nd', March 5, 1976, TNA FCO37/1791, 1; O'Neill to Pumphrey, 'Jaguars for India', March 3, 1976, TNA FCO37/1791, 1.

[110] FCO to UKE Islamabad, 'Jaguars and India', May 6, 1976, TNA FCO37/1791.

than domestic, consumption.[111] The Islamabad Embassy's Commercial Secretary Harry Turner sought out Khan to collect more intelligence about the massive expansion plan. Khan appealed to his visitor's commercial instincts by outlining the opportunities available for British business in the expanding world of Pakistani atomic power. Khan suggested that British tenders for parts of the reprocessing plant and the new power reactor at Chashma would be welcomed.[112] Turner passed this intelligence to the Department of Trade (DoT) in Whitehall with the recommendation that the state-owned British Nuclear Fuels Limited (BNFL) contact Khan to arrange exploratory meetings.[113] The FCO intervened and wrote to Turner about potential constraints on British involvement in the nuclear power programme, enclosing a recent statement made by Callaghan in answer to a written parliamentary question. Callaghan had emphasised that when considering the export of sensitive technology, non-proliferation issues must always be a top priority for the British government.[114]

Other factors influenced British reticence to get too involved in Pakistan's nuclear affairs. The Labour government was in the midst of secretly undertaking a reversal of its previous policy not to develop new weapons, a policy that had shielded it from criticism at the 1975 NPT Review Conference. As the United States was pressuring Pakistan, London authorised the conversion of Chapel Cross nuclear power station in Dumfriesshire to a tritium production plant for a new generation of nuclear warheads. This, and the improvements to the Polaris missile system being planned under the codename *Chevaline*,

---

[111] Turner to Ashwood, 'Pakistan's Nuclear Power Programme', March 25, 1976, TNA FCO96/575.

[112] 'British Embassy–Commercial Section–Islamabad: Visit Report', April 1, 1976, TNA FCO96/575, 1–2.

[113] Turner to Ashwood, 'Pakistan's Nuclear Power Programme', April 5, 1976, TNA FCO96/575.

[114] James Callaghan, Written Answer, 'Nuclear Equipment and Technology', March 31, 1976, *Hansard Online* hansard.millbanksystems.com/written_answers/1976/mar/31/nuclear-equipment-and-technology (accessed April 21, 2013); Robin Fearn, Transcript, interview by Malcolm McBain, November 18. 2002, BDOHP, www.chu.cam.ac.uk/archives/collections/BDOHP/Fearn.pdf (accessed May 3, 2013), 29.

was hidden from parliamentary scrutiny by disguising the new developments as 'maintenance'.[115]

Not only were upgraded nuclear weapons under development, but plans were in train to build a massive new reprocessing facility, the Thermal Oxide Reprocessing Plant (THORP) at Windscale in Cumbria. The apparent disregard for the wider implications of this new facility was challenged in the House of Commons in late 1976. The reasons for objections were many and varied, from the risk of radioactive pollution to the competence of the planning officers who judged the application to build the plant.[116] Conservative MP Nigel Forman, in a speech lambasting the government's desire for a reprocessing plant, linked the Pakistan situation directly to Western commercial desires, stating:

> [T]here is the distinct danger that this lethal technology will spread to Brazil and Pakistan, unless steps can be taken at the highest international level to undo the damage already done by cut-throat commercial competition with its ruthless disregard for the long-term survival chances of mankind—and I put it no lower than that. All this adds up to a disturbing picture of a world drunk on the prospect of nuclear power and desperate for a slice of the Faustian bargain so thoughtlessly entered into by the Americans, the British and the Canadians more than 30 years ago.[117]

This questioning—coupled to internal debates—led to a lengthy public enquiry and, in 1977, collision with the Carter administration's non-proliferation policy objectives.[118] Cabinet Secretary John Hunt highlighted the international implications to Callaghan in late 1976. Reflecting

[115] Roger Ruston, *A Say in The End of the World: Morals and British Nuclear Weapons Policy, 1941–1987* (Oxford: Clarendon Press, 1989), 176–177.

[116] See questions by Nigel Forman, David Steel, and Robin Cook, November 3, 1976, *Hansard Online* hansard.millbanksystems.com/commons/1976/nov/03/windscale-nuclear-reprocessing (accessed April 25, 2013); David Penhaligon, Speech, December 14, 1976, *Hansard Online* hansard.millbanksystems.com/commons/1976/dec/14/nuclear-fuel-reprocessing-windscale (accessed April 25, 2013).

[117] Nigel Forman, Speech, December 20, 1976, *Hansard Online* hansard.millbanksystems.com/commons/1976/dec/20/nuclear-power (accessed April 25, 2013).

[118] William Walker, *Nuclear Entrapment: THORP and the Politics of Commitment* (London: Institute for Public Policy Research, 1999), 13–14.

on the "plutonium economy" concerns that Ford's nuclear policy speech had highlighted, Hunt cautioned that there were significant political and proliferation implications of embarking on the THORP project, implications that included spurring global proliferation and 'nuclear terrorism'. These were counterbalanced by the important investment and employment that would be generated by the project.[119] The Treasury argued that backing out of the one nuclear supply area where the UK had demonstrable advantage—the provision of reprocessing services—would lead to a loss of Britain's global credibility.[120] These arguments would reappear over the next few years, as alleged Western double standards came up repeatedly in the relationship with Pakistan, as did the specific case of British hypocrisy regarding reprocessing, a hypocrisy rooted in the conflict between wider political and narrower commercial interests.

By the end of May, the question of whether or not Britain should, or could, participate in Pakistan's nuclear industry led to a vigorous debate between the FCO, DoT, and the Department of Energy (DoE) that lasted until the end of the year. The DoE argued that it was doubtful whether Britain had anything to sell the Pakistanis anyway.[121] The FCO challenged this assertion, but pointed out that the political difficulties of selling nuclear equipment and services to Pakistan were the main issue.[122] Finally, officials at the Export Credits Guarantee Department pointed out that Pakistan's substantial debts to the UK were not conducive to Islamabad spending even more money on nuclear equipment.[123]

Michael Wilmshurst of the FCO's Joint Nuclear Section (JNS) also contacted Christopher 'Kit' Burdess at the Embassy in Islamabad. As the resident nuclear expert and a colleague of Harry Turner (who had started the ball rolling with his March and April missives), Burdess was asked to keep an eye on the Embassy's commercial department and to make sure

---

[119] Hunt to Callaghan, 'Windscale: Nuclear Waste Proposal', December 15, 1976, TNA Records of the Prime Minister's Office (hereafter PREM) 16/1059, 1.

[120] Barnett to Callaghan, 'Windscale—Nuclear Waste Proposals', December 6, 1976, TNA PREM16/1059.

[121] Chew to Hamilton, 'Pakistan's Nuclear Power Programme', June 4, 1976, TNA FCO96/575; Holt to Hamilton, 'Pakistan', June 16, 1976, TNA FCO96/575.

[122] Cox to Broughton, 'Nuclear Power in Pakistan', June 23, 1976, TNA FCO96/575.

[123] Coggins, ECGD Memo, June 16, 1976, TNA FCO96/575.

his consular colleagues were aware of the situation's political implications.[124] Wilmshurst acknowledged the potential importance of British nuclear sales, but Pakistan's non-adherence to the NPT was problematic. Moreover, he added:

> Our doubts about the peaceful intentions of the Pakistan nuclear programme must make us very wary of any cooperation. Should the Pakistan programme at a later stage become blatantly directed at achieving a military nuclear capability, even if our contribution had not in practice assisted this, we should be liable to embarrassing allegations of collaboration and irresponsibility. Regional considerations too dictate a policy of caution towards nuclear cooperation with Pakistan; we must be careful to consider possible repercussions, not only in India but also, for example, in Iran.[125]

By July, Pakistani representatives were enquiring with BNFL about "projected plans for Pakistan" but it was Burdess in Islamabad who encapsulated the problems in the first of two letters he submitted on July 13.[126] He boiled the situation down to three questions: what was politically acceptable for Britain to export? What would UK suppliers actually be able to provide? What credit terms could the UK offer Pakistan?[127] In his second letter, much less widely circulated around government departments, the diplomat went into more detail about intelligence pointing to Pakistani nuclear ambitions. Burdess did not want the situation to evolve the same way as the embarrassing Pakistani interest in Jaguar, where Islamabad felt it had been encouraged, while Britain had little intention of allowing them to purchase the aircraft.[128]

Burdess ominously speculated about Pakistani intentions. There was the chance that if ostensibly civilian equipment was offered and then

---

[124] Wilmshurst to Burdess, 'Nuclear Power in Pakistan', June 23, 1976, TNA FCO96/575, 2.

[125] Ibid., 1.

[126] Bourke to Wilmshurst, 'Nuclear Energy in Pakistan', July 15, 1976, TNA FCO96/575.

[127] Burdess to Wilmshurst, 'Nuclear Power in Pakistan', Letter 1, July 13, 1976, TNA FCO96/575, 1.

[128] Burdess to Wilmshurst, 'Nuclear Power In Pakistan', Letter 2, July 13, 1976, TNA FCO96/575, 1.

withdrawn, there could be considerable—and public—political embarrassment.[129] Australian diplomats had passed on rumours that Britain had already assisted in the design and construction of a small reprocessing facility in Pakistan. It was later confirmed that this facility dated from the 1960s and Britain had only contributed some consultancy for an "active chemistry" laboratory.[130] Burdess' Australian source also vouchsafed that D.A.V. Fischer, the Director of External Relations at the IAEA, had suggested that Pakistan's motives were "clearly dishonourable".[131] All of this pointed to a murky and complicated picture of what was actually happening on the sub-continent. In response to this, Wilmshurst reiterated the centrality to British non-proliferation policy of Callaghan's March 31 statement about sensitive technologies. In light of this, it was vital not to raise unrealistic expectations in Islamabad.[132] Indeed, the Callaghan statement became the "basic political framework for our nuclear exports to all countries".[133]

The MoD had also weighed into the debate over exports to Pakistan, suggesting that the possibility of Pakistan reaching for nuclear weapons could not be discounted. There was perceived to be minimal scope for any nuclear exports to Pakistan, especially with the looming shadow of the French reprocessing plant deal. "If this deal comes off," argued the MoD's Dennis Fakley, "it is difficult to see how any significant nuclear co-operation with Pakistan could be politically justifiable. Safeguards apart, Pakistan will have an indigenous capability to produce plutonium for explosive device purposes and I do not believe we should help them with this capability by, for example, providing fuel fabrication services."[134]

Regardless of how justifiable participation in the Pakistani nuclear programme was, there was a scramble amongst European nuclear suppliers

[129] Burdess to Wilmshurst, Letter 2, July 13, 1976, 1.

[130] Maclean to Wilmshurst, Letter, August 4, 1976, TNA FCO96/575, 1.

[131] Burdess to Wilmshurst, Letter 2, July 13, 1976, 2.

[132] Wilmshurst to Burdess, 'Nuclear Power in Pakistan', July 23, 1976, TNA FCO96/575, 1.

[133] Bourke to Broughton, 'Nuclear Power in Pakistan', August 9, 1976, TNA FCO96/575.

[134] Fakley to Bourke, 'Nuclear Exports to Pakistan', August 11, 1976, TNA FCO96/575.

to carry out feasibility studies to see if they could profit from Bhutto's nuclear bonanza.[135] There was the problem that Britain did indeed have little to sell on the global nuclear market. Back home on leave from Islamabad, Turner was summoned to the FCO and after a briefing on nuclear matters agreed that pressing for nuclear business in Pakistan was pointless.[136] There was also the matter of Pakistani enquiries via a third party (presumed to be the Belgian company Belgonucléaire) about CANDU-type reactor fuel.[137] The Canadians, who had originally supplied the reactor and its fuel, were engaged in heated discussions with Islamabad about safeguards, discussions that ended with the complete suspension of Canadian nuclear cooperation with Pakistan in December of 1976. After lengthy discussions, it was decided that BNFL should not respond to further enquiries for reactor fuel from Pakistan.[138]

In the end, the decision not to participate in the Pakistani nuclear programme was, for Britain, a complex network of Callaghan's stated non-proliferation ideals, suspicion about Pakistan intentions, fear of the loss of international credibility, and the bald fact that the British nuclear industry had little to sell Islamabad. By October the FCO was advising Burdess that London strongly suspected Pakistan wished to acquire a nuclear explosive capability. Furthermore, Pakistan was seen as such a bad economic risk that sales must also be declined on those grounds. That aside, Martin Bourke concluded, "Even if these difficulties did not exist we would be reluctant on proliferation grounds to sell anything but the most innocuous items to the Pakistanis."[139] Despite some further desultory communications, this missive represented a final decision on

---

[135] Burdess to Bourke, 'Pakistan Nuclear Power Programme', September 14, 1976, TNA FCO96/575.

[136] Holt, File Note, October 15, 1976, TNA FCO96/575.

[137] Delooze to Holt, Letter, October 12, 1976, TNA FCO96/575; Makepeace to Wilmshurst, 'Supply of CANDU-type Fuel to Pakistan', November 2, 1976, TNA FCO96/575; the CANDU reactor was the CANadian Deuterium-Uranium nuclear reactor, the type used at the sole Pakistani commercial reactor, located in Karachi (KANUPP).

[138] Wilmshurst to Herzig, 'Supply of CANDU Fuel for Pakistan', November 15, 1976, TNA FCO96/575.

[139] Bourke to Burdess, 'Nuclear Power in Pakistan', October 27, 1976, TNA FCO96/575, 1–2.

British involvement in the reprocessing plant, coinciding as it did with Ford's policy statement on October 28. What Bourke, Burdess, and their colleagues could not possibly know was that just over a month later, "innocuous items" would open a whole new chapter in the story.

DISCOVERING THE CLANDESTINE PROGRAMME

Late 1976 and early 1977 saw increasing awareness within the British government that the Pakistanis were following another route to the bomb. Officials came to the realisation that there was another, more secretive, strand to the Pakistani programme, one that tapped into the currents of globalisation, economic competition, and the ambiguity of international nuclear trade standards. It all began with seemingly innocent electrical components ordered by West German engineering consultants.

The first indications of a clandestine uranium enrichment programme came when BNFL contacted the DoE to inquire if there were any non-proliferation issues regarding items called inverters, electrical components that controlled high-speed alternating current (AC) electric motors. In this case, BNFL had received an inquiry from E.S. Harris, the managing director of Emerson Electric Industrial Controls (EEIC) in Swindon, Wiltshire. Harris had been contacted by TEAM Industries of Leonberg, West Germany, a company working as consultants on behalf of the government ordnance factory in Rawalpindi, Pakistan.[140] Harris believed that the inverters were for controlling a centrifuge system, although TEAM had not confirmed this to him.[141] In his submissions to BNFL and the DoE, Harris had noted that his was a small company and such a contract provided £400,000 worth of vital business, thus preventing redundancies. Michael James, the official in the DoE's Atomic Energy Division dealing with the situation, was convinced that the most likely use for the inverters was in a centrifuge system "with military purposes in mind". James suggested, however, that getting hold of such equipment was not that

---

[140] James to Butler, 'Centrifuge Equipment: Possible Export of Invertor to Pakistan', December 3, 1976, TNA FCO96/575, 1; Feroz Hassan Khan, *Eating Grass: The Making of the Pakistani Bomb* (Stanford: Stanford University Press, 2012), 169.

[141] James to Butler, December 3, 1976, 1.

difficult and if EEIC did not supply them, another European supplier would.[142] The FCO's Michael Wilmshurst made urgent enquiries with Bonn to see if FRG non-proliferation restrictions prevented the export of these items from Germany to Pakistan and suggested informing ministers of this new development in the Pakistani saga.[143]

This first-ever indication of Pakistan's enrichment programme set off a debate that brought together BNFL, the DoE, DoT, FCO, and MoD. It was assumed that BNFL had the expertise to judge whether or not these inverters were suitable for gas centrifuges, having purchased similar items from EEIC in the past. Other officials within the DoE urged that the case be taken "very seriously unless or until we are reassured by the informa-tion as to end use".[144] Not all government departments shared this view. The MoD's Denis Fakley was less convinced that there were sinister implications to the inverter order. Although he agreed with the enquiries being made in West Germany, he suggested that it could be the case that "this affair is wholly innocent" and that James had made more of the available evidence than was warranted.[145] James argued that it was the managing director of EEIC himself who had made the link between the Pakistani order and uranium enrichment centrifuges and that, after examining the specifications given by Harris, BNFL agreed that their only use could be in a centrifuge (and expressed disbelief that nothing could be done to address the export issue).[146] In light of this, Fakley agreed that action should be taken if it were shown that TEAM

[142] Ibid., 2.

[143] Wilmshurst to Cromartie, 'Nuclear Suppliers: Exports to Pakistan', December 3, 1976, TNA FCO96/575; James to Butler, December 3, 1976, 2–3.

[144] Nichols to Herzig, 'Centrifuge Equipment: Possible Export of Invertor to Pakistan', December 6, 1976, TNA FCO96/575; Herzig to Butler, 'Centrifuge Equipment: Possible Export of Invertor to Pakistan', December 7, 1976, TNA FCO96/575.

[145] Fakley to Wilmshurst, 'Nuclear Supplies: Exports to Pakistan', December 9, 1976, TNA FCO96/575.

[146] James to Fakley, 'Export of Invertors to Pakistan', December 14, 1976, TNA FCO96/575; Coleman to Turner, Letter, December 7, 1976, Records of the Department of Energy (hereafter EG) 8/269; James to Butler, 'Centrifuge Equipment', December 1, 1976, TNA EG8/269, 1.

Industries were involved in helping to build a uranium enrichment plant for Pakistan.[147]

On December 20, James offered an update and a scathing assessment of the British intelligence community. He reiterated that BNFL had verbally confirmed that the inverters were of a type intended for centrifuge power supplies and the DoT had informed him such products would come under the Export of Goods (Control) Order when it was revised in two or three months. Any further action would have to be taken at the ministerial level and the political consequences of such action required very careful assessment.[148] James then commented, "I find it very strange if the Pakistanis really are building a centrifuge plant with military intentions in Rawalpindi, this is the first we have heard about it. Do we not have some sort of Intelligence Service or are they all writing spy novels?" The DoE was already preparing to assign blame elsewhere if the entire fracas turned out to be nothing more than a mistake: "We should look pretty silly if we prevented the export of these invertors on our present assumptions, only to find they really had been intended for a paper mill or something equally innocent and that BNFL had got the story wrong."[149] Though a seemingly throwaway comment, this particular statement is significant. It illustrates the confusion over inverters and—at least in part—explains why it took so long for London to take serious action against the Pakistani enrichment programme. As will be demonstrated, it took almost two years for sustained diplomatic action to take place. This was partially because of the situation's opacity—James described the situation as confused, noting that, "we are not clear what is going on"—and an unwillingness to suffer public embarrassment by creating a diplomatic 'scene'.[150]

During his visit to Pakistan in January of 1977, Callaghan hardly mentioned the nuclear issue and did not touch on British suspicions about the enrichment programme. Agha Shahi was interested in discussing

---

[147] Fakley to James, 'Export of Invertors to Pakistan', December 17, 1976, TNA FCO96/575.

[148] James to Butler, 'Centrifuge Equipment: Possible Export of Invertor to Pakistan', December 20, 1976, TNA FCO96/575.

[149] Ibid.

[150] James to Ellerton, 'Proposed Export of Invertors to Pakistan', December 22, 1976, TNA FCO96/575.

the issue of nuclear security guarantees for Pakistan, a topic that still had currency, even though two and a half years had passed since the Indian test. In his discussions with Callaghan, the Pakistani Army Chief, General Muhammad Zia ul-Haq, had seemingly expressed a vague willingness to adhere to the NPT and there were suggestions that Pakistani officials might visit London to discuss disarmament and security issues.[151]

In the new year, Ian Cromartie at Britain's Embassy in Bonn finally got back to Michael Wilmshurst regarding the situation's German aspects. Cromartie urged Whitehall to be as full and open as possible in order that his discussions with the *Auswärtiges Amt* (the West German foreign ministry) could be structured to avoid any future repercussions.[152] Cromartie was instructed to proceed in his discussion with the West Germans in as informative a manner as possible.[153] The position of inverters as regards the NSG and Zangger Committee 'trigger lists' started to dominate, with the assertion that if the items were on the list, the strictures applied to the FRG just as much as they did to the UK.[154] Even more evidence came to light when a British employee at the multinational URENCO uranium enrichment plant at Almelo in the Netherlands indicated that the Pakistanis had requested ten kilogrammes of uranium hexafluoride (the form of uranium used in the enrichment process) and had been buying substantial quantities of hexafluoride-resistant valves from a Swiss firm.[155] On the day that Jimmy Carter was being inaugurated as President, Cromartie wrote back to Wilmshurst confirming that his West German counterparts would get back to him with all possible speed.[156] They did not.

[151] O'Neill to Mallaby, 'Pakistan: Disarmament', January 17, 1977, TNA FCO37/2112.

[152] Cromartie to Wilmshurst, 'Nuclear Suppliers: Exports to Pakistan', January 5, 1977, TNA FCO96/728.

[153] Wilmshurst to Cromartie, 'Nuclear Suppliers: Exports to Pakistan', January 11, 1977, TNA FCO96/728.

[154] James to Wilmshurst, 'Proposed Export of Invertors to Pakistan', January 5, 1977, TNA FCO96/728.

[155] Brown to James, 'Proposed Export of Invertors to Pakistan', January 7, 1977, TNA FCO96/728.

[156] Cromartie to Wilmshurst, 'Nuclear Suppliers: Exports to Pakistan', January 20, 1977, TNA FCO96/728.

Although taking place below ministerial level and involving a collection of civil servants, junior diplomats, and businesspeople, this brief episode from early December 1976 to January 1977 is significant in the story of British and American involvement with the Pakistani nuclear weapons programme. It demonstrated there were two strands directed at obtaining materials to make a nuclear bomb. It also illustrates the situation's confusion and murkiness at this stage. Many works on the subject have—without having access to these new documents—suggested incompetence or conspiracy (and at times a combination of the two) as the reasons for Western inaction over the Pakistani enrichment programme.[157] In reality, the reasons for a lack of real action on the Pakistani nuclear issue were multifaceted and multilayered. Domestic politics, the brutal economic environment of the 1970s, interdepartmental politics, fear of public embarrassment, and the multinational nature of the emergent Pakistani programme all contributed to the lack of immediate response. Suggesting that there was a complicated conspiracy at work overlooks the situation's *actual* complexity.

CONCLUSION

The months from the signing of the French-Pakistani nuclear processing plant deal to Jimmy Carter's inauguration were a time of confusion and change. The complexity of international nuclear diplomacy and the difficulty of reconciling the needs of American and Pakistani domestic and foreign policy made it an intensely complicated time for policymakers. Electoral and congressional pressure forced Ford and Kissinger to reassess US non-proliferation policy, as perceptions in Washington and London hardened around the belief that Pakistan *was* pursuing the nuclear weapons option. Yet, there was little practical communication between the USA and the UK on ways and means to combat the problem. The Americans unsuccessfully attempted to bribe or coerce Bhutto with arms sales. Britain stubbornly pursued the Indian Jaguar deal, but chose not to seek commercial advantage from the Pakistani reprocessing plant deal when it became apparent that any involvement would be politically untenable and limited in profitability. The period was also revelatory in that,

---

[157] Frantz and Collins, Krosney and Weissman, Levy and Scott-Clark, and Venter all combine these two explanations to a greater or lesser extent.

with remarkable suddenness, it became apparent to British officials that they were dealing with two Pakistani efforts to acquire fissile materials. For all concerned, 1976 was indeed the end of the "first 'easy' phase" and the beginning of what promised to be a challenging period in international nuclear relations.

CHAPTER 4

"The omens are scarcely encouraging" Jimmy Carter, Nuclear Reprocessing, and the Clandestine Programme, February 1977 to March 1978

By the middle of 1977, things looked bleak for American and British efforts to discourage Pakistan's nuclear quest. Robin Fearn—British *chargé d'affaires* in Islamabad—commented that "the omens are scarcely encouraging".[1] Fearn's take on the situation was prescient as US-Pakistani relations dropped to their lowest ebb and British efforts to exercise influence came to naught. The year was also one of change in political leadership in the USA and Pakistan. Yet, despite personnel changes, these thirteen months were at the same time a story of continuity in policy rather than dramatic transformation.

Jimmy Carter came to the presidency promising to address human rights, the superpower arms race, and nuclear proliferation. His stance on proliferation created tensions between the United States and its European allies. The new President's pronouncements on non-proliferation were little different to his predecessor's late-period change of heart, but involved a jarring change in style that presaged a move away from stereotypical Cold

---

[1] United Kingdom Embassy (hereafter UKE) Islamabad to FCO, 'Pakistan: Reprocessing Plant: Nye's Visit', August 9, 1977, The National Archives of the UK (hereafter TNA) Records of the Foreign and Commonwealth Office (hereafter FCO) 96/728, 2.

© The Author(s) 2017
M.M. Craig, *America, Britain and Pakistan's Nuclear Weapons Programme, 1974–1980*, Security, Conflict and Cooperation in the Contemporary World, DOI 10.1007/978-3-319-51880-0_4

War bombast.[2] Carter's far-reaching, globalised policies, which tapped into transnational themes, impacted the foreign and domestic nuclear interests of states like the UK, France, and West Germany with little prior consultation.

Regarding Pakistan, efforts to halt or retard the nuclear programme faced multiple barriers. Although Washington and London hoped that a change of government in Islamabad might usher in a regime more pliable on nuclear issues, the July 1977 switch from Bhutto's civilian rule to General Muhammad Zia ul-Haq's military administration saw continuity in Pakistan's nuclear quest. Carter, in attempting to balance non-proliferation and conventional arms control policies, only succeeded in driving Pakistan further away from the United States. The one partial non-proliferation success was the French acceptance that non-proliferation mattered more than the reprocessing plant contract, although this was also mired in problems as Valery Giscard d'Estaing's government took their time to go public with its decision and break their contract with Pakistan.

Carter's non-proliferation strategy and its interactions with Pakistan faced a British government demonstrating that domestic interests influenced responses to the proliferation problem. Even though James Callaghan's government had a greater commitment to non-proliferation than other Western European states, arms sales and the avowed British requirement for a lucrative domestic nuclear reprocessing industry were the most prominent examples of economic imperatives interfering with non-proliferation policy. The Callaghan government's attitude towards British domestic reprocessing illustrated a division between the nuclear haves and the have-nots, as Foreign Secretary David Owen justified Britain's Thermal Oxide Reprocessing Plant (THORP) on the grounds of economic and energy needs that were identical to Islamabad's justifications.

Significantly for the future of non-proliferation policy, new evidence came to light about the late-1976 discovery of Pakistan's global, clandestine purchasing programme. The British government did not lack a willingness to tackle this issue, despite its effects on UK businesses. Regardless,

---

[2] William Glenn Gray, 'Commercial Liberties and Nuclear Anxieties: The US-German Feud Over Brazil, 1975–7', *International History Review*, 34:3 (2012), 468; Daniel J. Sargent, *A Superpower Transformed: The Remaking of American Foreign Relations in the 1970s* (Oxford: Oxford University Press, 2015), 229–30.

efforts to combat the purchasing programme were hampered by the vagueness of existing international proliferation controls, the dubious legal basis of action to prevent suspicious exports, and the reluctance of fellow supplier states to inhibit a lucrative global trade. So, by March 1978—when London officially alerted Washington to the existence of Pakistan's clandestine uranium enrichment scheme—the "omens" for future anti-proliferation activities focused on the sub-continent were indeed "scarcely encouraging".

Enter Carter

Jimmy Carter came to the presidency having made non-proliferation a key plank of his foreign policy platform. A devout born-again Southern Baptist, the new President sought to bring morality back to the centre of a human-rights focused US foreign policy, a foreign policy that had to grapple with an increasingly interdependent world.[3] Carter also had more direct nuclear experience than any other president, having served as a junior officer in Admiral Hyman Rickover's pioneering atomic-powered submarines. In an inaugural address involving frequent Biblical quotations and a refreshing lack of the Cold War bombast that had characterised such occasions since 1949, Carter vouchsafed that his ultimate goal was "the elimination of all nuclear weapons from this Earth".[4]

Carter attempted to govern as both a moralist and a realist, identifying US national interests with moral righteousness.[5] Unlike late-period Ford, forced towards non-proliferation by congressional and electoral pressure, Carter founded his passion for non-proliferation in a vision of a moral

[3] Scott Kaufman, *Plans Unraveled: The Foreign Policy of the Carter Administration* (DeKalb: Northern Illinois University Press, 2008), 11–17; Sargent, *A Superpower Transformed*, 230.

[4] Carter, 'Inaugural Address', January 20, 1977, *The American Presidency Project*, Public Papers of the Presidents: Jimmy Carter (hereafter PPPJC), www.presidency. ucsb.edu/ws/?pid=6575 (accessed July 16, 2013).

[5] Andrew Preston, *Sword of the Spirit, Shield of Faith: Religion in American War and Diplomacy* (New York: Anchor Books, 2012), 575; Betty Glad, *An Outsider in the White House: Jimmy Carter, His Advisors, and the Making of American Foreign Policy* (Ithaca: Cornell University Press, 2009), 281.

foreign policy driven by a personal commitment to human rights. Human rights, arms control, restraint in conventional arms sales, and non-proliferation were all tied together. Spending on conventional and nuclear weapons meant less for states to spend on curing social ills such as poverty, which in turn led to frustration and instability. Reducing arms expenditure and arresting the spread of nuclear weapons would result in a safer and more peaceful world, with more money to spend on combating societal problems.[6] Reflecting on India's 1974 test, Carter noted that he "wanted to do everything possible to prevent this capability from spreading to any additional nations".[7] While Carter's motivations may have been different from Ford's, the ends and means were broadly similar. The fundamentals of Carter's proliferation policy were little different from the Fri Report (when Carter took office, Fri became ERDA's Chief Administrator) and Ford's October 1976 statement: a halt to domestic reprocessing; increased pressure on other states to halt their own reprocessing plans; greater international debate about the nuclear fuel cycle; and closer attention to global means of curbing access to sensitive technologies. However, the emphasis on morality and humanitarianism created more problems than it solved, with human rights and nuclear proliferation issues antagonising enemies and allies alike.[8]

Although Carter drove the non-proliferation policy agenda, he was surrounded by advisers—such as Deputy Assistant Secretary of State for Nuclear Energy and Energy Technology Affairs Louis V. Nosenzo—who had been sidelined during the Nixon and Ford years, but now found themselves working for a President espousing a genuine enthusiasm for non-proliferation.[9] Problems that plagued Carter's time in office arose from the administration's very start. Disorganisation in decision-making, a managerial style that often initiated policies without a full understanding of their ramifications, and an—at times—rancorous relationship between his two key foreign policy advisers, Secretary of State Cyrus Vance and

[6] Kaufman, *Plans Unraveled*, 14–15.

[7] Jimmy Carter, *Keeping Faith: Memoirs of a President* (London: Collins, 1982), 215–216.

[8] Preston, *Sword of the Spirit*, 576.

[9] Michael J. Brenner, *Nuclear Power and Non-proliferation: The Remaking of U.S. Policy* (Cambridge: Cambridge University Press, 1981), 123–124.

National Security Adviser Zbigniew Brzezinski, led to confusion, misunderstanding, and—in some cases—disaster.[10]

Pakistan was on Carter's radar from day one, but there was little change from Ford's approach during the new administration's first three months. The twin measures of exerting pressure on France to cancel the reprocessing plant deal and using conventional arms sales to persuade Zulfikar Ali Bhutto to give up his nuclear ambitions remained in place as the administration formulated broader non-proliferation strategies. The use of conventional arms as an inducement for Bhutto not to pursue his nuclear aspirations represented a major flaw in Carter's policy platform. The administration wanted to prevent proliferation *and* reduce sales of advanced weapons to the developing world. In the case of Pakistan, either non-proliferation would have to be subordinated to arms sales restraint or vice versa. Lucy Benson (Under Secretary of State for Security Assistance) made the point to her British counterparts that the administration was unsure of how to achieve the desired compartmentalisation of Pakistan's nuclear weapons programme and conventional arms supplies.[11] The State Department favoured approaching Bhutto before the March Pakistani elections, but Vance decided to wait until it was clear whether Bhutto would remain in office or that Washington would have to deal with a new regime.[12]

On Pakistan's attempt to purchase the core of a nuclear fuel cycle, Vance's subordinates viewed the reprocessing plant deal's completion as increasingly unlikely, and argued that the emphasis should be on how to mollify Bhutto and persuade him that he would never have the facilities.[13] The French government bore significant responsibility for this reduced

---

[10] Burton I. Kaufman and Scott Kaufman, *The Presidency of James Earl Carter*, 2nd edition, revised (Lawrence: University Press of Kansas, 2006), 46–47.

[11] UKE Washington to FCO, 'US Nuclear Non-proliferation Policy', February 9, 1977, TNA FCO37/2066, 1–2.

[12] Action Memorandum to the Secretary, 'Further Steps on Brazil and Pakistan', January 28, 1977, United States National Archives and Records Administration (hereafter NARA); Record Group 59: General Records of the Department of State (hereafter RG59); Records of Anthony Lake (hereafter RAL), Box 17, Pakistan, 11; Vance to United States Embassy (hereafter USE) Paris, 'Letter to Foreign Minister', February 14, 1977, NARA, RG59, RAL, Box 17, Pakistan, 3.

[13] Action Memorandum to the Secretary, January 28, 1977, 2.

chance of completion, having concluded that the deal's proliferation risks were too great.[14] Giscard commented that the entire deal was a "bad mistake" but did not want to lose face by succumbing to American pressure.[15] Giscard feared that Washington would state publicly that a French policy change was the result of US influence on Paris, affecting his domestic political standing at a critical time.[16]

Vance wrote to his French counterpart, Louis de Guiringaud, in February seeking confirmation of cancellation. Not only would the deal's abandonment be a major step forward for sub-continental non-proliferation efforts, it would boost the chances of stopping the huge FRG-Brazil nuclear deal.[17] The new Arms Control and Disarmament Agency (ACDA) Director Paul Warnke pointed out to Deputy Secretary of State Warren Christopher the connections between the two nuclear deals, stating that "unless we obtain some kind of gain [on the FRG-Brazil deal], our prospects of obtaining success will be substantially reduced in Pakistan (where, unlike Brazil, the evidence on current weapons intentions is unambiguous)".[18] The problem was not that the French were unwilling to cancel the deal—Giscard had made his feelings plain in private—but French and Pakistani sensitivities required the final cancellation's discreet handling. French ministers had repeatedly emphasised the sanctity of a contract made in good faith and Bhutto had vigorously underscored his commitment to the reprocessing plant.

The disputed March elections in Pakistan caused Bhutto—at least in his diplomatic communications—to adopt a more conciliatory attitude on nuclear issues, prompting discussions within the Carter administration

---

[14] Gray, 'Commercial Liberties', 468.

[15] Brzezinski to Carter, 'Info Items', March 12, 1977, Jimmy Carter Presidential Library (hereafter JCPL), Remote Archives Capture system (hereafter RAC) NLC-1-1-2-67-4, 1; USE Paris to State, 'Elysee on Schmidt Visit', February 8, 1977, NARA, RG59, RAL, Box 17, Pakistan.

[16] Brzezinski to Carter, 'NSC Weekly Report #24', February 17, 1977, JCPL, RAC NLC-28-63-3-1-8, 4.

[17] Vance to USE Paris, February 14, 1977, 1–3.

[18] Cyrus Vance, *Hard Choices: Critical Years in America's Foreign Policy* (New York: Simon & Shuster, 1983), 43; Warnke to Christopher, 'Memorandum for the Deputy Secretary: The FRG-Brazil Deal', March 25, 1977, JCPL, RAC NLC-132-36-3-2-0, 1.

regarding the Ford-era offer of advanced A-7 attack aircraft to Islamabad.[19] The elections had become mired in accusations of vote rigging, religious intolerance, and political violence. Bhutto attempted to woo Maulana Mufti Mahmood, the multiparty Pakistan National Alliance's powerful conservative clerical leader, to try and avert strikes and violent protests.[20] The embattled Prime Minister also announced the enforcement of sharia law and declared prohibitions on alcohol and gambling to quell calls for change emanating from the more radical Islamic elements of Pakistani society.[21] With Bhutto vulnerable to domestic opponents, Brzezinski criticised a State Department plan for a wide-ranging package of economic and military incentives—including substituting the advanced A-7s for far less capable F-5E fighters and perhaps a few ageing A-4 fighter-bombers—as threatening Carter's policies on arms sales and South Asia. The National Security Adviser contended that rewarding Pakistan would set a "tempting precedent" for other proliferators, but an overly tough stand with Bhutto might well cause him to adopt an even more "strident nuclear policy", damaging Carter's international non-proliferation objectives. Brzezinski recommended entering into negotiations demanding Pakistani nuclear restraint, gradually offering incentives if it became apparent that Bhutto "would not fold".[22] Carter favoured Brzezinski's stance and demanded a minimal arms package, opposed the State Department's suggestion of a substantial aid programme, and disputed Christopher's suggestion that the USA could finance the Pakistani purchase of a French nuclear reactor.[23]

[19] Brzezinski to Carter, 'Evening Report', March 10, 1977, JCPL, RAC NLC-1-1-2-54-8.

[20] Husain Haqqani, *Pakistan: Between Mosque and Military* (Washington D.C.: Carnegie Endowment for International Peace, 2005), 121.

[21] Ibid., 123.

[22] Brzezinski to Carter, 'Pakistan: Reprocessing and Arms Sales Negotiations', April 1977 (exact date unknown), JCPL, Records of the National Security Staff, North-South, Box 95, Pakistan 4/77–12/78 (hereafter NSS-NS Box 95), 2.

[23] Christopher to Carter, 'Reprocessing Negotiations With Pakistan: A Negotiating Strategy', April 2, 1977, National Security Archive Electronic Briefing Book (hereafter NSAEBB) 'The United States and Pakistan's Quest for the Bomb', (hereafter USPQB), www.gwu.edu/~nsarchiv/nukevault/ebb333 (accessed July 16, 2013), Doc.2, 2–4. Carter's thoughts are outlined in his handwritten marginalia.

While the administration was getting to grips with Pakistan, Carter unveiled his global non-proliferation policy. On April 7 he argued that while nuclear power was a vital resource against the background of the oil crisis, there were dangerous consequences to the spread of certain technologies. Carter called for international efforts to curb the spread of plutonium producing facilities and uranium enrichment technology, although the focus was firmly on the former. Where Ford had opted for a three-year moratorium on US reprocessing, Carter announced an indefinite deferral of US commercial reprocessing, expansion of enriched uranium production to meet expected global demands, legislative steps to permit foreign nuclear fuel supply contracts, and international discussions on energy needs. The President pointed out that he was in no way attempting to impose his will on France, Japan, the UK, and West Germany, which had already embarked on extensive reprocessing projects and which did not have ready access to domestic supplies of oil. This approval of European and Japanese reprocessing surprised Carter's advisers, whom the President had not informed about this aspect of his speech. Finally, Carter called for the establishment of an evaluation programme looking at all aspects of the nuclear fuel cycle, a call that led to the International Nuclear Fuel Cycle Evaluation Program (INFCE).[24]

Carter's dramatic emphasis on reprocessing implicitly made the situation in Pakistan part of a major, global policy announcement. In the post-speech press conference, a journalist pointedly asked if "some of the smaller nations that are now seeking reprocessing technology are doing so in order to attain nuclear weapon capability as well as or in addition to meeting their legitimate energy needs?" Carter was circumspect, reiterating that his policies were aimed at stopping other nations copying India, where civilian atomic supplies had been used to create a nuclear explosive.

---

[24] Carter, 'Nuclear Power Policy Remarks and a Question and Answer Session With Reporters on Decisions Following a Review of U.S. Policy', April 7, 1977, PPPJC, www.presidency.ucsb.edu/ws/index.php?pid=7315 (accessed July 4, 2013); Brzezinksi recalls that Carter intended to announce the new policy two weeks earlier, but that he persuaded the President that the consequences of such an announcement—with only a few hours notice for allies—would be a foreign relations disaster. See Zbigniew Brzezinski, *Power and Principle: Memoirs of the National Security Adviser, 1977–1981* (London: Weidenfeld and Nicholson, 1983), 130–131.

Carter's presumption was that if the global community only used light water reactors and low-enriched uranium fuel, and agreed to restrictions on breeder reactor and reprocessing technology, it would be much harder to divert nuclear materials to military projects.[25] Furthermore, the President had broadened the debate from a discussion of nuclear *exports* to fundamental energy strategies, thus confronting a far greater number of interested parties.[26] Confusion ensued when Robert Fri (Acting ERDA Director) and Joseph Nye (Chair of the National Security Council group on nuclear non-proliferation) attempted to 'clarify' the President's statements. Contrary to the impression given by Carter, Fri and Nye claimed that the President was referring to domestic considerations, the international aspects not having been fully agreed. Their 'clarifications' did anything but, producing further confusion and consternation amongst foreign observers who were attempting to decipher Carter's policy.[27]

After placating nuclear suppliers by his willingness to allow continued reprocessing in those countries that already had facilities, Carter executed a *volte-face* on April 27 with the first draft of the Nuclear Non-proliferation Policy Act (NNPA). At the controversy's heart was the proposal that the United States should be able to renegotiate existing nuclear supply contracts and retain the right of approval over the foreign reprocessing of US-origin fuel.[28] This placed commercial reprocessing facilities—such as the proposed British THORP and French UP_3 facilities—in danger, imperilling reprocessing contracts with nations using US-supplied fuel, as there was now no guarantee that Carter would permit that fuel's transfer for recycling. The primary reason for the controversial THORP's proposed construction was to reprocess foreign spent fuels

---

[25] Ronald E. Powaski, *March to Armageddon: The United States and the Nuclear Arms Race, 1939 to the Present* (Oxford: Oxford University Press, 1987), 182.

[26] Gray, 'Commercial Liberties', 464.

[27] J. Samuel Walker, 'Nuclear Power and Nonproliferation: The Controversy Over Nuclear Exports, 1974–1980', *Diplomatic History*, 25:2 (2001), 238.

[28] 'Nuclear Non-proliferation Fact Sheet on the Proposed Nuclear Non-proliferation Policy Act of 1977', April 27, 1977, PPPJC, www.presidency.ucsb.edu/ws/index.php?pid=7409 (accessed July 9, 2013); Carter, 'Nuclear Non-proliferation–Message to Congress', April 27, 1977, PPPJC, www.presidency.ucsb.edu/ws/index.php?pid=7408 (accessed July 9, 2013).

and operate as a revenue earner.[29] Carter's reservation of dramatic interventionist powers in the NNPA thus posed a direct threat to British economic and energy policy.

The Carter administration's first three months of non-proliferation activity highlight several factors that became hallmarks of the President's time in office. The tension between the State Department and Brzezinski was already becoming apparent as Carter adopted the latter's line on negotiations with Pakistan. More broadly, foreign enthusiasm for non-proliferation initiatives foundered on the rocks of the sudden and jarring shift outlined in the draft NNPA. One of the countries most worried by the proposals was America's closest ally in dealing with the Pakistani proliferation problem, the UK.

THE UK AND PAKISTANI REPROCESSING

The Foreign and Commonwealth Office (FCO) was undergoing a transition of its own as Ford gave way to Carter. On February 13 Foreign Secretary Crosland suffered a massive stroke, dying in hospital six days later. The surprising choice as his replacement was David Owen, a committed Atlanticist from the party's right and, at thirty-eight, the youngest foreign secretary since Anthony Eden in 1935.[30] Superiors, colleagues, and friends saw him as energetic, imaginative, and full of initiative, but imperious and difficult to work with.[31] Christopher Mallaby, Arms Control and Disarmament Department (ACDD) Chief during Owen's time at the FCO, recalled Owen displaying great enthusiasm for arms control and non-proliferation matters, issues that the new Foreign Secretary saw as vote winners for Labour and a means of increasing his personal popularity

---

[29] William Walker, 'Destination Unknown: Rokkasho and the International Future of Nuclear Reprocessing', *International Affairs*, 82:4 (2006), 745.

[30] Rhiannon Vickers, *The Labour Party and the World, Volume 2: Labour's Foreign Policy 1951–2009* (Manchester: Manchester University Press, 2011), 112.

[31] James Callaghan, *Time and Chance* (London: Collins, 1987), 448; Paul Lever, interview by Malcolm McBain, November 7, 2011, British Diplomatic Oral History Programme (hereafter BDOHP), www.chu.cam.ac.uk/archives/collec tions/BDOHP/Lever.pdf (accessed July 10, 2013), 8; David Gillmore, interview by Jane Barder, March 17, 1996, BDOHP, www.chu.cam.ac.uk/archives/collec tions/BDOHP/Gillmore.pdf (accessed July 10, 2013), 11.

and standing.[32] Owen was concerned about proliferation, particularly in the cases of Pakistan and South Africa.[33] By 1977 British public activism on nuclear matters was reinvigorated after languishing in the doldrums during the early 1970s, with the Campaign for Nuclear Disarmament (CND) calling attention to the dangers of proliferation and Friends of the Earth protesting against the THORP.[34] There were inherent tensions in Owen's stance, including the matter of balancing popular arms control measures and the NPT's developing world signatories' demands for genuine progress in arms reduction, with the requirements of British defence and foreign policy. This balancing act became apparent in the British approach to the Pakistani proliferation problem, where UK domestic, commercial, and foreign policies were in constant tension.

After Carter's inauguration several US delegations travelled to London. Vance, Warnke, and Vice President Walter Mondale all journeyed to meet Callaghan and his ministers to discuss non-proliferation matters.[35] Mondale homed in on Pakistan and Brazil as pressing problems, and outlined American willingness to give up commercial advantage to combat the proliferation problem. Callaghan was well aware of this position from Carter's campaign speeches and shared the American viewpoint. The Prime Minister was "frankly terrified about what was happening on nuclear proliferation". Mondale agreed and noted Carter's great interest in following through on his early non-proliferation comments.[36]

[32] Christopher Mallaby, interview by John Hutson, December 17, 1997, BDOHP, www.chu.cam.ac.uk/archives/collections/BDOHP/Mallaby.pdf (accessed July 10, 2013), 11–12.

[33] David Owen, *Time to Declare* (London: Penguin, 1992), 336.

[34] Lawrence S. Wittner, *The Struggle Against the Bomb, Volume 3: Toward Nuclear Abolition* (Stanford: Stanford University Press, 2003), 22.

[35] John C. Edmonds, interview by Malcolm McBain, May 21, 2009, BDOHP, www.chu.cam.ac.uk/archives/collections/BDOHP/Edmonds.pdf (accessed May 3, 2013), 25–26. There also existed a flourishing relationship with Carter's pick as US ambassador to the UK, Kingman Brewster. See Alex Spelling, 'Ambassadors Richardson, Armstrong, and Brewster, 1975–81', in Alison R. Holmes and J. Simon Rofe, *The Embassy in Grosvenor Square: American Ambassadors to the United Kingdom, 1938–2008* (Basingstoke: Palgrave Macmillan, 2012), 201–209.

[36] 'Extract from PM's Meeting with US Vice President Mondale, 27-1-77', January 27, 1977, TNA Records of the Prime Minister's Office (hereafter PREM)16/1182, 1.

As the new administration in Washington put its campaign promises into action, London was still dealing with the intertwined issues of the French reprocessing plant, inverter exports, and Jaguar sales to India. The ACDD briefed Owen on the Pakistani and Brazilian situations during his first day in office, suggesting that Britain should offer support for the USA, while being careful not to offend any of the four main parties involved in these nuclear deals. Regardless of the diplomacy's public or private nature, the fundamental UK interest was that Islamabad never got the reprocessing plant.[37]

British unwillingness to become overtly involved in resolving the reprocessing plant problem was a constant throughout 1977 and into 1978. Despite a flood of information, Whitehall preferred to let the USA continue to take the lead. British Ambassador to the United States Peter Ramsbotham made it clear to Warren Christopher that, although US and UK non-proliferation policies were in alignment, he preferred that secret contacts between London, Paris, and Bonn not become public.[38] Meanwhile, the FCO's Michael Wilmshurst opined that leaking suspicions about Pakistani nuclear weapons intentions might persuade the French (thought to be lacking compelling commercial reasons for pursuing the reprocessing plant) to finally cancel the contract.[39]

Information from Islamabad indicated that the reprocessing plant's cancellation could benefit British commercial interests in Pakistan. The FCO's South Asia Department (SAD) was curious about connections between the deal and the building of a huge truck manufacturing facility by the French company Saviem that might threaten the British firm Bedford, which did well from exports to Pakistan.[40] The disquieting news from Britain's Paris Embassy was that the French government had agreed to part-fund the new truck plant's construction. John Macrae in Paris doubted that there was a *direct* tie-up between the nuclear and automotive

[37] Edmonds to Moberly, 'The France-Pakistan and FRG-Brazil Reprocessing Deals', January 21, 1977, TNA FCO37/2066, 1.

[38] UKE Washington to FCO, 'US Nuclear Non-proliferation Policy', February 9, 1977, TNA FCO37/2066, 1–2.

[39] Wilmshurst to Burdess, 'Pakistan's Nuclear Activities: Safeguards', February 16, 1977, TNA FCO37/2066, 2.

[40] Fursland to Macrae, 'Pakistan: French Nuclear and Vehicle Exports', January 31, 1977, TNA FCO37/2066.

facilities, but believed that the reprocessing plant was *generally* beneficial to French commercial interests in Pakistan. The reprocessing plant's cancellation could, Macrae suggested, negatively influence wider French interests.[41] Thus, there was a specific British interest in the Franco-Pakistani deal's cancellation in order to preserve markets for UK manufacturers.

Several factors inhibited British government action on the issue, despite this obvious self-interest in France's ending the reprocessing agreement. Bhutto showed no signs of giving up on acquiring the facilities, while in London the FCO worried that Islamabad might view overt British attempts to intercede as US influence by stealth. Kit Burdess at the Embassy in Islamabad contended that Bhutto had done nothing to alter the Pakistani population's belief that the reprocessing plant was intimately associated with a national nuclear weapons programme.[42] Burdess argued that, for Pakistan, the nuclear effort's long-term political consequences were unclear, although Western pressure on nuclear issues could tilt Islamabad more towards the "Islamic world" and China.[43] The Pakistani press reported that France had committed itself to honouring the reprocessing plant contract, while, in the aftermath of the disputed March elections, Bhutto reiterated his commitment to the nuclear programme but indicated a willingness to discuss matters with the United States and France.[44] US diplomats paid heed to such reports, suggesting that Bhutto would do anything to keep the plant. This information prompted the FCO to question exactly what—other than sophisticated military equipment— would persuade Bhutto to give up his nuclear ambitions.[45] Diplomats also

[41] Macrae to Fursland, 'Pakistan: French Nuclear and Vehicle Exports', March 16, 1977, TNA FCO37/2066.

[42] Burdess to Fursland, 'Pakistan's Nuclear Intentions', February 19, 1977, TNA FCO37/2066, 2.

[43] Ibid., 3.

[44] Burdess to Findlay, 'Press Reports of French Government Statements on Reprocessing Plant Deal', February 24, 1977, TNA FCO96/728; Burdess to Wilmshurst, 'Work on Chashma Nuclear Complex at Kundian (including French reprocessing plant)', March 31, 1977, TNA FCO37/2066, 1–2; UKE Islamabad to FCO, 'Pakistan Elections', March 11, 1977, TNA FCO96/728, 2.

[45] Burdess to Fursland, 'Pakistan and the US', March 26, 1977, TNA FCO37/2066, 1; Fursland to Burdess, 'Pakistan and the US', April 7, 1977, TNA FCO37/2066.

worried about the UK getting too involved in the reprocessing fracas. Robin Fearn argued that any overt British efforts to intervene would be seen as "a stalking horse for the Americans and [the Pakistani government] would be disinclined to treat us as serious interlocutors".[46] Fearn doubted that London could exert any leverage on Islamabad and endorsed the FCO recommendation that the Americans or the NSG should take the lead.[47] Indo-US relations were warming in the wake of Indira Gandhi's electoral defeat and the state of emergency's end.[48] Kit Burdess claimed that India's return to being the "world's largest democracy" might make it difficult to overlook human-rights and other violations in Pakistan.[49]

Confusing messages were emanating from Paris, making the situation even more opaque. Diplomatic discussions failed to show a change in French attitudes towards the Pakistani deal as the Quai d'Orsay argued that, despite the nuclear weapons implications, the Pakistanis were justified in buying the facility.[50] Burdess speculated that the Carter administration was delaying taking any decision due to the deteriorating situation in Pakistan, suggesting that a change of regime might be on the cards and that the Americans may be waiting for a more pliable administration to come to power in Islamabad.[51]

THE UK AND CARTER'S AGENDA

While Pakistan remained high on the FCO's list of priorities, the government as a whole strove to understand the implications for Britain of Carter's stand on non-proliferation. Callaghan ordered Owen to form a

---

[46] Fearn to Wilmshurst, 'Pakistan: Nuclear Intentions', April 1, 1977, TNA FCO37/2066, 1.

[47] Ibid., 2.

[48] Dennis Kux, *Disenchanted Allies: The United States and Pakistan, 1947–2000* (Washington D.C.: Woodrow Wilson Center Press, 2001), 234–235.

[49] Burdess to Fursland, March 26, 1977, 2.

[50] Wright to Fursland, 'France and Pakistan', March 16, 1977, TNA FCO37/2066.

[51] Burdess to Fursland, 'The US and the French Reprocessing Plant', April 12, 1977, TNA FCO37/2066, 2; 'Conclusions of a Meeting of the Cabinet held at 10 Downing Street' April 21, 1977, TNA Records of the Cabinet Office (hereafter CAB) 128/61 Meetings 1–22, 3.

high-level ministerial group to assess non-proliferation issues. The Prime Minister instructed the group—which included Owen and the Secretaries of State for Energy, Industry, Environment, and Defence—"to keep under review problems arising from the transfer of civil nuclear plant and technology, to consider their implications including the need to avoid the proliferation of nuclear weapons, and to report to the Ministerial Committee on Energy (ENM)".[52] The new GEN 74 committee's first task was to evaluate and provide feedback on Carter's upcoming nuclear statement.

GEN 74 examined the likely outcomes of Carter's review, briefed by an interdepartmental committee of advisers on the commonality of non-proliferation concerns between the USA and the UK *and* the major differences between the two nations. The FCO argued that Britain could not wholeheartedly support any policy that attacked the UK's reprocessing programme because of British energy needs, environmental concerns, technical aspects of Britain's nuclear power programme, and worries that a general moratorium on reprocessing would *drive* further proliferation.[53] For one, the fuel element design of the British Magnox nuclear power stations meant that reprocessing or safe disposal would have to take place within a few years of removing the elements from the reactors. GEN 74 also contended that the transfer of reprocessing technology to Brazil and Pakistan dominated American thinking.[54]

Shortly after GEN 74's formation, the State Department gave the British, French, and West German governments three days to respond to a draft of Carter's proposed April 7 nuclear policy statement.[55] This sudden request for input was not entirely surprising to London, as Carter

---

[52] Callaghan to Owen, Memorandum, March 21, 1977, TNA FCO66/934, 1.

[53] Ministerial Group on Non-proliferation (hereafter GEN 74), 'US Non-proliferation Policy: Note by the Minister of State, Foreign and Commonwealth Office', March 28, 1977, TNA CAB130/963, 1–2.

[54] GEN 74, 'US Non-proliferation Policy', March 28, 1977, TNA CAB130/963, 1.

[55] GEN 74, 'US Message on Non-proliferation: Note by the Foreign and Commonwealth Office', March 30, 1977, TNA CAB130/963, 1. According to Brzezinski, this brief notice was an improvement on the original plan to give only a few hours' notice to the nations of Western Europe and Japan (Brzezinski, 130–131).

had warned Callaghan that it might be on the horizon during the latter's visit to the USA in early March.[56] The West Germans refused to comment on such a short timescale and hurried through the dispatch of key blueprints to Brazil before the United States could impose a moratorium on reprocessing plant exports.[57]

Whitehall's most serious worry was that American efforts to curb reprocessing could have damaging effects on Britain's nuclear industry and public opinion. If Carter were to ban the reprocessing of US-origin nuclear fuels, this would destroy the commercial basis for the THORP. The ongoing public enquiry into the plant was also at a delicate stage and an American announcement regarding reprocessing could seriously affect British public opinion.[58] The UK sought to move the United States away from what Owen and his colleagues saw as a rigid stance on reprocessing. This was not simply a domestic issue for Britain, and it was pointed out to Carter's people that in the 1960s the USA had raised expectations in the developing world about the benefits of nuclear technology. These expectations required careful management lest Carter's changes in policy accelerate, rather than retard, proliferation by reducing access to reprocessing services, thus forcing states that had invested in nuclear power to set up their own facilities.[59]

After Carter's April 7 speech, London welcomed the new policy while protecting UK interests. The official public response articulated broad support for US non-proliferation policy, while off the record reactions conveyed reservations. Officially, the UK would play an "active and constructive" role in global non-proliferation activity and press comment stated that the government "intends to give non-proliferation consideration [sic] their full weight, together with commercial considerations".[60]

[56] GEN 74, 'Minutes of a meeting held in Conference Room A, Cabinet Office' March 31, 1977, TNA CAB130/963, 7.

[57] Gray, 'Commercial Liberties', 462–467.

[58] GEN 74, 'US Message on Non-proliferation: Note by the Foreign and Commonwealth Office, Annex 'B'", March 30, 1977, TNA CAB130/963, 1–2.

[59] 'Minutes of a meeting', March 31, 1977, 7–8.

[60] 'Draft No.10 Statement in Reaction to President Carter's Non-proliferation Statement', April 7, 1977, TNA FCO66/934, 1.

Off the record, the policy required further study.[61] The government argued that there were strong non-proliferation, economic, and environmental reasons for offering reprocessing services, statements that must be seen in the context of the massive THORP project.[62] Callaghan demanded a more in-depth analysis of how far the UK could identify itself "with all or any aspects of President Carter's policy without endangering our commercial interests and, in so far as our commercial interests are endangered, what is at stake for us".[63]

GEN 74 meetings and briefings emphasised the fissures between the USA and the UK. Ministers had hoped that ground had been gained after the feedback to the Carter administration before the April 7 announcement, but after Carter's confusing introduction of the draft NNPA at the end of April, GEN 74 participants felt such optimism was groundless. The FCO believed the upcoming May 7–8 Group of Seven industrialised nations (G7) summit in London could be a forum for engaging with Carter to persuade him to change tack.[64] British analysts argued—presciently as it turned out in the case of Pakistan—that it was the cheaper, smaller, and easier to build gas centrifuge uranium enrichment technology that posed the greater global proliferation risk, contrary to the American position where reprocessing was the *bête noire* of a proliferation-free world.[65] Indeed, it was centrifuge-based enrichment and not giant, industrial-scale processes, such as reprocessing, that invalidated the technology-based non-proliferation controls that were the

---

[61] Work on reprocessing and breeder reactor technologies had accelerated in the 1960s and early 1970s when estimates suggested that global uranium supplies were running out. Carter's speech drew upon the more positive estimates of uranium supplies, especially those offered by the 1977 Ford-MITRE study.

[62] GEN 74, 'Agenda Item 1: World Uranium Supply and Demand', April 27, 1977, TNA FCO66/935, 2; 'Draft No.10 Statement', April 7, 1977, 2.

[63] Rose to Moberly, 'President Carter's Statement on Non-proliferation', April 25, 1977, TNA FCO66/934.

[64] GEN 74, 'Nuclear proliferation Implications of the Reprocessing of Irradiated Fuel and the Storage of Spent Fuel Elements', May 2, 1977, TNA FCO66/935.

[65] GEN 74, 'Nuclear proliferation Implications of the Enrichment of Nuclear Fuel', April 27, 1977, TNA CAB130/963, 1–3.

NPT's, NSG's, and Carter's anti-proliferation strategy.[66] A centrifuge plant was exactly the kind of nuclear facility that more junior figures in the British government suspected Pakistan was creating. Owen suggested a negotiating strategy that was predicated on British commercial interests, but couched within a framework of supporting non-proliferation. Rather than emphasising economic interests, Owen suggested, "we might point out that the provision of reprocessing services by nuclear weapons states could have the non-proliferation benefit of reducing the incentives for non-nuclear weapons states to acquire this capability".[67] With the THORP in the pipeline, the nuclear weapon states in question obviously included Britain.

Changes in the US attitude towards reprocessing had serious commercial implications for the UK. British attitudes towards the issue—and the Pakistani case in particular—were based on a strategy designed to avoid damaging the UK's economic interests. Aside from the fact that British analysts suggested that reprocessing technology was "a door that was already jammed open", US plans to have oversight over the reprocessing of US-origin fuel threatened the entire THORP programme. If the THORP did not go ahead, the economic implications were vast: loss of a multimillion pound Japanese contract; loss of future contracts; loss of wider infrastructural investment; loss of thousands of new jobs; and irretrievable damage being done to the financial position of British Nuclear Fuels Limited (BNFL).[68] This was to inform Callaghan's position at the G7 summit.[69] The Prime Minister was briefed to display a positive attitude towards Carter's non-proliferation agenda while making it clear that further study of both the proposals and their underlying assumptions was required. The FCO argued that a small, expert group should undertake such a study before the organising of the proposed INFCE project even got underway.[70]

---

[66] R. Scott Kemp, 'The End of Manhattan: How the Gas Centrifuge Changed the Quest for Nuclear Weapons', *Technology and Culture*, 53:2 (2012), 274–75, 298.

[67] GEN 74, 'Minutes of a Meeting Held in Conference Room C, Cabinet Office', May 2, 1977, TNA CAB130/963, 4.

[68] GEN 74, 'The Implications of President Carter's Statement on Non-proliferation', April 29, 1977, TNA CAB130/963, 4–5.

[69] 'Downing Street Summit–7–8 May1977: Non-proliferation; Brief by Foreign and Commonwealth Office', May 5, 1977, TNA FCO66/935, 6.

[70] Ibid., 3.

Carter arrived in the UK on May 6, and Callaghan saw the President as a man he could get on with, in part because of their shared naval experiences and Baptist upbringings.[71] The Prime Minister believed that Carter was a foreign policy neophyte who would need to turn to a wiser, more experienced head of state for advice.[72] Callaghan rather parochially hoped that he could use this rapport to act as an intermediary between the President and other European leaders, particularly the West German Chancellor Helmut Schmidt, whose relationship with Carter became increasingly fractious.[73] Callaghan had, with the consent of Giscard and Schmidt, persuaded Carter to add non-proliferation to the summit agenda.

Before the summit, the State Department briefed Carter on the appropriate approach to his fellow leaders, appealing to Carter the pragmatic politician rather than Carter the moralist, urging him to see the European and Japanese points of view.[74] The President still saw European anger over reprocessing as rooted in ruffled national pride over perceived US intrusion into their domestic and economic affairs.[75] In viewing the allied positions through this particular lens, Carter failed to recognise that states such as Britain might have genuine economic *and* physical needs for reprocessing. However, State Department appeals to reason prevailed and Carter was accommodating, reasonable, and pragmatic during the summit. Nowhere was this more apparent than in the summit's non-proliferation session.[76]

After testy exchanges between Schmidt, Giscard, and Canadian Prime Minister Pierre Trudeau, the British delegation spoke up. Owen supported the INFCE project but formally called for the prior investigation by a small group of experts that had been suggested before the summit, and reiterated the distinct problems Britain faced in dealing with spent

[71] Jimmy Carter, *White House Diary* (New York: Farrar, Straus, and Giroux, 2010), 46.

[72] Kenneth O. Morgan, *Callaghan: A Life* (Oxford: Oxford University Press, 1997), 590–591.

[73] Kaufman, *Plans Unraveled*, 110.

[74] Brenner, *Nuclear Power*, 164.

[75] Carter, *White House Diary*, 48.

[76] Owen, *Time to Declare*, 320.

nuclear fuel.[77] Callaghan observed that negative international perceptions of Carter's policies stemmed from the view that policy was being made because America was wealthy and energy-rich status, and not because Carter was concerned about proliferation, thus making the President's life more difficult. The Prime Minister accepted that Carter's stance was founded in a deep concern about nuclear proliferation, but it was important not to give the appearance of attempting to deprive other countries of nuclear technology's full benefits.[78] On the meeting's margins, the FCO's Patrick Moberly spoke to Joe Nye regarding worries that a non-proliferation communiqué from the summit might make it look to the rest of the world that INFCE was being fixed by a cartel of advanced nuclear supplier nations.[79]

Notwithstanding the debate's—at times—acrimonious nature, Callaghan recounted in his memoirs that it eased bilateral difficulties, lessened misunderstandings, and found "a useful method of carrying forward policy on a matter of vital importance to the world's future".[80] INFCE had been broadly agreed upon and the proposal for a smaller, two-month preparatory study was welcomed. Particularly apparent were the differing perceptions of what posed the greater proliferation danger. In Carter's eyes reprocessing and the plutonium economy were the main threats. For Britain, more invested than the United States in reprocessing for domestic needs and as a source of revenue, centrifuge-based uranium enrichment programmes were far more dangerous. The GEN 74 pre-summit briefings starkly illustrate the point made by British analysts that giant industrial reprocessing mega-projects, such as the THORP, were not the way forward for developing states seeking weapons-grade fissile material. However, there was also a British interest in de-prioritising reprocessing because of the economically important THORP. It was because of centrifuge-based enrichment that countries such as the PRC and Pakistan could pursue (and in both cases gain) nuclear capability. The Carter administration's near obsession with

---

[77] 'Note of the Second Session of the Downing Street Summit Conference', May 7, 1977, TNA PREM16/1223, 13.
[78] Ibid., 14.
[79] 'Downing Street Summit: Non-proliferation', May 11, 1977, TNA FCO66/935.
[80] Callaghan, *Time and Chance*, 484.

reprocessing as a source of fissile material would—over the coming eighteen months—cause mounting evidence of Pakistan's clandestine enrichment programme to be ignored, even when allies in London pointed out the existence of that programme. Britain's discoveries about inverters pointed at the real direction in which Pakistan was heading to obtain the bomb.

FROM BHUTTO TO ZIA

One month after the G7 summit, on July 5, the Pakistani military, under the command of army Chief of Staff General Muhammad Zia ul-Haq, ousted Bhutto, promising to hold new elections within ninety days. Military rule eventually lasted eleven years. Zia was a different sort of leader to Bhutto, seen by British diplomats as taking action because of the incumbent government's corruption.[81] Like Carter, Zia was a man of profound religious faith, envisaging Pakistan as an ideological state founded on Islamic principles, where religion was a key part of civil and political life. He dreamt of Pakistan holding a leadership position within the Islamic world, a dream that would come to encompass the nuclear weapons programme.[82]

Before the coup there had been further American efforts to gain concessions on the reprocessing plant. Despite Bhutto indicating his willingness to take part in constructive negotiations, US-Pakistani relations were sliding towards their lowest ebb yet. Bhutto accused the United States of conspiring against him by interfering in the election and again claimed that the Soviets were attempting to penetrate South Asia.[83] Although Bhutto made conciliatory noises in late May, Peter Constable—*chargé d'affaires* at the US Embassy in Islamabad—expressed doubts about his sincerity.[84] Such doubts were proved right in a meeting on the margins

---

[81] Robin Fearn, interview by Malcolm McBain, November 18, 2002, BDOHP, www.chu.cam.ac.uk/archives/collections/BDOHP/Fearn.pdf (accessed July 17, 2013), 25.

[82] Haqqani, *Pakistan*, 131–136.

[83] Bhutto to Vance, Letter, May 3, 1977, JCPL, Records of the Office of the National Security Adviser (Carter Administration), 1977–1981, Zbigniew Brzezinski's Country Files: Pakistan, (hereafter Brzezinski: Pakistan), Box 59.

[84] Memorandum, May 24, 1977, JCPL, RAC NLC-1-2-7-19-1.

of the May Conference on International Economic Cooperation (CIEC) in Paris, when Vance avoided open confrontation with Aziz Ahmed over the nuclear issue. Ahmed, briefed by Bhutto, accused the Americans of electoral interference, of lacking commitment to Pakistan, and of discriminating against his nation's nuclear energy programme. Vance backed away from the allegation of electoral interference, but sought to emphasise that the United States had itself chosen to forego reprocessing in favour of "less dangerous alternatives".[85] Consequently, Joe Nye was anxious not to give the impression that any French decision was the result of American pressure. The Quai d'Orsay was quiet, even going so far as to rebuff British enquiries about the reprocessing plant deal's status.[86] The opacity of Paris' position and the Pakistani accusations of electoral interference all served to increase the murkiness surrounding the reprocessing plant.

Relations between Washington and Islamabad were deteriorating because of several factors. The two main contributors to this decline were mounting US suspicions that the Pakistanis had military intentions for their nuclear programme and Bhutto's belief that America was conspiring with his political opponents to remove him from power. By the beginning of June Carter had caused outrage in Pakistan by rescinding Kissinger's offer of A-7 aircraft. This came about because of mounting evidence that Pakistan was purchasing the reprocessing plant as part of a nuclear weapons programme and Carter's evolving policies on conventional arms sales. These policies were given concrete form by Presidential Directive 13 (PD-13), which established a policy of conventional arms sales restraint and—significantly for the Pakistan situation—stated, "the United States will not be the first supplier to introduce into a region an advanced weapons system which creates a new or significantly higher combat capability".[87] The A-7 announcement caused a furore not only because of its nature, but also because the information release was poorly handled. Carter had made the decision on April 9 in a marginal note for Brzezinski, who had held on to the information for over a month.

---

[85] USDEL Secretary in Paris to White House, 'Evening Report for the President', May 31, 1977, JCPL, RAC NLC-128-12-8-20-6.

[86] Macrae to Wilmshurst, 'Reprocessing Plant for Pakistan', June 6, 1977, TNA FCO96/728.

[87] Presidential Directive/NSC-13, 'Conventional Arms Transfer Policy', May 13, 1977, JCPL, PD-PRM, www.jimmycarterlibrary.gov/documents/pddirectives/pd13.pdf (accessed July 21, 2013), 2.

There was, it transpired, never any formal Presidential notification of the decision and leaks to the media forced the administration's official announcement.[88] Carter demanded that Vance explain how the debacle had occurred and it transpired that a State Department press officer released the information before the Department of Defense and NSC's official decision on the announcement's timing.[89]

The A-7 offer's withdrawal was a strand of a much harder-line stance on the Pakistani nuclear programme that attempted to tread a fine line between a lack of action and the imposition of congressionally mandated sanctions. There was, however, considerable confusion over the application of various legal restrictions on aid to near-nuclear nations. Newspapers had carried stories about technology transfers from France to Pakistan, giving rise to administration worries about a run-in with congressional representatives curious as to why the administration had not applied the Symington Amendment, terminating all economic assistance.[90] Government legal advisers had stated that the Amendment need not be applied, because of "negotiations in good faith" being conducted between the USA, France, and Pakistan.[91] The State Department wanted to discuss the issue with key congressional figures to avoid the Symington Amendment's immediate, and potentially damaging, imposition.[92]

[88] Aaron to Thornton, 'Aircraft Purchase by Pakistan', June 24, 1977, JCPL, White House Central Files, Subject File, CO119: Pakistan, Box CO48.

[89] 'Meeting with the Democratic Members of the Arkansas, Illinois, Missouri, Kentucky, Texas, and Oklahoma Congressional Delegations', June 9, 1977, JCPL, Records of the Office of the Staff Secretary, Presidential Files, Box 30, 1.

[90] 'Pakistan Sticks to French Nuclear Deal', WP, January 4, 1977, A9; 'U.S., Pakistan Plan Talks to Resolve Dispute Over Nuclear Reprocessing', WP, March 25, 1977, A2; 'U.S.-Pakistan Rift On Atom Fuel Grows', NYT, May 8, 1977, 9; 'France Denies Shift On Pakistan A-Plant', NYT, June 2, 1977, 11.

[91] Atherton and Bennet to Christopher, 'Pakistan's Purchase of a Nuclear Fuel Reprocessing Plant: The Symington Amendment and Consultations with Congress', June 23, 1977, USPQB, Doc.3, 1–2.

[92] Ibid., 2; There was considerable misunderstanding in the correspondence regarding whether the Symington or Glenn Amendments applied. This confusion has percolated into the literature on non-proliferation. Mitchell Reiss succinctly clarifies the situation, noting that technically the Glenn Amendment is the appropriate citation for either reprocessing or uranium enrichment technology transfers.

Vance decided to wait for six to eight weeks before making a decision on the Amendment, while briefing key congressional figures on the situation and attempting to obtain concessions from Pakistan's new military government.[93] Vance argued that overenthusiasm in the law's application—which Pakistan would interpret as a "slap" at the new government—could retard efforts to successfully conclude the nuclear negotiations.[94] Intelligence indicated that Islamabad might be attempting to obtain nuclear raw material and technology supplies from China, Niger (who, as a majority Islamic country, became more prominent in 1979), and South Africa.[95] It was clear to Vance that Pakistan could do nothing with uranium supplies unless it had access to fuel fabrication services, services that the French had declined to offer. Here was a key pressure point in the non-proliferation campaign: "By constraining Pakistani access to nuclear fuel services wherever possible, pressure can be built up to encourage Pakistan to adopt and follow responsible non-proliferation policies, including cancellation or indefinite deferral of its reprocessing project."[96] Vance argued that, in a climate of warming Sino-US relations, the PRC could be a useful ally in Pakistani-focused non-proliferation activity.[97] When visiting Beijing Vance noted that nuclear matters were the main obstacle to good relations with Pakistan. The hope was that a new, post-Bhutto government would set aside reprocessing ambitions in favour of a nuclear fuel supply agreement with the USA.[98] In the interim Carter had agreed to authorise in principle the cash sale of forty F-5E fighter aircraft to Pakistan, a move

See Mitchell Reiss, *Bridled Ambition: Why Countries Constrain Their Nuclear Capabilities* (Washington, D.C.: Woodrow Wilson Center Press, 1995), 214 fn14.

[93] Vance to Carter, Memorandum, July 9, 1977, JCPL, RAC NLC-128-12-10-1-5, 1.

[94] Ibid., 2.

[95] Gelb to Vance, 'Nuclear Safeguards–Pakistan, South Africa, China', July 12, 1977, NSAEBB, 'China, Pakistan, and the Bomb' (hereafter CP&B), www.gwu.edu/~nsarchiv/NSAEBB/NSAEBB114 (accessed July 21, 2013), Doc.4.

[96] Vance to Brzezinski, 'Nuclear Safeguards–Pakistan, South Africa, China', July 14, 1977, CP&B, Doc.4, 2.

[97] Ibid., 3.

[98] Memcon, August 22, 1977, *Foreign Relations of the United States 1977–1980* (hereafter *FRUS 77–80*), Vol.XIII, China, Doc.47, 153–154.

intended to dampen Islamabad's demands for the more powerful and advanced A-7.[99]

Joe Nye was dispatched to Islamabad at the end of July to engage in a two-pronged assault on Pakistani nuclear ambitions: to negotiate a delay in the reprocessing plant deal; and to get Pakistan to join the inaugural INFCE meeting due to take place on October 19–21.[100] Nye's mission used coercion, rather than bribery, but with little expectation of demonstrable results.[101] Like Kissinger before him, Nye hid behind Congress, telling Agha Shahi and Munir Khan that unless they withdrew from the reprocessing plant deal, the USA would have no legal option but to cut off economic assistance. As Dennis Kux contends, the fact that Pakistan had agreed to special international safeguards on its nuclear facilities made little impression on the US nuclear specialists. Shahi and Khan informed Nye that the reprocessing plant would go ahead and the American left Islamabad with nothing to show for his efforts.[102] Robin Fearn was provoked to comment: "On the evidence of the Nye visit, there is little scope for further US pressure until after October [the proposed timing of Pakistani elections]... but the omens are scarcely encouraging."[103]

With these omens hanging in the air, Carter decided in September to cut off economic aid without invoking the Symington or Glenn Amendments, a decision prompted by Nye's fruitless mission. The CIA had argued that success in obtaining conventional military aid *or* gaining nuclear capability might reduce Pakistan's determination to acquire the other. However, the administration's non-proliferation and conventional arms control policies frustrated both objectives simultaneously. Thus, the Carter administration undermined its own 'sticks and carrots'. The intelligence agency suggested that Pakistani anger might translate into encouragement—within the limits necessary to maintain the support of

[99] Brzezinski to Vance, 'Military Assistance for Pakistan', July 10, 1977, JCPL, Brzezinski: Pakistan, Box 59.

[100] Christopher to Carter, Memorandum, August 12, 1977, JCPL, RAC NLC-128-12-11-10-3, 2.

[101] UKE Islamabad to FCO, 'Pakistan: Reprocessing Plant: Nye's Visit', August 9, 1977, TNA FCO96/728, 1.

[102] Kux, *Disenchanted Allies*, 237.

[103] UKE Islamabad to FCO, August 9, 1977, 2.

friendly Muslim states—for the more outspoken developing world nations to reject the whole spectrum of US policies, including human rights.[104] The US government cut-off of aid was enforced without any formal announcement in the hope that the diplomatic fallout could be minimised. This was a forlorn hope as US-Pakistani relations collapsed.

In light of this collapse Arthur Hummel in Islamabad argued that, after Zia's postponement of elections, the USA needed to rethink its Pakistan strategy. The Ambassador asked permission to engage in private, back-channel discussions with the Pakistani military leader.[105] Hummel also called for the aid embargo's lifting, as the French government was attempting to work out how it could gracefully back out of the reprocessing contract. For Hummel this represented the "virtual achievement of our objective of assuring that the existing contract for a reprocessing plant in Pakistan will not be carried out".[106] Hummel argued that US legislation had not been violated and that to perpetuate a tough stance on aid only served to further damage US-Pakistani relations.[107] Hummel's counterpart in Paris, Arthur Hartman, expressed reservations about an early end to the aid cut-off, arguing that it would send the wrong message and make it appear to the French that the administration was not serious about non-proliferation. Hartman contended that there would be a negative domestic impact, with Congress assuming that a French decision had been made on the reprocessing plant that tallied with US objectives, when that was clearly not the case.[108]

While Washington digested Hummel's and Hartman's analyses, it was clear that Pakistan's new military government had lost none of the Bhutto-era desire for the nuclear option. Brzezinski observed that Pakistani diplomats were trying to reduce the upcoming inaugural INFCE organising conference's impact by positioning it as a "vehicle to

---

[104] CIA, 'Possible Collision Points for US-LDC Relations', August 19, 1977, JCPL, RAC NLC-31-53-2-6-1, 4.

[105] White House Memorandum, October 5, 1977, JCPL, RAC NLC-1-4-1-21-2.

[106] Hummel to Vance, 'Economic Assistance to Pakistan', October 14, 1977, USPQB, Doc.4, 1.

[107] Ibid., 2–3.

[108] Hartman to Vance, 'Economic Assistance to Pakistan', October 18, 1977, USPQB, Doc.4, 1–2.

threaten the agreement [between France and Pakistan]". Furthermore, Brzezinski argued that Islamabad had made it clear that "the issue has become one of national sovereignty and honor, making it impossible for any Pakistani government or leader to give up the plant or to acquiesce in non-implementation of the agreement with France".[109] Pakistan had been reticent to get involved in the INFCE discussions, claiming that the "interim" government could not take any major decisions on the nuclear issue.[110] A lack of progress on the reprocessing plant negotiations put Zia under pressure from Bhutto's Pakistan People's Party (PPP). While Bhutto languished in a Rawalpindi jail, his party claimed that the Martial Law Administrator had secretly "knuckled under" to the Americans. This was, Foreign Secretary Shahanawaz stated, a major reason why Pakistan could not participate in INFCE at this time, despite the fact that they were the only country that had replied in the negative.[111] Zia's refusal to send delegates to the opening INFCE meetings was, for British diplomats, not entirely unexpected. The Pakistanis were seen as making political capital from the reprocessing plant and INFCE situations, claiming that "industrialised countries are intent on hindering third world development" and that the way to control proliferation lay in "negotiated and enlightened self-restraint rather than manipulating or withholding nuclear technology".[112]

In the meantime Zia had rejected a French proposal to restructure the reprocessing plant to make its output useless for weapons production (so-called co-processing). This rejection confirmed to US observers that the plant was part of a nuclear weapons programme. It also persuaded the French, who promised not to supply any further "sensitive" components to their customer and sought ways to abrogate the contract. According to Alfred Atherton and George Vest it was vital, if for nothing more than French domestic stability, that news of this change in posture not be

---

[109] Brzezinski to Carter, 'Information Items', October 12, 1977, JCPL, RAC NLC-1-4-1-41-0, 1.

[110] Hummel to Vance, 'GOP Participation in INFCEP Organising Conference Unlikely', September 21, 1977, National Security Archive Nuclear Non-proliferation Unpublished Collection, Box 6 (hereafter NSANNUC Box 6), 1.

[111] Ibid.

[112] Burdess to Wilmshurst, 'Pakistan Nuclear Affairs', October 3, 1977, TNA FCO96/728, 1–2.

leaked.[113] The Pakistani domestic situation was also volatile, making it difficult for the Zia administration to renege on the deal. Likewise, for the State Department, immediate discussions with Congress on a resumption of aid were impossible because of France's and Pakistan's publicly stated intentions to go through with the deal. Such was Carter's desire to pass the NNPA that anything that could harm the fragile relationship between administration and Congress must be kept under wraps.[114] Because of this Vance instructed Hummel to open a channel to Zia in order to maintain good relations, but observed that nothing could be done on aid for at least two to three weeks. Hummel was also asked to emphasise US sensitivity to Pakistani security concerns, express support of a South Asian joint declaration on nuclear weapons, explain the ramifications of US laws (such as the Symington and Glenn Amendments), and to reiterate the US position's domestic context to Zia.[115] Vance instructed Hartman to keep pressure on the French, and to remind Giscard's government of their statements that, if Pakistan rejected the restructuring offer, France would quit the contract.[116]

While Pakistan was kept under pressure, Carter was about to inaugurate a cornerstone of his wider anti-proliferation policy, the INFCE conference. American scrutiny of British nuclear policy before the INFCE meeting highlights the contours of agreement and disagreement between the USA and the UK. US diplomats in London argued that Callaghan's government were sceptical about technological solutions to the proliferation problem and comfortable with far more ambiguity in non-proliferation policy. A key factor was the importance that Britain attached to reprocessing—typified by the THORP—as a vital part of its energy and economic plans. Despite all of this US diplomats did not question Britain's dedication to the cause of non-proliferation and contended that the UK was ready to act with the USA as long as its national

[113] Atherton and Vest to Vance, 'The Nuclear Reprocessing Issue with Pakistan and France: Whether to Resume Aid to Pakistan', October 18, 1977, USPQB, Doc.4, 2.

[114] Ibid., 3.

[115] Vance to Hummel, 'Reprocessing Plant and the Question of Aid', October 18, 1977, USPQB, Doc.4, 1–2.

[116] Vance to Hartman, 'French/Pakistan Reprocessing Deal', October 18, 1977, USPQB, Doc.4, 2.

interests were not threatened.[117] Three weeks later Callaghan admitted to visiting Congressmen that UK signals on reprocessing might be confusing. According to the recollections of Callaghan's guests, the Prime Minister stressed that he "would be ready to meet with the US if UK actions seemed to stand in the way of arrangements following from INFCE", but he firmly believed that there was plenty of time to ensure that non-proliferation took hold.[118]

Prior to the INFCE meeting, British ministers and officials expressed concerns about the conference, and about non-proliferation in general. Owen contended that the US non-proliferation gaze was almost wholly focused on INFCE, an undertaking that the United Kingdom must not allow the USA to dominate.[119] Owen outlined the British government's non-proliferation stance in a warmly praised speech to the Royal Institute of International Affairs. Speaking before a receptive audience, he argued that Britain had a major voice, significant standing, and a duty to be involved in international nuclear affairs.[120] Using a memorable turn of phrase, Owen stated that when it came to nuclear proliferation, "there are savage risks in doing nothing".[121] Deliberately publicising British commercial interests, Owen made the case for a strong, well-regulated reprocessing industry, an industry that would, of course, be dominated by countries such as Britain and France.[122] For the GEN 74 meeting on May 20 FCO analysts made it clear that the government should use the fuel cycle evaluation discussions as a means to defend British commercial interests by ensuring it could enter into international reprocessing contracts. British officials had concluded that selling reprocessing services was

[117] USE London to Vance, 'British Nuclear Policy on the Eve of the INFCE Meeting', October 18, 1977, NSANNUC Box 6, 1–5.

[118] USE London to Vance, 'Primin Callaghan's Meeting with CODEL Ribicoff/ Bellmon', November 11, 1977, JCPL, RAC NLC-10-16-5-24-2, 2.

[119] Palliser to Hunt, 'Downing Street Summit: Non-proliferation', May 11, 1977, TNA FCO66/935.

[120] 'Speech Prepared for Delivery by the Foreign and Commonwealth Secretary to the Royal Institute of International Affairs', May 19, 1977, TNA PREM16/ 1182, 1–2.

[121] '"Need to act" on A-power', TG, May 20, 1977, 8.

[122] 'Speech Prepared for Delivery by the Foreign and Commonwealth Secretary', May 19, 1977, 10.

more profitable than selling the technology itself. Furthermore, if the UK were to join INFCE, then the discussions should be based on the UK view of the nuclear world's reality, rather than any US theorising.[123]

The inaugural INFCE conference took place on October 19–21 in Washington, and Carter opened the discussions with a broad reflection on its aims and purposes, and his hopes for a robust but productive discussion of the nuclear fuel cycle's global dimensions.[124] Joe Nye, writing for *Foreign Affairs* in 1978, characterised INFCE as an opportunity for "the supplier countries and the consumer countries [to] come together to study the technical and institutional problems of organizing the nuclear fuel cycle in ways which provide energy without providing weapons".[125] The first meeting's final communiqué noted that non-proliferation action should be taken without jeopardising international energy supplies and that the energy needs of developing countries needed special consideration.[126] After the conference Pakistan acceded to US demands and joined INFCE on a limited basis.[127]

REPROCESSING, INVERTERS, AND JAGUARS

Concurrent with American wrangling over the reprocessing plant, arms sales, and Pakistan's involvement in INFCE, Britain's government combated *and* provoked Pakistan's nuclear programme. There were ongoing British efforts to ascertain the inverter order's nature and its place in a clandestine enrichment programme, but a constellation of problems meant that efforts to hinder Pakistani purchasing achieved little during

---

[123] 'Ministerial Group on Non-proliferation: 20 May: Item 1: Study Group on Fuel Cycle Evaluation', May 19, 1977, TNA FCO66/935, 1–2.

[124] Carter, 'International Nuclear Fuel Cycle Evaluation, Remarks at the First Plenary Session of the Organising Conference', October 19, 1977, PPPJC, http://www.presidency.ucsb.edu/ws/index.php?pid=6809&st=&st1 (accessed July 25, 2013).

[125] Joseph S. Nye, 'Nonproliferation: A Long-Term Strategy', *Foreign Affairs* 56 (1978), 615.

[126] R. Skjoldebrand, 'The International Nuclear Fuel Cycle Evaluation–INFCE', *IAEA Bulletin*, 22:2, 30–33.

[127] USE Islamabad to Vance, 'Pakistan Chooses INFCE Committees', November 1977, NSANNUC, Box 6.

this period. The ongoing British attempts to negotiate a massive arms deal with India and the continuing debates over the THORP negatively influenced non-proliferation efforts. These two strands were the mainstay of UK involvement in the Pakistani nuclear problem through the second half of 1977 and into 1978.

The main British worry throughout this period remained the Emerson Electrical Industrial Controls (EEIC) inverter order and its international dimensions. Ian Cromartie's belief that the FRG's government would quickly respond to news that inverters were being funnelled to Pakistan by a West German consultancy was tragically optimistic. By the end of February the *Auswärtiges Amt* (Federal Foreign Office) and the *Bundesministerium für Forschung und Technologie* (Federal Ministry of Research and Technology) had vaguely noted that they could find no grounds for believing the inverters were intended for a centrifuge enrichment plant.[128] The confusion and inertia surrounding the inverter sale meant that Whitehall took no action by the time Islamabad accepted the EEIC tender in the middle of March.[129] The FCO thought that stopping the export was impossible, but all the interested departments thought it worthwhile getting Cromartie in Bonn to again contact the *Auswärtiges Amt*.[130]

Throughout April and May Cromartie kept pressure on the West Germans with the aim of gaining a definitive decision on whether or not the inverters fell under German export restrictions.[131] By June he had lost patience with Georg-Heinrich von Neubronner (his contact in the *Auswärtiges Amt*) whom he described as "stonewalling", and took advantage of an absence to approach Werner Rouget, Neubronner's head of department, who appeared to take the case more seriously.

---

[128] Cromartie to Wilmshurst, 'Nuclear Suppliers: Exports to Pakistan', February 24, 1977, TNA FCO96/728.

[129] Makepeace to Wilmshurst, 'Export of Invertors to Pakistan', March 14, 1977, TNA FCO96/728, 1.

[130] Wilmshurst to James, 'Nuclear Suppliers: Exports to Pakistan', March 14, 1977, TNA FCO96/728; James to Wilmshurst, 'Nuclear Suppliers: Exports to Pakistan', March 15, 1977, TNA FCO96/728.

[131] Makepeace to McCulloch, 'Nuclear Suppliers: Exports to Pakistan', May 9, 1977, TNA FCO96/728; James to Wilmshurst, 'Proposed Export of Invertors to Pakistan', May 23, 1977, TNA FCO96/728.

Despite this approach, Cromartie was pessimistic about a speedy answer from the Germans.[132]

By September the *Bundesamt für Gewerbliche Wirtschaft* (Federal Office for Industrial Economics, the department that made decisions on export controls) informed Cromartie that, although they now believed the inverters were indeed for use in an enrichment facility, the items in question were not covered by German export laws because they were produced and exported by a British company.[133] Although the West Germans had, in Cromartie's words, "put the ball firmly back in our court", Bonn felt that London had discovered a loophole in the nuclear suppliers guidelines that required discussion.[134]

After a hiatus of nearly nine months, debate in London was reinvigorated by Bonn's disavowal of any responsibility for the inverter order. The first shipment of inverters was due to be exported in September, placing the government in a legal quandary. If the inverters were trigger listable, the export should be stopped, an action that would leave the government open to a compensation claim from EEIC.[135] Legal advisers regarded the entire situation as opaque, both in terms of the export chain *and* in the inverters' relationship to the NSG trigger list. Lawyers judged that EEIC were the supplier to the German firm TEAM Industries and that TEAM were supplying the Pakistani government, thus responsibility for the order lay with the West German government.[136] However, the advice was inconclusive and the corresponding adviser suggested obtaining more technical expertise from BNFL and the Department of Energy (DoE).[137]

By September 30 the DoE were determined to have the export stopped, while leaving EEIC "in the best possible position" to make a claim for

---

[132] Cromartie to Wilmshurst, 'Nuclear Suppliers Exports to Pakistan', June 10, 1977, TNA FCO96/728.

[133] Cromartie to Wilmshurst, 'Nuclear Suppliers: Exports to Pakistan', September 8, 1977, TNA FCO96/728, 1.

[134] Ibid., 2.

[135] Falconer to Berman, 'Proposed Export of Invertors to Pakistan', September 13, 1977, TNA FCO96/728, 1.

[136] Berman to Falconer, 'Proposed Export of Invertors to Pakistan', September 14, 1977, TNA FCO96/728, 2–3.

[137] Berman to Falconer, September 14, 1977, 3.

compensation to the Pakistani, rather than the British, government.[138] However, the export went through, determination to prevent the dispatch to Rawalpindi hamstrung by a dubious legal case and the NSG trigger list's inadequate nature. With the next batch due to go towards the end of November, the DoE wondered about asking Islamabad about end uses. Problematically, this could result in a situation where the government must either believe a hypothetical Pakistani statement that the inverters were not for enrichment centrifuges, or create a diplomatic incident by accusing Islamabad of lying.[139] A situation where EEIC might have a moral or legal claim for compensation from the British government was something best avoided, even if TEAM Industries turned out to be "men of straw".[140] In opposition to the DoE stance, the DoT remained sceptical about the inverters' end uses, but the consensus between interested parties in the FCO, Ministry of Defence (MoD), DoE, and BNFL was that they *could* be used to drive uranium enrichment centrifuges.[141] If the inverters were covered by the NSG trigger list, the West German government would need to be informed and was thought likely to blame the UK for not stopping the export. An interdepartmental meeting suggested reminding Bonn that their failure to respond to British enquiries for nine months left "insufficient time for mutual consideration of the trigger list definition, and for executive action here if the export was to be prevented".[142] The eventual decision was that there were very poor grounds for stopping the export because of an insufficient legal basis and the FCO opinion that inverters were *not* on the NSG trigger list. A cross-departmental decision was taken to look at modifying export controls to catch such items in the future, to inform the FRG and the Embassy in Islamabad of the background, and to ask for reports of any information that might clarify the inverter's end uses.[143]

---

[138] Butler to Ellerton, 'Centrifuge Equipment: Export of Invertors to Pakistan', September 27, 1977, TNA FCO96/728.

[139] James to Butler et al., 'Centrifuge Equipment: Export of Invertors to Pakistan', September 28, 1977, TNA FCO96/728, 1–3.

[140] Ibid., 5.

[141] 'Possible Export of Invertors to Pakistan', September 30, 1977, TNA FCO96/728, 1.

[142] Ibid., 4.

[143] Ibid., 5–6.

London found itself caught between an obvious case of proliferation, a dubious legal case for stopping an export, and the inadequacies of existing international non-proliferation agreements. In this instance, the government erred on the side of caution. This was only the beginning. Over the next few years, Britain—unlike other supplier states such as West Germany—took a strong stance on the export of 'grey area' technology, repeatedly upgraded export restrictions, and placed considerable diplomatic pressure on other nuclear technology supplier states.

The conclusions reached in September did not curtail further discussion. The UK informed the USA for the first time about the suspected clandestine Pakistani efforts. Over dinner in Vienna Michael Wilmshurst informally acquainted a surprised Lou Nosenzo with the case as an example of the complexities inherent in interpreting the NSG trigger list. Nosenzo claimed this was the first time he had heard suggestions that Pakistan had embarked on a centrifuge programme, and advised Wilmshurst that he would investigate further upon his return to Washington and pass on any US information.[144] This dinner conversation is doubly significant. It is the first documentary evidence showing British communication with the United States about the clandestine programme. More importantly, archival research has not revealed evidence of American reaction to the news. Indeed, US officials would not respond to the Pakistani uranium enrichment programme until the late summer of 1978.

The late September decisions did little to retard the export of inverters to Pakistan. Bonn was adamant that they held no responsibility for the export and that British arguments in favour of new restrictions would have to be particularly convincing. Michael Wilmshurst angrily noted on the dispatch, "Indeed—the FRG kept us waiting 9 months for an answer!"[145] By this point the MoD had made further expert enquiries and had concluded that the *only* possible use for the inverters was in a centrifuge-based, uranium enrichment programme. MoD analysts believed that each inverter was able to control a cascade of 1,000 centrifuges, with the total order able to drive enough machines to produce enriched uranium for ten or more nuclear weapons per year. Reflecting on long-held suspicions

---

[144] Wilmshurst to Bourke, 'Exports of Invertors to Pakistan', October 4, 1977, TNA FCO96/728.

[145] Cromartie to Wilmshurst, 'Export of Invertors to Pakistan', October 5, 1977, TNA FCO96/728.

that Pakistan was aiming for nuclear weapons capability, the MoD's Dr. M. H. Dean stated: "We regard this development as a most disturbing one. Our advice is to stop the export if at all possible."[146] Dean's advice was too little, too late.

There were further inconclusive exchanges between the FCO and DoE centred on an exchange of suppliers notes—written assurances as to the end uses of the inverters—with Islamabad. The Pakistanis might well lie about end uses, the FCO's Martin Bourke noted, but at least their response would be on record, hedging against future embarrassment.[147] The DoE's Michael James supported an exchange of suppliers' notes, but argued that such action should only be taken if it was possible to prevent exports.[148] James put the ball back in the FCO's court, causing confusion for Bourke, who noted, "Surely it is D/En [Department of Energy] who make the policy decision on whether or not to export[?]"[149] Wilmshurst argued that although stopping the export was nearly impossible,

> that is not really the point of my suggestion that we try such an Exchange of Notes; the point is that if we propose such an exchange and the Pakistanis accepted, our requirements would be met; if they refused, we should then have good grounds for informing the Nuclear Suppliers Group that the refusal suggests that the Pakistanis might be building an illicit enrichment plant.[150]

After this exchange, discussion of the inverter problem petered out until it dramatically resurfaced in March 1978. The entire episode—from discovery in late 1976 to the failure to prevent the export one year later—is significant for the story of British and American attempts to frustrate Pakistani nuclear aspirations. The widely accepted narrative on the

[146] Dean to Wilmshurst, 'Export of Inverters to Pakistan', October 14, 1977, TNA FCO96/728.

[147] Bourke to Butler, 'Export of Invertors to Pakistan', October 17, 1977, TNA FCO96/728, 1–2.

[148] James to Bourke, 'Export of Invertors to Pakistan', November 1, 1977, TNA FCO96/728.

[149] Marginal Note, James to Bourke, November 1, 1977.

[150] Wilmshurst to James, 'Export of Invertors to Pakistan', November 16, 1977, TNA FCO96/728.

Pakistani enrichment programme—popularised in the work of Armstrong and Trento, Frantz and Collins, Weissman and Krosney, and others—posits that Pakistan was able to build its enrichment facilities due to the incompetence, blindness, or willingness on the part of the United Kingdom and the United States to allow Pakistani nuclear weapons development.[151] This narrative does not benefit from the evidence that the British government was deeply concerned about a Pakistani uranium enrichment programme and took the prospect extremely seriously. The issues in play were not incompetence, blindness, or an obscure desire to permit nuclear proliferation, but the NSG trigger list's confused and incomplete nature, the complexities of international diplomacy regarding such a sensitive subject as proliferation, the fear of being blamed for damaging the economic prospects of British companies, and the difficulty of being certain of the inverters' end uses. This made for a situation where the British government found itself in an extremely difficult position. The difficulty would not ease as more evidence of clandestine Pakistani purchasing came to light. However, Britain would take a leading role in diplomatically combating the Pakistani enrichment programme but—as had been the case in 1977 and early 1978—would find itself hampered by the non-proliferation stances of other states and the complexities of maintaining a relationship with Islamabad while secretly trying to frustrate Pakistan's national nuclear ambitions.

Britain continued its attempts to juggle commercial and non-proliferation priorities when reinvigorated Jaguar negotiations came under increasing pressure from the United States. Discussions at the British High Commission in New Delhi underscored that there was potential for a Jaguar sale to India *and* Pakistan. Although the Indians had repeatedly delayed negotiations, Britain saw them as the better commercial prospect. British diplomats who discussed the issue in New Delhi argued that efforts to stimulate Pakistani interest, limited though they were, should be concealed from the Indians.[152] During US-UK discussions in Islamabad that covered arms sales and the nuclear issue, Hummel attempted to pressure the British delegation into not selling Jaguar to either India or Pakistan. The British

---

[151] For example, see Weissman and Krosney, 174–194, and Frantz and Collins, 83–90.

[152] Pakenham to O'Neill, 'Jaguars for India and Pakistan', October 6, 1977, TNA FCO37/1971.

side, knowing Carter's position on conventional arms sales, argued that Jaguar was not an improvement on weapons already in the region.[153] When Vance briefed Carter for an upcoming visit to India, British arms sales loomed large on the agenda. Following on from the US policy of restraint in arms sales to South Asia—typified by the A-7 decision—Vance urged Carter to request that India show similar restraint in its own arms purchases. Tacitly, this meant Jaguar.[154]

On December 2, in Washington, the Pakistani nuclear programme and British arms sales collided again when American officials implied that the British sale of Jaguar to India would push Pakistan even further down the nuclear path.[155] Richard Ericsson (of the State Department's Bureau of Politico-Military Affairs) outlined the difficult American decision to halt the A-7 sale and the serious domestic consequences this entailed for the administration and the manufacturer. Jaguar, the Americans argued, was superior to any other aircraft in the region and an Indian purchase could provoke Pakistan to "escalation and an arms race".[156] Back in London the SAD outlined the compelling reasons for ignoring the US protestations; if the sale did not go ahead, aircraft production lines may have to close and Britain's overall aircraft sales strategy would suffer a setback. In terms of foreign affairs India would perceive this as a judgement on them and, more widely, it would appear as if the British government had capitulated in the face of American demands.[157]

As the State Department maintained pressure on the FCO the British delegation's leader Donald Murray questioned the American assertion that selling Jaguar to India would push Pakistan further towards the nuclear option and challenged the US view that Jaguar represented a vast improvement in Indian capabilities. Murray characterised US opposition as being within the context of an outmoded view of the sub-continent

---

[153] Cortazzi to O'Neill, 'Call on US Ambassador in Islamabad', October 12, 1977, TNA FCO96/728, 1.

[154] Vance to Carter, 'Your Visit to India, November 27–29', November 1977, JCPL, RAC NLC-10-16-5-24-2, 13–14.

[155] 'Record of a Meeting Held at the State Department at 11.30 on December 2, 1977', December 6, 1977, TNA FCO96/728, 3.

[156] Ibid., 7–9.

[157] O'Neill to Murray, 'Sales of Jaguar Aircraft to India', December 19, 1977, TNA FCO37/1971, 2–4.

and because of the A-7 manufacturers' fury at the cancellation. There was also a European dimension. If Britain pulled out Murray considered it likely that "the unscrupulous French" would pick up the contract with their Mirage aircraft.[158] If Britain withdrew from the competition someone else was bound to pick up the sale. The government would therefore be depriving the country of valuable export earnings while doing nothing to alter the situation on the sub-continent.[159] David Owen accepted the arguments for the sale and on December 21 dispatched a note to his subordinates: "I agree completely. Push ahead with this sale. If we can land it, I will be pleased."[160] The US State Department was most certainly not pleased. On the very last day of December 1977 it was noted that the Indian purchase of Jaguar would "cause considerable alarm in Pakistan and could set off an arms race in South Asia".[161]

The Indian Jaguar contract and its connections to the Pakistani nuclear programme provide an illustrative case of British willingness to participate in non-proliferation activity coming up against the hard realities of an economy in crisis. Given David Owen's personal liking for arms control and non-proliferation policies, the case demonstrates that the economic and political need for commercial gains could override anti-nuclear considerations. When major British commercial interests remained unthreatened—as with the inverter situation—the UK was strongly anti-proliferation. Where commercial considerations above a certain magnitude were concerned—such as the THORP or the Jaguar contract—the UK government subordinated non-proliferation to the wellbeing of the British economy. The Jaguar case in particular also illustrates London's and Washington's quite different perceptions of the situation. The State Department believed that selling a supposedly nuclear-capable, advanced strike aircraft to India would alarm the Pakistanis so much that they would be even more determined to gain nuclear capability as a hedge against

[158] Murray to O'Neill, 'Sales of Jaguar Aircraft to India', December 20, 1977, TNA FCO37/1971.

[159] Luard to Owen, 'Sale of Jaguar Aircraft to India', December 20, 1977, TNA FCO37/1971.

[160] Wall to O'Neill, 'Sale of Jaguar Aircraft to India', December 21, 1977, TNA FCO37/1971.

[161] State Department Bureau of Intelligence and Research, 'Analysis', December 31, 1977, JCPL, RAC NLC-4-7-1-16-2, 1.

increased Indian military superiority. The FCO, from Owen on down, remained unconvinced by this argument and—because of economic imperatives—compartmentalised arms sales to India and the Pakistani nuclear problem. Furthermore, this demonstrates the lack of influence that the United States could bring to bear on its key ally when hundreds of millions of pounds of foreign investment were at stake. The situation would continue to evolve into 1978, becoming a source of increasing frustration for the United States and the cause of British resentment at American interference in its commercial affairs.

THE USA AND PAKISTAN AFTER THE INFCE MEETING

After the first INFCE meeting and Pakistan's decision to have a limited part in future discussions, there were significant further developments. Brzezinski agreed to a US abstention from a UN General Assembly (UNGA) vote on a South Asian Nuclear Weapon Free Zone (SANWFZ) proposal by Pakistan, a proposal that the Pakistanis had been pushing since the Indian test in 1974. Brzezinski's staff believed that the proposal was a Pakistani political ploy and a positive US vote could damage the warming US-Indian relationship and, furthermore, unless China was party to the SANWFZ, India would have nothing to do with it.[162] There was, however, a later shift in attitude during a further round of voting, when both the USA and UK voted in favour of the SANWFZ proposal, a move that was welcomed in Islamabad.[163]

By the end of 1977 the French government had—it appeared—finally agreed to abandon the Pakistani reprocessing deal. The situation was problematic for all concerned. Even though France had tentatively informed Pakistan and the USA that the deal was now off, the French did not want the news made public due to the adverse effect it might have on the Giscard government's chances in the March 1978 legislative elections. This placed the Carter administration in a difficult situation; the US government could now resume aid to Pakistan and patch up relations with Zia, but to do so would mean explaining the French cancellation to

[162] Thornton and Tuchman to Brzezinski, 'South Asian Nuclear Free Zone Vote in the UNGA', November 10, 1977, DDRS, DDRS-268103-i1-2.

[163] Burdess to Candlish, 'Pakistan's UN Resolution on South Asian Nuclear Weapon Free Zone', December 29, 1977, TNA FCO37/2066.

Congress, an action that would automatically mean that the information would become public.[164] Nye opined that such an eventuality might come to pass anyway, as the Pakistanis were apparently briefing their domestic press to expect an unfavourable French announcement.[165] The French decision was a victory of sorts for the United States. Even though the change in attitude could not be announced or used overtly, it was a worthwhile outcome to months of often painful and protracted diplomacy.

Just past the first anniversary of Carter's inauguration, Vance submitted a lengthy report on proliferation, asking the question, "where are we after the first year?"[166] There were some successes to reflect upon: the NSG trigger list's final publication in January; the welcome given to INFCE; and, most importantly, the increased prominence of non-proliferation as an international issue.[167] On the negative side, Vance expressed doubt about France's firmness regarding the decision on Pakistan and reflected upon the lack of progress regarding the FRG-Brazil nuclear deal. The Secretary of State contended that the new non-proliferation policy had left the USA isolated, arguing,

> We must recognise, however, that while we have sensitized the international community to the dangers of proliferation, we remain essentially isolated (with Canada and Australia) among the major industrialized states in questioning the inevitability of moving toward reprocessing and early commercialization of breeder technology. The prevailing attitude remains that non-proliferation goals can be pursued without conflict with perceived energy needs if reliance is placed on political and safeguards arrangements rather than limits on technology. The success of our policy will depend to a great extent on our ability to reconcile these differences.[168]

[164] Vance to Carter, 'Evening Report', December 27, 1977, JCPL, NSS-NS Box 95.

[165] Vance to USE Paris, 'Meeting with Ambassador de Laboulaye on Nuclear Issues', December 1977, JCPL RAC NLC-16-110-1-6-5, 5-6.

[166] Vance to Carter, 'Non-proliferation Policy Progress Report', February 26, 1978, JCPL RAC NLC-15-123-2-4-3, 1.

[167] Ibid.; The NSG trigger list was officially published in IAEA Information Circular (INFCIRC) 254 on January 11, 1978. See www.iaea.org/Publications/Documents/Infcircs/Others/infcirc254.shtml (accessed July 25, 2013).

[168] Vance to Carter, February 26, 1978, 2.

Again, a key figure in the Carter administration had failed to note the genuine economic concerns expressed by the UK and others surrounding abandoning reprocessing. Likewise, the focus remained on reprocessing and plutonium as the primary proliferation route. Once again, centrifuge-based programmes went unremarked. However, Vance's scepticism about the reprocessing plant was well founded. The issue would rumble on into 1978 as Pakistan attempted to keep the deal alive.

By March 1978, after being passed by overwhelming majorities in both houses, Carter signed into law the Nuclear Non-proliferation Act of 1978 with the intent of bringing to a halt "the spread of nuclear weapons capability while preserving the peaceful use of nuclear energy".[169] The NNPA placed strict criteria on US nuclear exports and required the renegotiation of existing agreements to bring them into line with the new policy. Any countries embarking on a nuclear weapons programme, assisting others to do so, or failing to submit to full-scope safeguards inspections would have all US cooperation terminated.[170] As J. Samuel Walker notes, the NNPA enshrined distrust of foreign countries when it came to non-proliferation matters.[171] In an extreme interpretation, the Act served warning to the nuclear-exporting states of Europe to bring their own nuclear export activities into line with the United States or face the consequences. For Britain the NNPA could theoretically precipitate an economic disaster by destroying the basis for the THORP.

BRITAIN, PAKISTAN, AND THE REPROCESSING ISSUE AFTER THE INFCE MEETING

Irrespective of Britain's own reprocessing ambitions, Callaghan's government—less well informed than the Americans about evolving French thinking—continued in its attempts to wheedle information on the Pakistani reprocessing plant out of the Quai d'Orsay. Callaghan's

---

[169] Wittner, *Struggle Against the Bomb*, 50; Carter, 'Nuclear Non-proliferation Act of 1978', March 10, 1978, PPPJC, www.presidency.ucsb.edu/ws/index.php?pid= 30475&st=reprocessing&st1=#axzz1muxPn8eP (accessed February 20, 2012).

[170] Peter A. Clausen, *Nonproliferation and the National Interest: America's Response to the Spread of Nuclear Weapons* (New York: HarperCollins, 1993), 135.

[171] J.S. Walker, 'Nuclear Power and Nonproliferation', 246.

impending visit to India and Pakistan necessitated more detailed information, even though the British Embassy in Paris had previously advised the FCO not to press the French government too hard on the reprocessing plant issue. With Zia having now consolidated his power, Whitehall wanted to know more about his attitudes towards the reprocessing plant and the French stance on the issue in order to assess how much scope Callaghan had for influencing Zia's position.[172] Responding to British enquiries, the French were coy. According to the Quai, the reprocessing deal was—by early December—in a state of "pause", but from the French point of view there was little diminution in Pakistani determination to go ahead with the reprocessing facility and little hope for Pakistani acceptance of full fuel cycle safeguards.[173] In private, French officials admitted to British diplomats that they found themselves in something of an embarrassing position, having decided that the reprocessing plant contract was a bad idea but not wanting to lose credibility by appearing to cave in to American pressure.[174] British consular officials in Paris believed that the French were doing their best to frustrate the contract, while in public Giscard's government was still committed to the deal and expressed faith in Pakistani assurances that the plant would not be used for military purposes.[175]

From Islamabad, Kit Burdess commented that US diplomats seemed resigned to the fact that there was no moving the Zia government on the reprocessing issue. French reluctance to take public action on the deal might, Burdess suggested, stem from wider commercial interests, such as the Saviem truck plant discussed earlier in the year.[176]

[172] Cortazzi to James, 'Pakistan Nuclear: French Reprocessing Plant', November 30, 1977, TNA FCO37/2066.

[173] James to Cortazzi, 'Pakistan Nuclear: French Reprocessing Plant', December 11, 1977, TNA FCO37/2066, 1–2.

[174] UKE Paris to FCO, 'French Sale of Nuclear Reprocessing Plant to Pakistan', January 9, 1978, TNA FCO96/822.

[175] Macrae to Wilmshurst, 'Supply of a Reprocessing Plant by France to Pakistan', January 13, 1978, TNA FCO96/822, 2; UKE Paris to FCO, 'MIPT', January 10, 1978, TNA FCO96/822.

[176] Burdess to Wilmshurst, 'Pakistan Nuclear Affairs–The Reprocessing Plant', November 14, 1977, TNA FCO96/728, 1.

Britain sprang to prominence in Pakistan with a raft of press stories that suggested the UK supported the wide availability of reprocessing plants. Headlines such as "UK Backs N-plant Stand by Pakistan" caused John Bushell to urgently request clarification from London. According to Bushell, the Pakistani media was reporting that the United Kingdom Atomic Energy Authority (UKAEA) had stated that withholding reprocessing facilities could not control proliferation and that reprocessing was an essential part of nuclear power programmes.[177] The FCO suspected that the mistaken headlines came from speeches by British officials, submissions to the Windscale enquiry, or from the UKAEA annual report.[178] Owen told Bushell that, if queried by Pakistani officials, the Ambassador should adopt the line that:

> the UK has had a large and sophisticated nuclear industry for over 25 years, of which reprocessing has formed a major part. As the UK is a nuclear weapon state proliferation dangers do not arise. Pakistan, on the other hand, has no such industrial or economic requirement for reprocessing, and she is not a nuclear weapon state. The circumstances are not, therefore, comparable.[179]

Owen's justifications highlight the problematic nature of British domestic nuclear policy when it came to dealing with the Pakistani situation and the stark division that arose between the nuclear haves and the have-nots. The arguments made by Pakistan in favour of having a domestic reprocessing capability—energy needs and the economic basis—were identical to the British justifications for the THORP. Hence, the British reticence to become deeply involved in anti-reprocessing plant diplomacy, even though Owen attempted to delineate a qualitative difference between British and Pakistani needs and capabilities, predicated on the fact that Britain was a nuclear weapon state. This conflicted with Carter's desire

---

[177] Bushell to FCO, 'UKAEA Report on Nuclear reprocessing', November 16, 1977, TNA FCO96/728.

[178] Jay to Bushell, 'UKAEA Report on Nuclear Reprocessing', November 18, 1977, TNA FCO96/728; Owen to Bushell, 'UKAEA Report on Nuclear Reprocessing', November 21, 1977, TNA FCO96/728; Owen to Bushell, 'UKAEA Report on Nuclear reprocessing', November 18, 1977, TNA FCO96/728.

[179] Ibid.

that every state should be a have not in terms of reprocessing, even though he had grudgingly conceded that states—such as Britain and France—with existing reprocessing facilities should be allowed to retain them.

Between January 11 and 13 Jim Callaghan visited Pakistan—the first visit by a major Western political leader since the military takeover—as part of his sub-continental tour. Just before Callaghan left New Delhi for Islamabad, Owen sent an urgent telegram advising that the Prime Minister should not raise the topic of reprocessing and instead try to persuade the Pakistanis to accept full fuel cycle safeguards on their nuclear facilities.[180] In the midst of the pomp, ceremony, and proclamations of mutual affection, Callaghan emphasised to his hosts UK concerns about democracy, non-proliferation, and—contrary to Owen's advice—the reprocessing plant.[181] During their private conversations Zia apparently stated his willingness to sign up to the NPT and to accept full safeguards, although Agha Shahi later struck a note of caution regarding this new-found enthusiasm for the treaty.[182] The most significant conversations took place outside the Callaghan–Zia meetings, when officials from both nations discussed the nuclear issue and the potential for formal talks in London. Agha Shahi was particularly interested in nuclear guarantees from nuclear weapon states to non-nuclear weapon states, and hoped for a sympathetic hearing from the British. The prospect of Anglo-Pakistani nuclear talks in London was guardedly welcomed and the FCO put plans in place for a meeting sometime in February to discuss security assurances and transfers of nuclear technology.[183]

The hastily convened nuclear talks took place in London on February 23. The discussions followed a familiar pattern, with the Pakistanis seeking nuclear security assurances against the Indian threat, and the British

[180] FCO to UKHC New Delhi, 'French Reprocessing Plant for Pakistan', January 10, 1978, TNA FCO96/822.

[181] UKE Islamabad to FCO, 'Prime Minister's Visit to Pakistan', January 16, 1978, TNA FCO96/822, 2.

[182] Cartledge to Prendergast, 'Prime Minister's Discussions in South Asia, 3–13 January, 1978: Nuclear Matters', January 17, 1978, TNA FCO96/822; UKE Islamabad to FCO, 'Anglo/Pakistani Talks on Non-proliferation and Disarmament', January 24, 1978, TNA FCO96/822.

[183] O'Neill to Mallaby, 'Pakistan: Disarmament', January 17, 1978, TNA FCO96/822; Bushell to FCO, January 24, 1978.

attempting to cajole their guests into providing some assuran
they would consider full fuel cycle safeguards or even sign u
NPT. Naiz Naik, Additional Foreign Secretary in the Pakistani Ministry
of Foreign Affairs, argued that security assurances were vital for the
"safety and security of the third world".[184] Neither side in the discussions mentioned the reprocessing plant or the clandestine uranium
enrichment programme. Despite mutual declarations of respect, these
talks were fruitless beyond a vague appearance that the Pakistanis might
be a little more flexible on the subject of security assurances and that
further talks at a higher level might take place in March.[185]

CONCLUSION

If there was one truism to come out of 1977 it was Robin Fearn's
statement that the "omens were scarcely encouraging" when it came to
Pakistan's nuclear weapons programme. For all the Carter administration's
and Callaghan government's efforts, there was little in the way of genuine
progress. New faces in Washington, London, and Islamabad had not
brought new solutions. One dim ray of light emerged on the reprocessing
front, with expressed French willingness to disengage from the Pakistani
deal. However, this was tempered by the Giscard government's lack of
enthusiasm for going public on their change of heart. The clandestine
enrichment programme's emergence in late 1976 and into 1977 represented a new strand in the tangled web of Pakistani nuclear efforts. Success
in hindering the Pakistani purchasing programme was lacking, despite
British attempts to comprehend and utilise the evidence provided by
Pakistani purchasing in the UK. The ability to use that evidence to move
the non-proliferation agenda forward collided with the difficulty of prov-
ing the intended end use of inverters, the dubiety of the legal case against
exports, and the unwillingness of fellow nuclear supplier states to support
Britain's efforts. While committed to non-proliferation, Callaghan's government had to juggle such global commitments with the harsh realities
of Britain's economy. British economic self-interest in the form of the

---

[184] 'Record of the Anglo/Pakistani Talks on Non-proliferation and Disarmament',
February 23, 1978, TNA FCO37/2112, 1.
[185] UKE Islamabad to FCO, 'Pakistan, India, and Nuclear Proliferation', March
13, 1978, TNA FCO37/2112.

THORP and the Indian Jaguar negotiations complicated non-proliferation activity on the sub-continent and created friction between the US and UK governments. In 1978 non-proliferation policy would see some success with the final cancellation of the Franco-Pakistani reprocessing plant contract. However, as reprocessing faded into the background, the clandestine enrichment programme became a much more significant proliferation threat.

CHAPTER 5

"We do find this statement of intentions to be disquieting" The US-UK Diplomatic Campaign Against Pakistan, March 1978 to December 1978

On November 1, 1978, US Deputy Secretary of State Warren Christopher expressed unease over Pakistani nuclear ambitions. From prison, Zulfikar Ali Bhutto had claimed that Pakistan was near to a nuclear breakthrough, stating, "We know that Israel and South Africa have full nuclear capability. The Christian, Jewish, and Hindu civilizations have this capability. The communist powers also possess it. Only the Islamic civilization was without it, but that position was about to change."[1] Reflecting on Bhutto's political testament, the number two man in the State Department stated that, "we do find this statement of intentions to be disquieting".[2] The Deputy Secretary of State had good reason for disquiet. Although there was a non-proliferation success in 1978 when the French government

---

[1] Zulfikar Ali Bhutto, *If I Am Assassinated*, www.bhutto.org/books-english.php (accessed May 1, 2013), 151, originally published by Vikas Press, New Delhi, 1979. From 1972–77, Bhutto made no mention of any Islamic dimensions to Pakistan's nuclear programme.

[2] Department of State (hereafter State) to United States Embassy (hereafter USE) London et al., 'Pakistani Reprocessing Plant', November 1, Jimmy Carter Presidential Library (hereafter JCPL), Remote Archives Capture system (hereafter RAC) NLC 16-114-1-9-8, 3.

© The Author(s) 2017
M.M. Craig, *America, Britain and Pakistan's Nuclear Weapons Programme, 1974–1980*, Security, Conflict and Cooperation in the Contemporary World, DOI 10.1007/978-3-319-51880-0_5

143

abandoned the Pakistani reprocessing plant deal, it became clear that Islamabad's nuclear ambitions were now focused on the clandestine enrichment programme.

With stronger evidence to hand, James Callaghan's government convinced Jimmy Carter's administration that Islamabad was pursuing a secret enrichment programme. Thus, both governments began a campaign conducted in the diplomatic shadows, aimed at marshalling a consensus amongst nuclear supplier states. At the same time, the British government continued to complicate sub-continental non-proliferation policy. Callaghan and his ministers remained determined to keep conventional arms sales and non-proliferation compartmentalised, in the hope of selling Jaguar strike aircraft to India, despite American and Pakistani concerns. The Carter administration—continuing to link arms sales to bribing or coercing General Muhammad Zia ul-Haq into giving up his atomic aspirations—made fruitless efforts to frustrate the deal. Britain's nuclear industry also affected non-proliferation action, as the decision to allow the building of the controversial Thermal Oxide Reprocessing Plant (THORP) weakened London's non-proliferation credibility.

Furthermore, statements made by Bhutto and Zia laid the foundations for public and private discussions about the seeming transnational, religious aspects of Pakistani nuclear aspirations that took place in 1979 and beyond. These provocative pronouncements spurred debate about proliferation in the Muslim world. Although the 'Islamic bomb' idea did nothing to change actual non-proliferation policy in 1978, it created the basis for a much larger and broader series of public and private discussions in the future.[3]

REPROCESSING, ARMS SALES, AND THE FOUNDATIONS OF THE 'ISLAMIC BOMB'

Despite Valery Giscard d'Estaing's indications during 1977 that France was considering quitting its reprocessing plant contract with Pakistan, the deal still vexed American and British politicians and diplomats. The

---

[3] The information on the 'Islamic bomb' issue in this book is in part derived from Malcolm M. Craig, '"Nuclear Sword of the Moslem World"?: The United States, Britain, Pakistan, and the "Islamic bomb", 1977–80', *International History Review*, 38:5 (2016), 857–879.

issue was complicated not simply by French recalcitrance, but also by the Carter administration's resentment of Britain's continued quest for a vastly increased reprocessing capability, by Islamabad's foregrounding of Britain's hypocrisy, and by the ongoing British efforts to sell Jaguar strike aircraft to India.

In early April Pakistan's media highlighted (prompted, Christopher 'Kit' Burdess of the British Embassy in Islamabad assumed, by their government) Britain's Janus-faced position on non-proliferation by running stories focusing on the "discriminatory" aspects of the recent parliamentary vote to allow the THORP's construction. The Pakistani press brought this up as a "contradiction in policy" when the UK and the USA were trying so hard to prevent Pakistan from acquiring reprocessing facilities.[4] During his January 1978 sub-continental tour, Callaghan had given Britain's critics ammunition by stating that, "I am not a fan of reprocessing myself. It won't help us in our task to rid the world of the threat of nuclear weapons."[5] The THORP decision reinforced the appearance of discrimination against the post-colonial world and the Pakistani reaction highlighted the inequalities inherent in anti-proliferation strategies. While Pakistan was subject to diplomatic initiatives to prevent it gaining reprocessing capability, the UK moved on with expanding its own facilities.

The State Department was anxious about how to deal with an aggrieved Pakistan following the French deal's eventual cancellation, and was worried that Pakistan might decide to "go it alone" with the plant.[6] US diplomats proposed a Franco-American solution that would offer Pakistan increased access to economically and developmentally helpful, but less sensitive, nuclear technology, such as reactors. This would also reassure Giscard's government, who were concerned about how the cancellation would affect Paris' credibility as a nuclear exporter.[7] From Islamabad Arthur Hummel

---

[4] Burdess to Wilmshurst, 'Pakistani Reaction to Commons Windscale Vote', April 6, 1978, The National Archives of the United Kingdom (hereafter TNA) Records of the Foreign and Commonwealth Office (hereafter FCO) 37/2112.

[5] 'Callaghan Presses Zia to Tow Line', The *Guardian* (hereafter *TG*), January 13, 1978, 7.

[6] State to USE Islamabad, Paris, and Tehran, 'Pakistan Reprocessing Issue', April 21, 1978, Digital National Security Archive (hereafter DNSA), NP01574.

[7] USE Paris to State, 'Pakistan Reprocessing Issue', April 21, 1978, DNSA, NP01575.

argued that such a plan would help counter accusations of nuclear discrimination against Pakistan and the developing world.[8]

At the end of April Zia demonstrated how angry he was when he connected the THORP, Pakistani nuclear aspirations, and discrimination against the Islamic world in an interview with American journalist Bernard Nossiter. The *Washington Post* gave the reprocessing issue a brief mention amongst a wider discussion of the General's Islamic faith.[9] Kit Burdess, analysing Pakistani media coverage of the interview, pointed out that there was a "clear plea to the USA to stop trying to block the reprocessing plant; and to the French to get on with supplying the plant as under the original project".[10] Zia obliquely referred to nuclear weapons, all the while assuring the world that Pakistan did not intend to build any. Burdess also highlighted remarks about discrimination against the Muslim world, observing that, "Zia emphasised that Brazil was going to get a reprocessing plant, the Indians would soon have a third one, and 'The Jews have got it [reprocessing technology]. Then why should Pakistan, which is considered part of the Muslim world, be deprived of this technology?'" It was these connections between nuclear power and the Islamic world that would come back to haunt the Pakistani leader by creating a propaganda problem for Islamabad, London, and Washington when the Islamic bomb paradigm became publicly prominent from 1979 onwards. Zia also played on American and European concerns regarding oil supplies, commenting that Pakistan needed energy security just as badly as the developed world states, so why should Pakistan not have a full nuclear fuel cycle?[11] Although not explicitly mentioned, Britain—with its recent vote on the THORP—was part of a perfidious developed world that sought to prevent Pakistan from obtaining facilities available to European states.

In the face of Zia's rhetoric, Washington and London attempted to mitigate an adverse Pakistani reaction when it became apparent that the

---

[8] USE Islamabad to State, 'Pakistan Reprocessing Issue', April 25, 1978, DNSA, NP01576.

[9] 'Pakistani Chief Denies Political Ambition', *Washington Post* (hereafter *WP*) May 1, 1978, A17.

[10] Burdess to Wilmshurst, 'Pakistan: Reprocessing Plant', May 15, 1978, TNA FCO96/822, 2.

[11] Ibid., 1.

French were preparing to cancel the reprocessing plant contract. Hummel contended that Pakistan would raise an "almighty fuss" and emphasised the perceived injustice of US pressure on France. The Ambassador argued that the Carter administration must go to considerable lengths to prove that it was not a "willing accessory or accomplice" to the cancellation.[12] Cyrus Vance (whom French Foreign Minister Louis de Guiringaud told that the deal was off) took Hummel's advice on May 18 when he outlined the two main US goals as being to guard against an "intemperate" reaction and to dissuade Pakistan from continuing its efforts to develop a nuclear capability.[13] Vance outlined various methods of achieving these goals, including a "tangible package of inducements" (which included restarting financial aid and military equipment sales), and a political dialogue to reassure Pakistan over its security concerns. Optimistically, Vance noted that the "reprocessing issue is behind us".[14]

In London the THORP was not the only major British commercial interest that adversely affected Britain's credibility on non-proliferation policy. Bilateral Anglo-American discussions once more highlighted that the US government saw arms sales as a non-proliferation tool, in contrast to a British government that wilfully ignored the connections between its arms sales to India and non-proliferation. London saw no problem in continuing to market the Jaguar aircraft to New Delhi, in selling a warplane known to have a tactical nuclear role to the nation whose conventional *and* nuclear superiority to Pakistan was *the* major motivating factor for Islamabad's atomic programme.

Callaghan's official visit to the sub-continent in January was focused on bolstering trade links, with Jaguar forming a major part.[15] Even though Jaguar and non-proliferation had to be discussed, the Cabinet Office and the Foreign and Commonwealth Office (FCO) were keen to make sure

---

[12] USE Islamabad to State, 'Pakistan, Iran, and Reprocessing', May 18, 1978, DNSA, NP01589.

[13] State to USE Islamabad, 'Reprocessing Issue', May 30, 1978, National Security Archive Electronic Briefing Book (hereafter NSAEBB) 'The United States and Pakistan's Quest for the Bomb' (hereafter USPQB), Doc.6, 1–2.

[14] Ibid.

[15] Vile to Cartledge, 'Prime Minister's Visit to Bangladesh, India, and Pakistan, Annex A', December 19, 1977, TNA Records of the Prime Minister's Office (hereafter PREM) 16/1308, 1.

that the two should never intersect.[16] In New Delhi Indian Prime Minister Morarji Desai equivocated, suggesting that—although he would shortly invite a British team to India for discussions—the aircraft might not be suitable for Indian needs. Callaghan argued that a combination of India's Soviet MiG-21 fighters and the Jaguar would provide all the defensive and offensive firepower India could need.[17] True to British government wishes, at no point did non-proliferation and Jaguar cross paths.

By July Washington became ever more exasperated by British intransigence over Jaguar. Joe Nye called on the FCO's Joint Nuclear Unit (JNU) chief Robert Alston to discuss Pakistan's nuclear ambitions and security anxieties. Islamabad had become increasingly nervous about its security situation, as the communist coup in neighbouring Afghanistan at the end of April saw the pro-Soviet—but factionalised and unstable—government of Nur Muhammad Taraki take control. Nye and his colleagues were worried that a fearful Pakistan might move away from the West, towards non-aligned status, by leaving the crumbling Central Treaty Organisation (CENTO), and go all out for nuclear weapons if not reassured. Nye wondered if, given Pakistan's sensitivity, the UK might not delay or halt the Jaguar negotiations.[18]

FCO officials thought the American attitude hypocritical, given that the USA was considering the sale of F-5E fighter aircraft to Pakistan in an attempt to assuage security concerns and ameliorate annoyance at the various non-proliferation efforts.[19] British intransigence regarding Jaguar—still in competition with French aircraft—went to the very top. Owen, despite his Atlanticism and close relationship with Vance, had committed himself to the deal in 1977 and saw no reason to delay or

[16] Hunt to Cartledge, Confidential Memo, December 9, 1977, TNA PREM16/1308.

[17] 'Note of a Conversation Between the Prime Minister and Mr Morarji Desai', January 7, 1978, TNA PREM16/1692, 2; 'Record of a Meeting Between the Prime Minister and the Prime Minister of India', January 7, 1978, TNA PREM16/1692, 4–5

[18] Alston to White, 'Pakistan Reprocessing Plant', July 3, 1978, TNA FCO96/823.

[19] Ibid. The F-5 was a lightweight fighter, the offer of which the Pakistanis declined, still desiring the more capable A-7.

halt it now.[20] Owen angrily commented: "We should look after our relations with the Pakistanis, the Americans can look after theirs."[21] Although British Ambassador to Pakistan John Bushell appreciated that the Jaguar negotiations were a complicating factor in British-Pakistani relations, he did not expect it to cause particular problems.[22] British determination was bolstered by word from New Delhi indicating that British procrastination might result in India making this valuable purchase from France.[23] A contract worth hundreds of millions of pounds would be a fillip to the ailing (and now nationalised) aircraft industry and a feather in the government's cap.

Much to Owen's annoyance, Carter personally attempted to discourage the Indian purchase. The President, in correspondence with Desai, emphasised his policy of restricting the supply of advanced weapons to the sub-continent, arguing that such supplies could precipitate an arms race and disrupt the delicate regional balance.[24] In addition to this presidential intervention, Vance also warned Indian Minister of External Affairs Atal Bihari Vajpayee that the acquisition could only have a harmful and destabilising effect on the region and potentially cause a rift in US-Indian relations.[25] The mood in Whitehall was not lightened when word filtered back to the FCO that senior State Department official Eric David Newsom had approached the British High Commission in New Delhi on a mission to measure Indian tolerance for the sale of American combat aircraft to Pakistan.[26] British officials recognised that there was genuine alarm in Islamabad regarding the Afghan coup, a

---

[20] Cyrus Vance, *Hard Choices: Critical Years in American Foreign Policy* (New York: Simon & Schuster, 1983), 262; United Kingdom High Commission (hereafter UKHC) New Delhi to FCO, 'French Nuclear Reprocessing Plant in Pakistan', July 17, 1978, TNA FCO96/823.

[21] 'French Nuclear Reprocessing Plant in Pakistan', July 17, 1978, TNA FCO96/823.

[22] United Kingdom Embassy (hereafter UKE) Islamabad to FCO, 'French Nuclear Reprocessing Plant', July 17, 1978, TNA FCO96/823, 1.

[23] UKHC New Delhi to FCO, July 17, 1978, 1.

[24] Carter to Desai, Letter, August 14, 1978, JCPL, RAC NLC 128-3-2-9-5, 4.

[25] Christopher to Carter, Memorandum, October 3, 1978, JCPL, RAC NLC 7-20-8-1-3, 2.

[26] UKHC New Delhi to FCO, July 17, 2.

circumstance not helped by perceptions of a relaxed American attitude towards Afghanistan and the ever-present anxieties about India.[27] British diplomats in Islamabad suggested that a way around the problem posed by Indian rearmament might be to resurrect the old idea of offering the aircraft for sale to India *and* Pakistan, thereby putting both sides on an equal footing.[28]

Owen took the opportunity to plead Britain's case to a wider audience at the 1978 G-7 summit in Bonn. At a quadripartite foreign ministers' meeting involving Owen, Vance, Louis de Guiringaud of France, and Hans-Dietrich Genscher of West Germany, the Foreign Secretary admitted that, yes, the situation had contributed to regional concerns, but that Britain had offered to sell the aircraft to India *and* Pakistan. The British government could not help it, Owen contended, if Pakistan was not interested in, or lacked the money for, Jaguar. According to the Foreign Secretary, the UK had taken a measured attitude, kept the Pakistanis informed, and believed that, in the final analysis, the sale would not be taken as a "tilt" against Pakistan.[29]

In the end American protestations were fruitless. Despite persistent delays and extensions to the deadline, in October 1978 Desai's government made the decision to buy the warplane to fulfil the deep strike role for the Indian Air Force (IAF).[30] The IAF would purchase forty complete aircraft—worth £260 million—from Britain, and receive a license to produce upwards of 150 more.[31] Callaghan was "naturally delighted" and

---

[27] UKE Islamabad to FCO, 'Your Telegram Number 654: Visit En Newsom', July 19, 1978, TNA FCO96/823, 1; Dennis Kux, *Disenchanted Allies: The United States and Pakistan, 1947–2000* (Washington D.C.: Woodrow Wilson Center Press, 2001), 237.

[28] UKE Islamabad to FCO, July 19, 1978, 2.

[29] 'Record of Meeting of Foreign Ministers', July 17, 1978, Thatcher MSS (Digital Collection, hereafter TMSS), www.margaretthatcher.org/document/111458 (accessed December 10, 2013), 4.

[30] Prendergast to Cartledge, 'Mr Desai's Visit: UK/Indian Agreements', April 26, 1978. TNA PREM 16/1694, 2; Beswick to Callaghan, Letter, October 25, 1978, TNA Records of the Ministry of Defence (hereafter DEFE) 13/1318; 'Britain wins big Delhi order for jet planes', The *Times* (hereafter *TT*), October 7, 1978, 5.

[31] 'Brief for the Call by the Indian Finance Minister on the Prime Minister', September 16/17, 1978, TNA PREM16/1556, 1.

hoped for a "deepening and widening" of Britain's collaboration with India.[32] By July 1979 the first aircraft had landed at Jamnagar airbase and the IAF entered a new era of sophistication.[33]

The tension surrounding the sale of these aircraft illustrates the way in which domestic economic needs transcended the global requirements of non-proliferation policy. The mid-year diplomatic arguments over the reprocessing plant also highlighted key differences between Britain and America. Britain—in approving the THORP and in its intransigent insistence on the sale of Jaguar to India—looked inward towards its own domestic economic difficulties, which were intertwined with issues exacerbated by the energy crises. Thus, Callaghan's government consistently compartmentalised the issues of the Jaguar deal and non-proliferation, despite the fact that Jaguar was *always* an impediment to a successful sub-continental non-proliferation policy.

American officials—from the President down—had argued for four years that the Jaguar deal was having an adverse effect on anti-proliferation action in the sub-continent. For the United States conventional arms sales were a tool of non-proliferation policy, as either a bribe or a means of coercion. Weapons were a means of reassuring Pakistan in the face of unstable or nuclear neighbours, with this reassurance a key theme in attempts to discourage Pakistan from pursuing its nuclear ambitions. If Washington could bolster Islamabad's conventional military forces and persuade the Indians not to rearm, logic dictated that there was less incentive for the pursuit of nuclear weapons. In contrast, London took a compartmentalised approach. Despite the arms control commitment of senior figures such as David Owen, the Jaguar deal's value (like that of the THORP project) was such that non-proliferation had to subordinate itself to economic necessity. The need to support an ailing British industry in a time of economic tumult trumped concerns about the spread of nuclear weapons in South Asia.

At the same time as American interference provoked Owen's wrath, Zia made comments that would partially lay the foundations of a decades-long media fascination. On July 17 the Pakistani leader stated in a Saudi Arabian newspaper interview that, "no Muslim country has any

---

[32] Callaghan to Desai, Letter, November 13, 1978, TNA PREM16/1556.

[33] Kapil Bhargava, 'Quarter Century of the Jaguar in India', www.bharat-rakshak. com/IAF/Aircraft/Current/607-Jaguar-25.html (accessed December 10, 2013).

(atomic arms). If Pakistan possesses such a weapon it would reinforce the power of the Muslim world."[34] The English-speaking media failed to cover the interview, even though Arthur Hummel assumed that the information had been passed to major news outlets. Hummel advocated a restrained US stance on Zia's remarks, seeing them as a "gaffe" that could—if the US or French governments publicised the comments—create domestic problems for Zia at an already difficult time.[35]

Such statements were not entirely new. India's 1974 nuclear test had coincided with changes in Pakistan catalysed by Bhutto's 'tilt' towards religious conservatism, a shift founded in his economic and security agendas rather than any deep-seated faith. The Organization of the Islamic Conference (OIC) had acclaimed Pakistan as a leading Muslim state and, after the explosion, Bhutto turned for financial aid to the countries that had so recently elevated his country to a position of esteem.[36] The move towards the increased significance of Islam in Pakistani political life had reinvigorated domestic debate about Pakistani national identity, tying that identity to Islamic traditions rather than geography or ethnicity.

The Pakistani government vaguely exploited international bonds of faith in post-test verbal attacks on India. In an announcement timed to coincide with Pakistani Independence Day on August 14, 1974, Pakistan Atomic Energy Commission (PAEC) Chief Munir Ahmad Khan contended that:

> The history of Muslims is full of instances which show that any threats to their existence brings out their inner strength of faith which unifies them and helps them surmount great difficulties. India's explosion was aimed at demoralizing us but it may serve as a jolt to awaken us from a long slumber so that we quickly summon and deploy all our moral, human, and material resources in defence of our country... It may also bring realization in other Muslim countries that India's ambitions extend far beyond the subcontinent and this threat to Pakistan is also a threat to them.[37]

[34] USE Islamabad to State, 'Nuclear Reprocessing', August 6, 1978, USPQB, Doc.11, 1.

[35] Ibid.

[36] Husain Haqqani, *Pakistan: Between Mosque and Military* (Washington D.C.: Carnegie Endowment for International Peace, 2005), 107–108.

[37] Munir Ahmad Khan, 'Challenge and Response', August 14, 1974, TNA FCO66/664, 1–2.

Here Pakistan's senior atomic scientist—a man cognisant of the drive for the bomb—Islamised the nuclear issue to elicit Muslim support. But although Washington and London had expressed mild concern since 1974 about "Arab funding" for Pakistan's nuclear efforts, it was only in 1978 that Bhutto and Zia explicitly connected their national atomic quest to the global Islamic community. In Bhutto's case, his comments were also an effort to elicit pan-Islamic support and preserve his own life.[38]

However, it was Zia's reinforcing the power statement, coupled to the blustering rhetoric in his April interview with Bernard Nossiter, to Bhutto's death cell testimony, and to the furore surrounding the, so-called, Khan Affair, which formed the basis for the media imbroglio surrounding the so-called 'Islamic bomb' that surfaced in 1979. Until 1979 such remarks would—in the West at least—remain solely the subject of governmental speculation. Zia's aim with his pronouncements was clear. Under pressure from Western states, the Pakistani leader, his subordinates, and religious allies sought to dissipate the American and British pressure on Pakistan's nuclear programme by changing matters from an *anti-Pakistani* issue to an *anti-Muslim* issue. By positioning Pakistan as a victim Zia hoped to leverage support from his co-religionists in the Middle East and South Asia. This quest for support from a faith community would give rise to a media outcry that the American, British, and Pakistani governments grappled with in 1979 and beyond.

Zia's statements were discussed at the G7 summit in Bonn, where Owen was pressing the case for selling Jaguar to India. De Guiringaud contended that such "disquieting" statements "exposed the Pakistani position to the public", and made it easier for France to justify delaying or cancelling the reprocessing plant. Regarding the Islamic aspect, Vance believed Zia's pronouncements were a bluff and was quick to point out that the quite different Islamic states of Saudi Arabia and Iran had urged Pakistan not to build a nuclear weapon.[39] Reflecting on non-proliferation in general, Vance was sceptical about the chances for truly controlling the spread of nuclear technology but—reflecting the Carter administration's

---

[38] Pervez Hoodbhoy, 'Iran, Saudi Arabia, Pakistan, and the "Islamic Bomb"', in Pervez Hoodbhoy (ed.), *Confronting the Bomb: Pakistani and Indian Scientists Speak Out* (Oxford: Oxford University Press, 2013), 152.

[39] 'Record of Meeting of Foreign Ministers', July 17, 1978, 5.

continued focus on reprocessing—contended that plutonium was still the prime concern.[40]

After the summit, Owen cabled Islamabad to see if his subordinates could expand upon Zia's statements.[41] Burdess responded on July 23, observing that the reprocessing plant had become a matter of national pride in Pakistan. The plant—and by extension the implicit nuclear weapons programme—had become solidly identified with Pakistani national identity and self-image.[42] In a well-publicised speech that Burdess was sure had government endorsement, prominent religious and political leader Maulana Mufti Mahmood spoke in "terms which left no doubt that he was referring [positively] to nuclear weapons (for Pakistan and the Muslim world)".[43] British Ambassador John Bushell supported Burdess' analysis, highlighting the Pakistani government's approved media comment that the reprocessing plant was "a matter of life and death for Pakistan". Furthermore, Zia had stated: "Not only the government but the people of Pakistan were committed to the procurement of [a] nuclear reprocessing plant."[44] Meantime, French and American pressure on Pakistan was ill received. André Jacomet, de Guiringaud's envoy, met with Zia's representatives on July 17 and 19, attempting to persuade the Pakistanis to agree to the plant's delay or cancellation, something they refused to commit to. Jacomet later related that his hosts reacted badly to this request, refusing to agree. The Frenchman then made a tense situation even worse when he attempted to use Zia's "power of the Muslim world" interview as evidence that Pakistan was planning to build a bomb and to proliferate.[45]

---

[40] Ibid., 8.

[41] FCO to UKE Islamabad, 'French Nuclear Reprocessing Plant', July 20, 1978, TNA FCO96/823.

[42] Burdess to Candlish, 'Pakistan Public and Political Attitudes to reprocessing and "The Bomb"', July 23, 1978, TNA FCO96/823, 1.

[43] UKE Islamabad to FCO, 'French Nuclear Reprocessing Plant', July 20, 1978, TNA FCO96/823, 1–2.

[44] UKE Islamabad to FCO, 'Pakistan's Reprocessing Plant', July 28, 1978, TNA FCO96/823.

[45] UKE Islamabad to FCO, 'French Nuclear Reprocessing Plant', July 20, 1978, TNA FCO96/823, 1–2.

However, the roots of the Islamic bomb were not just based on Bhutto and Zia's blustery rhetoric. Libyan leader Colonel Gaddafi's Lieutenant Abdessalam Jalloud visited Islamabad in mid-August, prompting worries about a Pakistani-Libyan nuclear alliance and cooperation in building a reprocessing plant, should the French deal be cancelled.[46] Moreover, at the end of September the JNU received a press agency release noting Pakistani-Libyan connections that it regarded as "well informed". The release drew on rumours of a triangular Chinese-Libyan-Pakistani relationship that Whitehall viewed as "a nasty idea".[47] The idea of Pakistani-Libyan nuclear cooperation was one of the main platforms on which later public narratives surrounding the Islamic bomb would be built.

As well as questions of pan-Islamic proliferation, policymakers in London and Washington also had to consider matters of Pakistani national pride. Senior US diplomats wanted to influence any post-cancellation Pakistani nuclear plans and gain assurances that the reprocessing plant was dead and buried. To restart the flow of aid Pakistan would publicly have to relinquish sovereign rights and subjugate itself to American dictates. The State Department strongly implied that if Pakistan abandoned its pursuit of the reprocessing plant and assured Washington of its good intentions, then the economic aid cut-off in September 1977 might be restored.[48] Zia's government was opposed to giving any assurances, stating that if they did wish to press ahead with a reprocessing plant, "it would not matter how many assurances [it] provided".[49] Hummel thought the requests unacceptably impinged on Pakistani sovereignty and argued that no government of Pakistan could survive the public outcry over such an assurance.[50]

Given the level of anti-American feeling welling up in the Pakistani media over the reprocessing plant, anxiety on both sides was unavoidable.

---

[46] UKE Islamabad to FCO, 'Further Visit of Libyan Vice-President to Pakistan: 15-17 August', August 17, 1978, TNA FCO96/823.

[47] 'Libya's Qadafi Seeks Nuclear Fuel Deal', September 29, 1978, TNA FCO96/824.

[48] State to USE Islamabad, 'Pak Ambassador's Call on Undersecretary Newsom', August 1, 1978, USPQB, Doc.8, 2.

[49] Ibid., 3.

[50] USE Islamabad to State, 'Pakistani Reprocessing Plant: USG Stipulation', August 6, 1978, USPQB, Doc.10, 1.

According to British diplomats in Islamabad, the Pakistani press and public believed that Washington was engaged in a conspiracy against the country. Consular staff viewed the situation as irrational, as the Pakistani press stressed the reprocessing plant's importance and denigrated the value of US aid, "which most serious officials must consider to be indispensable".[51] Here was a situation where outside observers contended that Pakistan should suppress national pride and self-image in order to gain Western aid. The embassy in Islamabad advised against becoming involved in the fracas, which might be counterproductive to British interests.[52] Ambassador Hummel—in a frank discussion with his British opposite number John Bushell (a discussion that Bushell advised should not reach other American or French ears)—was pessimistic. Hummel was certain that a final negative reply from Paris would lead to a Pakistani withdrawal from CENTO, something that Bushell had mooted at the end of July.[53] Hummel suggested that one way the UK could help ameliorate the effects of the anti-US line emanating from Zia's administration was through the BBC World Service's influence in Pakistan. The BBC could be used as a medium to clarify the French position regarding the plant, emphasising that the cancellation was the Giscard government's decision alone and not made under US pressure.[54] Finally, Bushell commented on the potential for a Pakistani move to non-aligned status or worse. The Pakistanis were, he commented, receptive to Soviet blandishments.[55]

By the beginning of September, American and French officials believed they had resolved the reprocessing issue, despite the fact that Zia and other senior officials were doing all they could to reinvigorate the ill-fated project by linking it to the Saviem truck plant deal.[56] After

---

[51] UKE Islamabad to FCO, 'Pakistan's Reprocessing Plant', August 9, 1978, TNA FCO96/823, 1.

[52] Ibid., 2.

[53] UKE Islamabad to FCO, 'French Reprocessing Plant', August 11, 1978, TNA FCO96/823, 2; UKE Islamabad to FCO, 'French Reprocessing Plant', July 31, 1978, TNA FCO96/823, 2; USE Paris to State, 'Surfacing of French-Pakistan Reprocessing Issue', August 25, 1978, DNSA, NP01605, 1–2.

[54] UKE Islamabad to FCO, August 11, 1978, 2.

[55] Ibid., 3.

[56] USE Paris to State, 'Next Steps on Pakistan Reprocessing Issue', September 1, 1978, NSAEBB 'Non-papers and Demarches' (hereafter NPAD) www2.gwu.

Jacomet's hapless mission Giscard had written to Zia indicating, in convoluted and obscure language, that once the International Fuel Cycle Evaluation (INFCE) studies had been completed, France was willing to renegotiate the deal. Hummel believed Giscard's message was tantamount to the contract's cancellation.[57] On August 23 Zia announced that the deal was off.[58] Pakistani commentary was initially mild, but became more heated when Agha Shahi remarked on the perfidy of "some of the nuclear supplier states and their allies [who] prevent the transfer of nuclear technology for peaceful purposes to Third World nations".[59] In September Zia repeated his earlier assertion that it was Washington's pressure on Paris that led to the cancellation and that he would continue to press Paris to meet its contractual obligations.[60] Burdess noted Zia's statement that Pakistan was "determined to acquire nuclear technology and hoped that with the help of Allah its efforts in this regard would be successful", had apparently convinced the French of Pakistan's desire to develop nuclear weapons.[61] British analyses suggested that the Pakistanis were unable or unwilling to recognise that remarks made by Zia—and not rumours—were the root cause of anxiety about a Pakistani bomb being developed for use by the wider Arab or Muslim world. Furthermore, Burdess suggested that it was the implicit connection between the reprocessing plant and nuclear weapons—and the position of nuclear weapons as a 'prestige' technological

edu/~nsarchiv/nukevault/ebb352/index.htm (accessed November 10, 2012), Doc.1, 2–3; USE Islamabad to State, 'Pakistan Reprocessing vs French Commercial Deals', September 3, 1978, NPAD, Doc.2, 1.

[57] USE Islamabad to State, 'Reprocessing Plant', August 20, 1978, USPQB, Doc.14, 2; USE Paris to State, 'Surfacing of French-Pakistani Reprocessing Issue', August 24, 1978, DNSA, NP01605, 1–2.

[58] 'Paris cancels nuclear deal with Pakistan', *TT*, August 24, 1978, 3.

[59] USE Islamabad to State, 'Press and GOP Reactions to Reprocessing Deal Cancellation Surprisingly Mild', August 30, 1978, DNSA, NP01607, 2.

[60] USE Paris to State, 'French Views on Pakistan Reprocessing Plant', August 25, 1978, USPQB, Doc.17; USE Islamabad to State, 'Ambassador's Talk with General Zia', September 5, 1978, NPAD, Doc.3, 1.

[61] UKE Islamabad to FCO, 'French Reprocessing Plant', August 24, 1978, TNA FCO96/823; Smith to Murray, 'French Nuclear Deal with Pakistan', August 25, 1978, TNA FCO37/2112, 1.

development—that made non-proliferation efforts seem like such an attack on Pakistani national image and identity.[62]

Now that there had been public acknowledgement that the reprocessing plant was—seemingly—dead (the French Council on Foreign Nuclear Policy having decided the entire thing was a bad idea), the issue entered a new phase when the US government sought assurances from Giscard's representatives that the deal was also buried.[63] With Washington considering the resumption of aid to Pakistan, and with key anti-proliferation figures in Congress to be persuaded of Islamabad's *bona fides*, the State Department thought an explicit commitment from Paris would be helpful.[64] Jacomet rebuffed the request when he met with Gerard Smith, Carter's roving ambassador for non-proliferation affairs. Jacomet felt that a public statement from Giscard would give Jacques Chirac—former prime minister and the French right's *de facto* leader—and the Gaullists ammunition to attack Giscard's beleaguered government.[65] Despite considerable pressure from Smith, Jacomet refused to offer any assurances that the Carter administration could present to Congress as part of their negotiations over aid to Pakistan.[66]

The State Department pressed ahead with consultations on Capitol Hill despite French recalcitrance. These discussions took place between Newsom, Nye, John Glenn (one of the Nuclear Non-proliferation Act's main congressional supporters), and Clement Zablocki (the House Foreign Affairs Committee chairman). The talking points covered a number of issues, including the threat to US regional interests by a "disintegrating or radicalized Pakistan".[67] The State Department

---

[62] Burdess to Candlish, 'Pakistan Nuclear Affairs: Visit by Mr Justice Fox', August 20, 1978, TNA FCO37/2112.

[63] Jeffery T. Richelson, *Spying on the Bomb: American nuclear intelligence from Nazi Germany to Iran and North Korea* (New York: W.W. Norton & Co., 2006), 329.

[64] State to USE Paris, 'Next Steps on Pakistan Reprocessing Deal', September 7, 1978, NPAD, Doc.4A, 2.

[65] USE Vienna to State, 'Smith-Jacomet Meeting: French Position on Cancellation of Pakistan Reprocessing Plant Contract', September 13, 1978, NPAD Doc.4B, 3.

[66] Ibid.

[67] State to USE Vienna, 'Congressional Consultations on Pakistan', September 15, 1978, NPAD, Doc.5, 3.

portrayed Pakistan as fearful of the unfriendly (India) or unstable (Afghanistan and Iran) states surrounding it and lacking confidence in the West's ability to provide security.[68] Newsom and Nye argued that economic aid and military sales could reassure Islamabad and encourage movement away from a bomb programme.[69] In any case, US officials saw Pakistan as having neither the technical prowess nor the industrial ability to complete the reprocessing plant without external aid. The inverter problem was alluded to, but given scant attention, as Carter administration officials believed that the technical challenges of a centrifuge-based enrichment programme were insurmountable for Pakistan, and that such ambitions could be defeated by strict supplier controls.[70] According to the two State Department officials, restoring good relations with Pakistan would give the administration a better chance of dealing with the nuclear problem.

France returned to the spotlight when stories appeared in the French press that the firm Robatel had received a government license to export centrifuges for a reprocessing plant.[71] Jacomet, who perceived a Gaullist hand in the affair, strenuously denied this.[72] Intelligence from a highly placed anonymous source in the Élysée Palace did, however, indicate that the Pakistanis were searching for suppliers who would allow them to complete the reprocessing plant on their own or with the help of another nation.[73] The Pakistanis hinted that they might turn to their ally China for aid.[74] The State Department had already sought and received "credible" assurances from Beijing that the Chinese did not intend to offer

[68] Ibid.

[69] Ibid., 3–4.

[70] Ibid., 4–5.

[71] State to USE Paris, 'French Export of Centrifuges for Pakistani Reprocessing Plant', September 19, 1978, NPAD, Doc.6, 1–2. These were not the same type of centrifuge that the Pakistanis were attempting to build.

[72] USE Paris to State, 'Update on French-Pakistani Reprocessing Situation', September 21, 1978, NPAD, Doc.7, 1.

[73] USE Paris to State, 'Élysée Views on Reprocessing Issues', September 23, 1978, NPAD, Doc.8, 1–2.

[74] Burdess to Candlish, 'Pakistan's Reprocessing Plant: Self Sufficiency or Chinese Help?', September 24, 1978, TNA FCO37/2113, 1.

Pakistan assistance to complete the reprocessing plant.[75] Likewise, Chinese diplomats "emphatically denied" nuclear cooperation with any other state to their British counterparts.[76]

In early October the United States again attempted to use arms sales and aid as a means of luring Islamabad away from 'the bomb'. Vance indicated to Agha Shahi that the administration was ready to resume financial aid programmes and consider military sales. Vance wanted Shahi to understand that this potential resumption of aid was on the basis that Pakistan did not intend to develop a nuclear weapons capability. The Secretary of State observed that recent reports had suggested that Pakistan was exploring "other means" of "completing the reprocessing plant or otherwise acquiring a nuclear option through indigenous efforts".[77] Shahi had asked about the possible sale of A-7 attack aircraft in light of India's intention to buy Jaguar, but Vance made it clear that he saw no reason for the Indian purchase.[78] The Jaguar sale might, Vance suggested, create a new spiral of arms acquisitions in the region. Subject to congressional approval, Carter had decided in July that he was still willing to sell Pakistan the forty to fifty less-capable F-5E warplanes that had been considered in 1977 to replace the Pakistan Air Force's Korean War vintage F-86 fighters.[79] The Pakistanis again brought up Jaguar during November meetings with a US delegation to Islamabad. Lucy Benson (Under Secretary of State for Arms Control and International Security Affairs) made it clear to her hosts that they would have to come up with a list of arms requirements that both met their needs *and* did not contravene Carter's stated policies on conventional arms restraint.[80]

---

[75] State to USE Islamabad, 'PRC Assistance to Pakistan in Reprocessing', August 25, 1978, USPQB, Doc.16, 2. The irony being that evidence of PRC involvement would later emerge.

[76] Burdess to Candlish, September 24, 1978.

[77] 'Memo to Chris from Steve enclosing edits to draft cable to Islamabad and "Evening Reading" reports to President Carter on Pakistan', October 4, 1978, USPQB, Doc.18, 8.

[78] Ibid., 7; Kux, *Disenchanted Allies*, 235.

[79] Brzezinski to Carter, 'F-5E's for Pakistan', July 7, 1978, JCPL, National Security Staff (hereafter NSS) Files, North/South, Box 95, Pakistan: 4/77-12/78.

[80] USE Islamabad to State, 'Under Secretary Benson's Meeting at Pak Defense Ministry', November 7, 1978, JCPL, NSS Files, North/South, Box 95, Pakistan: 4/77-12/78.

By this point US intelligence had caught up with what the British had been telling them for several months, passing their conclusions to Congress in diluted form. Senators and Representatives were informed that, although the non-proliferation problem with Pakistan had not been completely resolved, the restoration of normal relations would "best serve both our non-proliferation objectives and our interest in regional stability".[81] Newsom, Hummel, and Nye briefed Glenn two days later, discussing the possibility of an indigenous Pakistani programme. The team concluded that although Pakistan was clearly exploring multiple means of gaining access to weapons-grade nuclear material, these methods would take time and could be closely monitored.[82] The US government soon acquired further intelligence on Islamabad's intentions. Pakistani Ambassador Iqbal Ahmed Akhund had informed Arthur Hartman that Pakistan "had every intention of finishing the reprocessing plant on its own".[83] Hartman concluded that Akhund had all but admitted the nuclear programme's military nature, its function being to give "the Pakistani people, Indians, and others a perception of a Pakistani military capability".[84]

By November the Americans were engaged in the final preparations for a diplomatic offensive that the Carter administration hoped would finally bury the reprocessing plant. Influenced by the reports that Pakistan may have been seeking to complete the facility with foreign aid, they were determined to ensure the impossibility of such completion. The UK and France were the first foreign eyes to see the US démarche, provided with the communication before general circulation. Warren Christopher articulated serious concerns about Islamabad's reprocessing and enrichment programmes, and was desirous of continued dialogue with France over reprocessing.[85] The main American worry was the impact of Pakistani

---

[81] 'Memo to Chris from Steve', October 4, 1978, 8.

[82] 'Memorandum of Conversation: Consultations on Pakistan: details on Indigenous Nuclear Capabilities (Supplement to October 6, 1978 memcon prepared by Ambassador Hummel)', October 6, 1978, USPQB, Doc.19.

[83] USE Paris to State, 'Pakistan Ambassador to France Hard-Lines on Reprocessing Plant', October 21, 1978, DNSA, NP01612, 1.

[84] Ibid.

[85] State to USE London et al., November 1, 1.

nuclear ambitions on the South Asian geopolitical situation. In the Afghan Revolution's aftermath and with growing instability in Iran, Christopher deemed it "critical to stability in the region and to our non-proliferation objectives to inhibit Pakistan from moving closer to the threshold of nuclear explosive capability".[86]

Christopher's cable also alluded to Bhutto's inflammatory death cell testament alluding to the nuclear programme and the Islamic world. "We do not necessarily accept Mr. Bhutto's claims of imminent success in this field," stated Christopher, "but we do find this statement of intentions to be disquieting."[87] Even though Christopher found Bhutto's remarks "disquieting" and having "profound implications", the implied Islamic dimension did not change basic American assumptions about the reasoning behind the Pakistani programme. Christopher and his colleagues "were under no illusion that Pakistan's motivations or intentions have changed", with Islamabad's nuclear ambitions still perceived as a "national project".[88] Thus, Bhutto's and Zia's rhetoric about the Islamic bomb failed to alter American perceptions and policy regarding Pakistani intentions. For the Carter administration, while the idea of the Islamic bomb might have proved "disquieting", it did not change the fundamental assumptions about the national quality of Pakistan's bomb project and the dual issues of regional stability and a global non-proliferation policy.

The responses from London and Paris to Christopher's telegram indicated a lack of European unanimity over non-proliferation. The JNU's Robert Alston indicated there was a slight case of confusion over which nations might be able to supply nuclear components to Pakistan, observing that the recipients of the proposed American cable and Britain's parallel démarche on the clandestine enrichment programme (discussed later in this chapter) were different. Britain should thus be given time to send further cables so that both campaigns were fully coordinated.[89] Alston also commented that US allies such as Taiwan and South Korea—both countries that had pursued nuclear weapons programmes—might prove

---

[86] Ibid., 3.

[87] State to USE London et al., November 1, 1978, 3.

[88] Ibid., 4.

[89] USE London to State, 'Pakistani Reprocessing Plant', November 2, 1978, NPAD, Doc.10A, 1.

problematic in their reactions to a campaign of Pakistani nuclear containment.[90] In contrast, André Jacomet was more anxious about the situation's European dimension. Now assured that France's own domestic reprocessing plants were untouchable, Jacomet told the US consular staff that furnishing sensitive nuclear equipment to Pakistan would conflict with the French position on nuclear proliferation. According to Jacomet, France's most serious concerns were with Italy and the FRG, which did not care about French non-proliferation policy.[91] Jacomet was relieved that the US démarche did not mention the reprocessing plant deal's cancellation, as there was still a "great sensitivity to any direct linking on our part of US policies with the French decision to cancel the deal".[92]

When the finalised American démarche—influenced by the French reporting on Pakistani efforts to complete the reprocessing plant and US-UK intelligence on the clandestine enrichment programme—was finally sent out, it left its recipients in no doubt that Washington was serious about ensuring that Pakistan would not complete the reprocessing plant. Sent in Cyrus Vance's name, the cable outlined "deep concern at the highest levels" of the Carter administration, the communication sketched out American concerns in light of intelligence flowing into Washington.[93] Considering the increasing instability in the region—continued Pakistani-Indian tensions, the Afghan revolution, and the stirrings in Iran—it was vital to prevent Pakistan from gaining nuclear capability.[94] In a move that caused some degree of peevishness in London, the cable cited the UK as the source of "firm information" that Pakistan was *also* constructing a centrifuge-based uranium enrichment facility. Vance encouraged supplier governments to be watchful when it came to Pakistan's overt and covert attempts to acquire the means to complete their reprocessing facility.[95]

---

[90] Ibid., 2.

[91] USE Paris to State, 'US Demarches on Pakistani Reprocessing Plant', November 2, 1978, NPAD, Doc.10B, 1.

[92] Ibid., 2.

[93] State to USE London et al., 'US Demarche on Pakistani Reprocessing Plant', November 4, 1978, NPAD, Doc.12, 2.

[94] Ibid.

[95] Ibid., 4.

Responses to this major diplomatic initiative were mixed. Many were routine expressions of acknowledgement or willingness to cooperate. Some governments, such as the Canadians, Italians, and Japanese, expressed surprise that Pakistan was considering the reprocessing plant's completion.[96] Despite niggling doubts the Netherlands government regarded Bhutto's death cell testimony as simple boasting. Furthermore, the Dutch were concerned about the developing world's reaction should word of a Western offensive against Pakistan leak out. Nuclear power, officials in The Hague suggested, was a great status symbol for developing nations, as it represented the acme of industrial and technological achievement. Actions that created even the perception that Western nations were trying to prevent developing states from gaining access to nuclear technology presented a "psychological question" that the West would have to consider carefully.[97] Some US conclusions about Pakistani intentions and progress surprised the West German government and, like the Dutch, Bonn was concerned about the campaign's potential public relations implications. The last thing the *Auswärtiges Amt* wanted to see was a headline stating "US and FRG in Nuclear Boycott of Pakistan".[98]

Concern over adverse reactions to a boycott of Pakistan was echoed by American diplomats. Hummel was nervous about plans to inform the Indians about the diplomatic campaign, arguing that, if the press found out that the US government had been talking to India behind Pakistan's back, there could be a hostile reaction from Islamabad. The Ambassador argued that colluding with Western nations to prevent Pakistan from continuing with its bomb programme was one thing, colluding with the Indians would be quite another.[99] Vance contended that India already seemed aware of Bhutto's overblown claims of nuclear capability *and*

---

[96] USE Ottawa to State, 'US Demarche on Pakistani Reprocessing Plant', November 7, 1978, NPAD Doc.13F; USE Rome to State, 'US Demarche on Pakistani Reprocessing Plant', November 7, 1978, NPAD Doc.13C; USE Tokyo to State, 'US Demarche on Pakistani Reprocessing Plant', November 9, 1978, NPAD Doc.13I.

[97] USE The Hague to State, 'Demarche on Pakistani Reprocessing Plant', November 9, 1978, NPAD, Doc.13H.

[98] USE Bonn to State, 'US Demarche on Pakistani Reprocessing Plant', November 8, 1978, NPAD, Doc.13E.

[99] USE Islamabad to State, 'Achieving USG Nonproliferation Objectives in Pakistan', November 14, 1978, NPAD, Doc.15A.

the UK inverter story, but that US representatives in New Delhi had been instructed not to mention Pakistan in talks.[100] Robert Goheen, US Ambassador to India, reinforced Vance's statement. Goheen was sure that the Indians had known for some time about American suspicions, and the Indians themselves believed that Pakistan was only two or three years from producing a bomb.[101]

At the end of November the French reassured US representatives of their firmness regarding the reprocessing plant. Jacomet emphasised that, despite a continuing dialogue with Islamabad, there was no possibility of France ever helping to complete it.[102] He did mention that Pakistan had requested bilateral discussions regarding a "plutonium storage scheme", something that Gerard Smith found unusual if no plutonium was to be produced in Pakistan.[103]

For the Carter administration the manoeuvering against the Franco-Pakistani reprocessing plant deal represented a qualified non-proliferation success. Paris had finally cancelled the contract and there was a generally positive response from nuclear supplier nations regarding US moves to prevent the Pakistanis completing the plant with the aid of third parties. Despite the anti-reprocessing offensive's success, it was quite clear that Pakistani nuclear ambitions remained undiminished. A major factor in this realisation was the mounting evidence of another strand to the nuclear programme. It was this much more covert aspect that became the focus of anti-proliferation action on the sub-continent from late 1978 onwards, action led by Britain.

THE CLANDESTINE ENRICHMENT PROJECT

Parallel with American-led anti-reprocessing efforts, action directed against Pakistan's clandestine uranium enrichment programme gathered pace. From 1978 onwards it became apparent to Callaghan's government

---

[100] State to USE Islamabad, 'Achieving USG Non-proliferation Objectives in Pakistan', November 16, 1978, NPAD, Doc.15B.

[101] USE New Delhi to State, 'Achieving USG Nonproliferation Objectives in Pakistan', November 17, 1978, NPAD, Doc.15C.

[102] State to USE Paris, 'Pakistan Reprocessing Plant', November 22, 1978, NPAD, Doc.18, 2.

[103] Ibid., 3.

and then the Carter administration that the inverters ordered from EEIC by agents working on behalf of Pakistan represented only a fraction of a global plan to gain the capability to produce weapons-grade enriched uranium.

In March of 1978 Mr E. S. Harris, EEIC's managing director, contacted the DoE—as he had in 1976—passing on his suspicions about another request to tender that his company had received from Weargate, the shadowy supply agents acting on the Pakistani government's behalf. The request was for inverters worth £2 million, a significantly larger order than the small shipments dispatched to Rawalpindi in 1977. Again, Harris argued that the inverters were alarmingly similar to those used by British Nuclear Fuels Limited (BNFL) in their centrifuges at Capenhurst in Cheshire, despite Pakistani claims that the devices were for use in a textile mill.[104] The textile mill excuse was the same cover that had been used by Israel in answer to queries from the United States in the 1960s about their nuclear plant at Dimona.[105]

Once Harris informed the DoE about this new request, British non-proliferation and economic interests collided again. Considerable evidence and expert opinion was amassed from a diverse range of sources—including BNFL, the MoD, and the DoE—that the likely end use for the inverters was as components of a uranium enrichment plant.[106] Given the British economy's precarious state in 1978, the DoE's Barbara Parkin observed that this was a substantial order of great benefit to a British manufacturer.[107] EEIC had made representations to the government that, in the event of not being permitted to fulfil the order, the prospects for their workforce were bleak.[108] In the end, the government decided that the FCO—in consultation with the United States and other Nuclear Suppliers' Group (NSG) members—should take the lead, with FCO

---

[104] Parkin to Bourke, 'Export of Inverters to Pakistan', March 15, 1978, TNA FCO96/822.

[105] See Gerlini, 'Waiting for Dimona', 150.

[106] 'Note of a Meeting Held in the Department of Energy, Atomic Energy Division, on March 17, 1978 to Discuss a Possible Export Order for Inverters to Pakistan', March 23, 1978, TNA FCO96/822, 1.

[107] Parkin to Bourke, 'Export of Inverters to Pakistan', March 15, 1978, TNA FCO96/822.

[108] 'Note of a Meeting', March 23, 1978, 3.

enquiries informing a final decision on whether or not to allow the export.[109] When Agha Shahi visited London just after the March 18 announcement that a Lahore court had passed a death sentence on Bhutto, British suspicions about the clandestine programme were kept under wraps. This was in spite of discussions on nuclear security assurances, and attempts by Owen to persuade Shahi to at least agree to delaying acquisition of the reprocessing plant.[110]

At the end of March the FCO officially brought inverters to the US government's attention. The JNU's Martin Bourke wrote to the Washington Embassy's First Secretary, the experienced diplomat Michael Pakenham, asking for US input.[111] Bourke's main interest was whether the Americans felt that their export restrictions covered inverters and how these controls related to their 'machinery' for combating the proliferation problem. There was less incentive for Britain to halt the sale if Washington felt that their export restrictions did not cover inverters. Furthermore, the FCO tasked Kit Burdess in Islamabad with subtly finding out whether or not the Pakistanis were planning to build an enrichment plant.[112]

Other than a brief message from Pakenham on March 28 to the effect that inverters probably fell within the spirit, if not the letter, of US nonproliferation legislation, it was several weeks before the State Department offered a substantive response.[113] In the FCO's eyes that response was less than satisfactory. Pakenham related the opinion of Joe Nye, who questioned the "quality and dependability" of the British intelligence that indicated the inverters were for a nuclear enrichment facility.[114] US intelligence agencies had "not come up with any evidence that the Pakistanis

[109] Ibid.

[110] 'Record of a Meeting Between the Foreign and Commonwealth Secretary and Agha Shahi', March 20, 1978, TNA FCO96/822, 1. Callaghan and Owen both called upon Zia to rescind Bhutto's death sentence, to no avail.

[111] Bourke to Pakenham, 'Export by a UK Firm of Inverters to Pakistan', March 23, 1978, TNA FCO96/822.

[112] Ibid., 2.

[113] UKE Washington to FCO, 'Inverters for Pakistan: Bourke's Letter to Pakenham of March 23', March 28, 1978, TNA FCO96/822.

[114] Pakenham to Bourke, 'Export of Inverters to Pakistan', April 18, 1978, TNA FCO96/822, 1.

were organising themselves for the development of an enrichment capability".[115] Despite London's conviction—based on evidence that had
been amassed since late 1976—that the inverters *were* for a uranium
enrichment project, the State Department wanted to wait until there was
incontrovertible evidence of wrongdoing before they would step into the
murkiness of this "grey area".[116]

Between the request for American input and the unsatisfactory mid-
April response, intelligence continued to flow into Whitehall from the
British Embassy in Islamabad. Burdess submitted detailed appraisals of
Pakistani intentions, noting that he had heard nothing concrete, but
was unsurprised that Pakistan may have embarked on an enrichment programme, as this would probably be the only way they could get access to
enriched material without acceding to full-scope safeguards or the NPT.[117]
Furthermore, Burdess doubted the claim that the inverters were for a
textile mill, being familiar with the plant for which the inverters were
allegedly intended. He observed that the factory had no means of spinning
cloth and produced items from materials manufactured elsewhere in
Pakistan. Nonetheless, Burdess concluded that stopping the inverter export
at this point might do more harm than good by arousing Pakistani suspicions and forcing Islamabad to adopt even more secretive methods in
future. He felt that confronting the Pakistanis over these discrepancies
could only stop a covert enrichment programme if there was incontrovertible proof that the inverters were intended for an enrichment facility.[118]

In Washington, the CIA presented the limited fruits of its intelligence
efforts on the Pakistani bomb programme on April 26. From the unre-
dacted sections—and despite London alerting the US authorities to
Pakistan's covert centrifuge project—it appears that the analysis did not
consider a centrifuge-based enrichment programme.[119] The report

---

[115] Ibid.

[116] Ibid.

[117] Burdess to Bourke, 'Inverters for Pakistan', April 3, 1978, TNA FCO96/822.

[118] Ibid.

[119] CIA, 'Pakistan Nuclear Study', April 26, 1978, DNSA, WM00212. William
Burr opines that the redacted sections were on issues other than a potential
enrichment programme; see www.gwu.edu/~nsarchiv/nukevault/ebb333/
index.htm, notes on Doc.5.

focused on reprocessing as the route to a bomb and assumed—incorrectly and in the face of all the evidence—that, unlike Bhutto, Zia attached a low priority to the nuclear programme.[120] Following on from this apparent oversight, the CIA would continue to award little significance to the possibility of an enrichment programme for some months to come, as it focused on the reprocessing issue.

This opening period of US-UK communication on Pakistan's enrichment project saw a marked difference between British and American views of the situation. British officials recognised that the EEIC inverter order was potentially a serious non-proliferation regime breach and attempted to formulate ways to handle the situation. When the FCO informed its American counterpart, the State Department offered little more than bland platitudes. From Carter downwards, the US focus at this stage concentrated on reprocessing as the main proliferation danger. With American intelligence agencies offering little or no information on enrichment and unwilling to rely on evidence from the UK, Carter's team gave enrichment a low priority. Because American policy—typified by the NNPA—was directed so much towards the 'big ticket' elements of proliferation, such as reprocessing plants, the Carter administration failed—despite warnings—to see the dangers inherent in a smaller-scale, clandestine uranium enrichment programme. This highlighted a contradiction in the administration's non-proliferation policy. Despite their enthusiasm for non-proliferation efforts directed at reprocessing capabilities, the US government—not wishing, as Nye intimated, to fight a "war on two fronts" until there was definite proof of Pakistan duplicity—deprioritised the possibility of a secret Pakistani enrichment scheme.[121]

From late summer onwards EEIC was in the process of manufacturing the components since the UK export control list did not yet include inverters. Transatlantic discussions saw the UK taking the diplomatic lead in an effort to gain unanimity amongst potential suppliers as

---

[120] 'Pakistan Nuclear Study', April 26, 1978, 1.

[121] Carter's administration was not alone in having such a blind spot. British officials had, since the late 1960s, been warning their US counterparts about the proliferation dangers of gas centrifuge technology, to no avail. See John Krige, 'US Technological Superiority and the Special Nuclear Relationship: Contrasting British and US Policies for Controlling the Proliferation of Gas Centrifuge Enrichment', *International History Review*, 36:2 (2015), 230–251.

intelligence pointing towards a covert Pakistani bomb programme accumulated. By late October the campaign against the Pakistani bomb effort was in full swing. The United States continued to make efforts to ensure the reprocessing plant deal would not come back to life, while Britain carried out an interlocking campaign against the clandestine enrichment project.

Up to this point discussion of inverters had taken place behind closed doors, but in late July the issue became public when Labour MP Frank Allaun asked in parliament "[I]f consideration has been given to the purchase of variable frequency inverters, which are used in gas centrifuges to enrich uranium, by TEAM, a German company, for supply to Pakistan, in the light of the provisions of the non-proliferation treaty or other measures to prevent the spread of nuclear weapons?"[122] Allaun was one of the parliamentary Labour Party's staunchest anti-nuclear campaigners, a passionate CND supporter, and an organiser of the Aldermaston Marches against nuclear weapons. Tony Benn, the Secretary of State for Energy and one of Owen's fellow non-proliferation enthusiasts, had to provide a written reply. "Yes", he responded,

> However, equipment specially designed for use in a gas centrifuge installation was not at that time subject to the Export of Goods Control Order; such equipment is now subject to control. Any application to export equipment which is specially designed for uranium enrichment plants would be treated with full regard to our Non-Proliferation Treaty—NPT—and other international obligations, in accordance with the statement of my right hon. Friend the Prime Minister on 31st March 1976.[123]

[122] Frank Allaun MP, Question, July 21, 1978, *Hansard Online*, http://hansard. millbanksystems.com/written_answers/1978/jul/21/nuclear-power-variable-frequency#S5CV0954P0_19780721_CWA_167 (accessed November 8, 2012). How Allaun came to find out about the matter is a matter of considerable speculation. Weissman and Krosney argue that it was Ernst Pfiffl—a disillusioned Pakistani purchasing agent in Germany—who spread the word about Islamabad's activities. Allaun took the secret to his grave in 2002.

[123] Owen, *Time to Declare*, 320; Tony Benn, in response to Frank Allaun MP, July 21, 1978, *Hansard Online*, http://hansard.millbanksystems.com/written_answers/1978/jul/21/nuclear-power-variable-frequency#S5CV0954P0_19780721_CWA_167 (accessed November 8, 2012).

On July 26 Allaun again probed the issue, this time asking the Secretary of State for Trade, Edmund Dell, if "the supply to Pakistan by Emerson Industrial Controls of Swindon of equipment which will form part of that Government's technological capability to manufacture their own nuclear weapons was made with his approval". Dell responded bluntly "No. A licence to export such equipment would have been needed only if the goods involved had been subject to the Export of Goods (Control) Order 1970."[124] Although buried within the parliamentary record, the questions asked by Allaun would shortly bring the clandestine Pakistani programme into the public gaze.[125] Allaun continued to plague the government with questions on the Pakistani nuclear programme throughout 1978 and beyond. Existing histories of Pakistan's nuclear programme posit Allaun's inquisition as the moment at which the British government began to take the issue seriously. As the archival record demonstrates, however, serious concerns had been raised long before Allaun asked his questions.[126]

Before Britain's diplomatic campaign against the covert Pakistani purchasing programme got under way, there were discussions within the government over whether or not to control the export of inverters, and, if control was required, what form this should take. The Department of Trade (DoT) was keen not to place additional burdens on hard-pressed British exporters, but overall the feeling was that export restrictions *must* be toughened up in the face of a major non-proliferation test.[127] Additionally, the FCO decided not to seek "end use assurances" from Pakistan. David Hannay, head of the FCO's Energy Department, believed that the Pakistanis would say the inverters were not for use in a bomb

---

[124] Frank Allaun MP, Question, July 26, 1978, *Hansard Online*, http://hansard. millbanksystems.com/written_answers/1978/jul/26/nuclear-power-equip ment-exports#S5CV0954P0_19780726_CWA_322 (accessed on November 8, 2012).

[125] 'Power export faces ban', *TG*, October 6, 1978, 1; 'Nuclear trade loophole closed by Government', *TG*, November 23, 1978, 4.

[126] Armstrong and Trento (74–76), for example, mention the early March to July discussions, but attach no significance to them.

[127] Dunning to Parkin, 'Export of Inverters', August 18, 1978, TNA FCO96/823.

programme, requiring the British government to either allow the export or accuse the Pakistanis of lying. [128]

Tony Benn wrote to Owen on August 30 expressing his deep concern about the problem. "I believe," he stated in an echo of the US NNPA, "that we should do everything possible to prevent the spread of sensitive technology."[129] The problem was that the inverters themselves could be used for a range of industrial purposes, not just for uranium enrichment centrifuges. In a summary that broadly reflected interdepartmental and ministerial thinking, Benn's subordinates asserted that the non-proliferation value of controlling the exports—even though such controls were of little value should other supplier nations not be persuaded to do the same—outweighed the problems additional export controls would cause for British manufacturers.[130] Despite the tension that existed between the two men, Owen agreed with Benn's assessment and restated his conviction that Pakistan *was* pursuing a nuclear weapons programme based on enriching uranium.[131] The Foreign Secretary asserted that the burden on British exporters—in terms of delays and possible lost contracts—would just have to be borne so that Britain would be able to fulfil this "major priority of our non-proliferation policy".[132] In this case, Britain could subordinate economic concerns to non-proliferation. Enhanced export controls might cause inconvenience to manufacturers, but if the EEIC case was typical, the losses would only amount to a few million pounds. This contrasted with the Jaguar and THORP cases, when the sums of money involved were orders of magnitude greater. In those cases prioritising non-proliferation over billions of pounds of orders and investment was too great a burden for Callaghan's government to bear.

[128] Hannay to Gittelson, 'Export of Inverters to Pakistan', August 9, 1978, TNA FCO96/823.

[129] Benn to Owen, Letter, August 30, 1978, TNA FCO37/2113.

[130] Inclusion with August 30 Benn to Owen Letter, 'Export of Inverters', Date unknown, presumed August 1978, TNA FCO37/2113.

[131] On the tension between Benn and Owen, see David Hannay, transcript of interview by Malcolm McBain, July 22, 1999, British Diplomatic Oral History Programme (hereafter BDOHP), www.chu.cam.ac.uk/archives/collections/BDOHP/Hannay.pdf (accessed December 10, 2013), 13.

[132] Owen to Benn, 'Proposed Sale of Inverters to Pakistan', September 12, 1978, TNA FCO37/2113, 1.

By early October Allaun's questions in the Commons brought the entire affair out into the public domain. Newspapers local to the EEIC factory in Swindon started asking the government questions, receiving bland replies from the DoT that consideration was being given to controlling exports of "certain electrical equipment".[133] Weightier media voices were raised when the *Guardian* and the *Financial Times* ran stories on October 6 about the exports. The *Guardian* revealed that the government had been pressuring EEIC to drop the Pakistani deal and included disgruntled comments from senior management at the company. EEIC management had already protested in a confidential meeting with DoE representatives from the interdepartmental GEN 141 Official Group on the Export of Inverters, when Harris and his colleagues noted that if they were forced to cancel the order, inverters could easily be obtained from other suppliers in the USA, FRG, or Switzerland.[134] During the meeting DoE officials had asked EEIC personnel not to discuss the matter in public but, frustrated at the potential loss of business, they expressed their anger to the press.[135] EEIC's personnel manager, Alasdair Malpas (who had not been present at the confidential meeting), stated that,

> The Government has asked us not to make it. But we pointed out that we have no contract with Pakistan. Our contract is with an agent, and we might be sued for damages. In any case there would be no difficulty in getting the equipment elsewhere. There is nothing magic about it. Swindon should have the business rather than anyone else.[136]

With the issue now embarrassingly public, diplomatic and inter-alliance communications intensified in the run-up to Britain's campaign to alert nuclear supplier states to the danger posed by the clandestine programme.

---

[133] Owen to UKE Islamabad, 'Inverters for Pakistan', October 5, 1978, TNA FCO37/2113.

[134] 'Note of a Meeting Held with Emerson Electrical Industrial Controls', September 26, 1978, TNA Records of the Cabinet Office (hereafter CAB) 130/1052, 1.

[135] Ibid., 4; Harris to Herzig, 'Weargate Contract for Inverters', September 28, 1978, TNA FCO37/2113.

[136] 'Power Export Faces Ban', *TG*, October 6, 1978, 1; Owen to UKE Islamabad, 'Inverters for Pakistan', October 6, 1978, TNA FCO37/2113.

John Bushell in Islamabad astutely commented that the Pakistanis were intent on pursuing a nuclear programme regardless of the barriers put in their way. Furthermore, Bushell argued that, because of the evolving situation in Afghanistan, Britain should offer Pakistan any support it could.[137] In a US-UK meeting in London, Joe Nye circulated drawings showing exactly how inverters could fit into a centrifuge array. Tom Pickering noted that Bhutto's prison cell statements about imminent nuclear capability made it somewhat easier to take action publicly against Pakistan. The meeting's tenor was that any efforts to retard the clandestine programme must be made secretly to avoid the perception of discrimination or a "US crusade" against Pakistan.[138] It had also become apparent to the FCO that the Pakistanis were attempting to purchase not just inverters, but a wide range of equipment from Britain that could be used as part of the enrichment project.[139]

On October 26 the British government began its diplomatic campaign to halt, or at least retard, the clandestine Pakistani enrichment project. Before the campaign's initiation FCO officials had bilateral meetings with the West Germans, who had been so obstructive in 1976 and 1977. Again, there were vague promises to investigate the situation and get back to the British government.[140] When the full campaign started Owen dispatched three related telegrams laying out overall British aims, assessing Pakistani intentions, and providing speaking notes for consular representatives. These cables shared the case's basic facts: that an order had been placed with EEIC; that there were suspicions of a covert weapons programme; and that the immediate aim was to prevent Pakistan from acquiring inverters from other suppliers.[141] Not wanting to create the impression of an international crusade, Owen was keen to place these efforts within a

[137] UKE Islamabad to FCO, 'Pakistan–A Difficult Time Ahead', September 24, 1978, TNA FCO96/826, 5.

[138] 'Record of Discussions of Non-proliferation Issues in the State Dept.', October 6, 1978, TNA FCO96/824.

[139] Fletcher-Cooke to Candlish, 'Nuclear Exports to Pakistan', October 10, 1978, TNA FCO37/2113.

[140] 'Anglo-German Bilaterals: Bilateral Discussions on Nuclear and Non-proliferation Questions', October 19–20, 1978, TNA FCO96/825, 2–4.

[141] FCO to UKE Bonn et al., 'Nuclear Exports', October 26, 1978, TNA FCO96/825, 1.

wider non-proliferation context, as opposed to a campaign that singled out Pakistan. The Pakistanis were described as being "in a nervous state about their security situation" and it was necessary to avoid any "over-reaction which could itself have long-term destabilising consequences in the region". [142] There was no mention that Britain was contributing to these security fears by pressing ahead with the Jaguar sale. Suspicions about a covert enrichment programme were outlined, highlighting the various purchases that front organisations had been making and the unsafeguarded nature of domestic Pakistani uranium supplies.[143] Owen's speaking note for diplomatic representatives covered the key UK concerns and emphasised British status as a depository party to the NPT and core NSG member. Recipient governments were to be made aware that revised British export controls on inverters would come into force on November 9, and other nations were urged to take similar steps themselves.[144] Finally, Owen contacted his ambassadors in nuclear supplier countries instructing them that, if the American démarche on reprocessing were mentioned in their discussions on inverters, UK officials should confirm British support for the US approach.[145]

There were mixed reactions to the démarche. The Dutch government was concerned that it might have difficulty in obtaining information about inverter orders, while the governments of Australia, Canada, Italy, Spain, and Sweden agreed with the British approach and assessment.[146] The West Germans assured the British that inverters were already covered

[142] Ibid.

[143] FCO to UKE Bonn et al., 'Nuclear Exports', October 26, 1978, TNA FCO96/825.

[144] FCO to UKE Bonn et al., 'Nuclear Exports. Following Is Suggested Speaking Note', October 26, 1978, TNA FCO96/825.

[145] FCO to UKE Bonn et al., 'Nuclear Exports', November 7, 1978, TNA FCO96/825.

[146] UKE The Hague to FCO, 'Nuclear Exports', October 30, 1978, TNA FCO96/825; UKHC Canberra to FCO, 'Nuclear Exports', October 30, 1978, TNA FCO96/825; UKHC Ottawa to FCO, 'Nuclear Exports', October 30, 1978, TNA FCO96/825; UKE Rome to FCO, 'Nuclear Exports', October 31, 1978, TNA FCO96/825; USE Madrid to State, 'US Demarche [sic] on Pakistani Reprocessing Plant', November 13, 1978, NPAD, Doc.14; UKE Stockholm to FCO, 'Nuclear Exports', November 2, 1978, TNA FCO96/825.

by their export controls, that they agreed with the démarche's aims, and that a more considered reply was imminent.[147]

While the British démarche was making its way around the capitals of Europe, North America, and Australasia, US-UK cooperation continued apace. Returning from a tour of Afghanistan, Iran, and Pakistan, Jack Miklos (Deputy Assistant Secretary of State for the Near East & South Asia) confided to Deputy Under Secretary of State Hugh Cortazzi (who informed Owen) that the Pakistanis were lacking in self-confidence and that this lack of confidence might influence a move towards non-aligned status.[148] At the same time US embassies were instructed to emphasise that the Carter administration shared British concerns, and to be aware that the USA would shortly be sending out its own démarche on reprocessing.[149] Furthermore, the Americans passed classified intelligence to the UK that indicated Pakistan was apparently working on various technologies necessary for a nuclear bomb programme, such as high explosives, triggering systems, and laboratory-scale reprocessing operations.[150]

Differences between the British and American positions came to light over the inverter issue and its place in wider superpower relations. Back in early October Alston had submitted a list of countries that Pakistan might turn to if the EEIC order was turned down. Those countries fell into three categories: those that were definitely known to have the capacity to manufacture inverters (including France, the FRG, the Netherlands, and Switzerland); those that possessed adequate technology to manufacture them, but whose capability to do so was unknown (Austria, Australia, Canada, Japan, and Sweden); and those countries that were assumed to have the capacity to manufacture comparable items if they wished (such as Brazil, India, Iran, Israel, South Africa, and the USSR.)[151] These lists formed the basis of the countries approached in both the UK démarche on

---

[147] UKE Bonn to FCO, 'Nuclear Exports', November 2, 1978, TNA FCO96/825.

[148] FCO to UKE Islamabad, 'Afghanistan/Pakistan', October 31, 1978, TNA FCO96/825, 2.

[149] State to USE Bonn et al., 'UK Approach to Supplier Governments on Pakistan', November 1, 1978, NPAD, Doc.9.

[150] Weston to Cullimore and White, 'Inverters for Pakistan', November 3, 1978, TNA FCO96/825.

[151] FCO to UKE Washington, 'Inverters for Pakistan', October 5, 1978, TNA FCO37/2113, 1–2.

inverters and its US counterpart on reprocessing. After both these diplomatic efforts had begun the matter of approaching the USSR remained unsolved. The US government planned to have an informal discussion with the Soviets on the margins of a late November INFCE meeting.[152] Robert Alston was concerned about this and made his feelings known via the US Embassy in London. British officials had considered approaching the Soviets at the point of sending out their démarche, but had ruled out such action because they were "not certain [of the] Soviet's [sic] commitment to non-proliferation" and whether or not it would outweigh their "special political interests vis a vis Pakistan".[153] Alston suggested meeting with Nye in Vienna to discuss the matter more fully. In this meeting Nye noted that the Americans were "well satisfied" with the reactions to their own démarche and Alston went over the British doubts about an approach to the Soviets. The meeting's conclusion was that when Gerard Smith met with Soviet representatives, he would speak from the notes prepared in Washington (thereby excluding the British notes on their démarche) and offer no further elaboration on them.[154]

THE FALLOUT

The two-pronged diplomatic attack on the Pakistani nuclear programme did not end the issue. Despite this, Whitehall believed that the campaign had been a qualified success. The major nuclear supplier nations had been alerted to the situation and had reached a broad agreement on exports. At the same time the Pakistani government had not publicly attacked the UK in the way it had America over the reprocessing plant.[155] Washington was more sanguine about the evolving situation. US intelligence was now feeding back information that, despite the "best efforts" of nuclear supplier states, Pakistan had made great progress in

[152] State to USE London, 'Pakistan Proliferation Problem', November 18, 1978, NPAD, Doc.17A, 2.

[153] USE London to State, 'Pakistan Proliferation Problem', November 24, 1978, NPAD, Doc.17B.

[154] 'Pakistan', December 4, 1978, TNA FCO37/2114.

[155] 'Pakistan: Inverters', December 12, 1978, TNA FCO96/826.

pulling together the major components for a uranium enrichment programme.[156] Thinking in London paralleled these assessments, the FCO suggesting that Pakistan was "pressing ahead as rapidly as possible with the development of a fissile material production capability" and "continuing its attempts to acquire a nuclear fuel reprocessing facility that would produce plutonium".[157]

As December progressed a variety of international diplomatic sources backed up the US intelligence appraisal of the Pakistani situation. On December 18, Burdess reported to the FCO that he had met with an Australian diplomat whose name was only recorded as Baker. Baker had approached Burdess because of the US démarche on reprocessing, which mentioned the UK's "hard evidence" of Pakistani nuclear intentions. When Burdess outlined the British intelligence Baker commented that he had recently been to Kahuta, east of Rawalpindi, where there was a "frenzy" of construction. The Australian seemed to think (and Burdess was quick to disabuse him of the notion) that machinery was being installed at Kahuta with the compliance of BNFL. Having seen other nuclear facilities Baker was convinced that the facilities being constructed at Kahuta formed part of a weapons programme.[158] Baker's companion on his visit to Kahuta was French diplomat Jean Forlot, who fed information back to the US Embassy in Islamabad. Forlot came to the same conclusions as Baker—that work was proceeding swiftly on constructing some form of nuclear facility—and Arthur Hummel passed these suspicions back to the State Department.[159]

This diplomatic intelligence arrived on the back of statements by Zia a few days earlier that reemphasised the possibly pan-Islamic dimensions of Pakistani nuclear ambitions. In a speech inaugurating a conference on commercial and industrial matters, he called for technical and economic

---

[156] NIO for Nuclear Proliferation to DCI, 'Monthly Warning Report–Nuclear Proliferation', December 5, 1978, USPQB, Doc.21, 2.

[157] 'Pakistan: Nuclear Weapons Development', December 22, 1978, TNA FCO96/827. The second page of this document—which presumably contains intelligence information—is readacted.

[158] Burdess to Alston, 'Pakistan Nuclear Affairs', December 18, 1978, TNA FCO37/2114, 1.

[159] USE Islamabad to State, 'Discussion with French Official on Nuclear Matters', December 19, 1978, DNSA, NP01615, 1.

cooperation between Islamic countries that should not remain dependent on the West.[160] Burdess suggested that the pan-Islamic comments on the value of nuclear energy, and the pursuit of that technology, would be taken by Pakistanis as a reaffirmation of Pakistan's intention to develop a nuclear weapons capability.[161]

The increased volume of evidence signalling Islamabad's determination to build an enrichment plant persuaded the United States and the United Kingdom to plan another series of démarches aimed at shutting down Pakistani access to further supplies of centrifuge components. In a meeting with US officials British diplomats fretted about the political consequences and tried to reassure themselves that the USA and the UK were thinking in the same direction. The main danger, from the British point of view, was that if Pakistan perceived itself as the victim of further Western diplomatic pressure on an issue so identified with the nation, then a departure from CENTO and a move towards non-alignment became all the more likely.[162] The Americans thought the Pakistani attitude towards CENTO was more influenced by other external matters, such as the developing situation in Iran. The US government drew on its experiences in dissuading South Korea and Taiwan from pursuing indigenous nuclear weapons programmes, and thought that the tactics used there would have a similar effect in the case of Pakistan.[163] British scepticism about this was well founded. South Korea and Taiwan were two nations tied to America on a much more deep-seated and fundamental level than Pakistan ever was. Therefore the USA had much greater diplomatic weight and could assert considerably more pressure on them than it could on Pakistan. Furthermore, the South Korean desire to develop nuclear weapons was in direct reaction to the US desire to reduce troop numbers in the country, leaving Seoul fearful in the face of North Korea's huge army.[164] Whitehall

---

[160] Burdess to Alston, 'Pakistan Nuclear Affairs: General Zia's Public Comments', December 22, 1978, TNA FCO37/2114.

[161] Ibid.

[162] Weston to Alston, 'Pakistan Nuclear Developments', December 22, 1978, TNA FCO37/2114, 1.

[163] Ibid., 2.

[164] Peter A. Clausen, *Nonproliferation and the National Interest: America's Response to the Spread of Nuclear Weapons* (New York: HarperCollins, 1993), 140–141.

analysts suggested that it might be easier for Zia to quit the enrichment programme, as it had not been publicly announced, and therefore cancellation would not result in a loss of face.[165] By this juncture, the United States had been persuaded by its British ally and its own belated intelligence gathering that Pakistan was engaged in a covert attempt to enrich uranium. As the reprocessing plant problem faded away, it was the enrichment project—global in nature and much harder to combat with state-to-state diplomacy—that would occupy the attention of the US and UK governments.

Conclusion

In 1978 two strands of non-proliferation activity became intertwined. The Carter administration achieved a measure of non-proliferation success by realising its aim of preventing Pakistani acquisition of a nuclear reprocessing plant. However, just as the threat of a Pakistani bomb based on recovered plutonium was receding, the covert enrichment project became the focus of non-proliferation policy and action. This was a field in which Britain took the lead, attempting to marshal international opinion against Pakistan. Despite this, Callaghan's government also impeded non-proliferation by pressing ahead with the lucrative deal to sell Jaguar aircraft to India and, through the success of the parliamentary vote, to approve construction of the revenue-generating THORP. Because of this the Carter administration was frustrated by British insistence on prioritising commercial interests over non-proliferation. These cases illustrate the level at which the British government was willing to subordinate non-proliferation to economic self-interest. The clandestine enrichment programme's emergence as an international issue saw Callaghan's government quite willing to accept relatively small economic burdens in order to enforce anti-proliferation policy. When it came to multibillion-pound enterprises, such as Jaguar and the THORP, the economic turmoil of the 1970s forced Callaghan, Owen, Benn, and their colleagues to deprioritise non-proliferation. Regardless, the passing of these two commercial milestones allowed Britain to take a more active role in non-proliferation into 1979 with the government freed from two major economic constraints.

[165] Weston to Alston, December 22, 1978, 2.

The year 1978 also saw the foundations being laid for a major cult component in debates on Pakistan's nuclear aspirations. Bhutto's and Zia's remarks about the pan-Islamic nature of their atomic ambitions became a huge part of the public discussion of Pakistan, the Middle East, and nuclear weapons. Although the media and governments responded quite differently to the Islamic bomb, these responses would nonetheless help to shape the outcry over 'Islamic' nuclear proliferation that emerged over the next twelve months and into the decades beyond.

CHAPTER 6

"A dream of nightmare proportions" The 'Islamic bomb' and the 'Khan Affair', January 1979 to December 1979

At 5.12 am on December 18, 1979, Labour MP Tam Dalyell stood in the House of Commons and delivered a philippic against the Pakistani nuclear programme and the prospect of atomic proliferation in the Islamic world. "This is a spine-chilling prospect—a dream of nightmare proportions," he intoned. "Great Governments, such as those of the Soviet Union or the United States, can be counted upon to act with deliberation...but the bad dream come true of a Gadaffi [sic] bomb or an ayatollah bomb is altogether different."[1] Dalyell's apocalyptic, orientalist speech, which starkly differentiated between 'responsible' states and the 'irresponsible' Islamic world brought to a close a year in which Pakistani nuclear aspirations had gone from a modest media story to front-page news.

Revelations about the 'Islamic bomb' and Abdul Qadeer (A.Q.) Khan's role in appropriating gas centrifuge designs from Europe ensured that the religious and clandestine elements of Pakistani nuclear ambitions became the dominant public narratives. Within these narratives, there was a strand of alarmism around the Islamic bomb's impact on global proliferation. 'Proliferation

[1] Tam Dalyell, House of Commons Debate, December 18, 1979, *Hansard Online*, http://hansard.millbanksystems.com/commons/1979/dec/18/joint-centrifuge-project-almelo#column_555 (accessed May10, 2013).

© The Author(s) 2017
M.M. Craig, *America, Britain and Pakistan's Nuclear Weapons Programme, 1974–1980*, Security, Conflict and Cooperation in the Contemporary World, DOI 10.1007/978-3-319-51880-0_6

cascade' alarmism was nothing new and was normally voiced in terms of security concerns and national prestige. This was proliferation of a different kind, founded in the cultural connections between disparate states.

This chapter and Chapter 7 examine in detail the critical year of 1979, looking at the various strands that were intertwined around the issue of non-proliferation. It was a year that saw the public emergence of the Islamic bomb issue and ended with the Soviet invasion of Afghanistan. Both these issues were intertwined and extended into the next year and beyond, the latter necessitating the creation of a broad-based alliance of US-aligned and Islamic states to combat the USSR in a reinvigorated Cold War.

THE 'ISLAMIC BOMB' IN PUBLIC

Although the concept of the Islamic bomb—a nuclear weapon transcending state boundaries and spanning a transnational religious community—had appeared in Pakistani and Western governmental comment since 1974, it was only in 1979 that the issue burst into the media, increasing public pressure on policymakers in Washington and London but failing to influence policy. The Iranian revolution of February 1979—a popular uprising that caught the Carter administration off guard and drove the Shah, America's staunchest ally in the Islamic Middle East, from power—dramatically brought to Western public attention a new form of political, puritanical Islamic radicalism. While sometimes characterised as anti-modern, Ayatollah Khomeini's new republic was by no means anti-technology. Khomeini emphasised that Muslims had to improve their access to technology and harness science in the service of Islam.[2] The later hostage crisis also negatively influenced the American public's perceptions of Islam.[3] In Pakistan, Zia gave the Muslim faith an ever more central political role as he strove to strengthen his rule and avoid revolutionary "excesses" of the kind seen in Iran.[4]

---

[2] Odd Arne Westad, *The Global Cold War: Third World Interventions and the Making of Our Times* (Cambridge: Cambridge University Press, 2005), 295–296.

[3] Kambiz GhaneaBassiri, *A History of Islam in America* (Cambridge: Cambridge University Press, 2010), 303; Fawaz Gerges, *America and Political Islam: Clash of Cultures or Clash of Interests?* (Cambridge: Cambridge University Press, 1999), 7–8.

[4] Lawrence Freedman, *A Choice of Enemies: America Confronts the Middle East* (New York: PublicAffairs, 2008), 86; Giles Kepel, *Jihad: The Trail of Political Islam* (London: I.B. Tauris, 2004), 98.

It was in this atmosphere—in which Islam appeared to be growing in power—that the Islamic bomb surfaced. Driven by Pakistani rhetoric and media revelations about the nuclear programme, Muammar Gadaffi's anti-Americanism, and an increasingly assertive Islamic world, in 1979 the fear of Muslim nuclear weapons became a key component of public debate on Pakistani atomic aspirations.

A pivotal moment was the West German ZDF television channel's March 1979 exposure of A.Q. Khan's 1975 theft of centrifuge designs from the British-Dutch-German URENCO uranium enrichment facility in the Netherlands. The documentary wove together Islam, Khan, Middle Eastern conflict, and Western efforts to disrupt the Pakistani nuclear programme into a web of speculation and recrimination.[5] The ZDF broadcast proposed that "radical Arab countries", like Libya, "whose hatred of Israel and all those who desire peace in the Middle East is well known" financed Pakistan's bomb programme.[6]

From the ZDF programme onwards, the Islamic bomb became a significant media trope. The *New York Times* and *Washington Post* repeatedly placed Pakistan's ambitions within a pan-Islamic context.[7] In June broadcaster Walter Cronkite introduced a CBS report on 'The Pakistani-Islamic Bomb' that painted an apocalyptic picture of Middle Eastern nuclear warfare. "Reliable" informants led reporter Bill McLaughlin to contend that "Libya wants it [a Pakistani nuclear weapon] to be the nuclear sword of the Moslem world. And Pakistan not only has close relations with Libya, it is also deeply committed to the Palestine

---

[5] 'Pakistan: Nuclear', April 2, 1979, The National Archives of the UK (hereafter TNA) Records of the Foreign and Commonwealth Office (hereafter FCO) 96/950; Translated transcript of ZDF broadcast, appended to Carter to Granger, 'Pakistan', April 20, 1979, TNA FCO 37/2203.

[6] Translated transcript of ZDF broadcast April 20, 1979, 2–3. For a recent in-depth study of Libya's nuclear programme, see Malfrid Braut-Hegghammer, *Unclear Physics: Why Iraq and Libya Failed to Build Nuclear Weapons* (Ithaca: Cornell University Press, 2016), 127–217.

[7] 'Pakistan Denies it is developing Nuclear Arms', *Washington Post* (hereafter *WP*), April 9, 1979, front page; 'How Pakistan Ran the Nuke Round the End', *New York Times* (hereafter *NYT*), April 29, 1979, E5; 'Arms Sales to Pakistan Urged to Stave Off A-Bomb There', *WP*, August 6, 1979, A7; 'Pakistan: The Quest for Atomic Bomb [sic]', *WP*, August 27, 1979, A1.

Liberation Organisation."[8] Observers in Washington and London concluded that McLaughlin's intelligence sources were probably Israeli, while the Pakistani Embassy in Washington responded furiously to the story, disavowing the Libyan connection.[9]

In Britain, the media and MPs speculated about Muslim nuclear proliferation. The liberal *Guardian* unquestioningly referenced an Islamic bomb in multiple stories on the Pakistani programme and the Khan imbroglio.[10] The magazine *8 Days*, published in London, funded by the Emirati Ambassador, and distributed around the Middle East and South Asia, produced a lengthy article on Khan. Correct in some respects, inaccurate in others, it concluded with further comment on the Pakistani bomb's pan-Islamic nature.[11] The publicity prompted MPs to ask awkward questions in parliament. Labour's Leo Abse, Frank Allaun, Bob Cryer, Jim Marshall, David Stoddart, and Tam Dalyell queried British involvement in the scandal and government approaches to the Pakistani problem.[12] Abse was most vocal initially, asking new Prime Minister Margaret Thatcher if she was aware that the Khan Affair had brought the possibility of an Islamic bomb, which would subvert the Western position with Middle Eastern oil producers, that much

[8] Transcript of CBS Evening News, 'Special Report on the Pakistani-Islamic Bomb, part 1', broadcast June 11, 1979, TNA FCO96/956, 3; Transcript of CBS Evening News, 'Special Report on the Pakistani-Islamic Bomb, part 2', broadcast June 12, 1969, TNA FCO96/956, 2.

[9] Pakenham to Alston, 'Nuclear Pakistan', June 25, 1979, TNA FCO37/2206; Embassy of Pakistan, Washington D.C., Press Release, June 13, 1979, TNA FCO96/956, 1. In truth, Libya did provide money to Pakistan on the basis of the relationship between Gaddafi and Bhutto. This relationship waned after 1977 when Zia took power, and there is no evidence to suggest that sensitive transfers took place during the 1970s or that there was a deep-seated 'Islamic' dimension to the relationship. See Braut-Hegghammer, *Unclear Physics*, 159.

[10] 'Zia uninterested in N-weapons, says Desai', The *Guardian* (hereafter *TG*), June 22, 1979, 6; 'Dutch step up inquiry after security slip which 'gave hydrogen bomb to Pakistan'', *TG*, June 22, 1979, 6; 'Security breach 'cover-up' at uranium plant', *TG*, June 29, 1979, 3.

[11] 'How Pakistan Fooled the World and Got the Bomb', *8 Days*, June 23, 1979, 13.

[12] *TG*, June 22, 1979; *TG*, June 29, 1979.

closer?[13] Thatcher remained silent regarding Islamic nuclear weapons posing a threat to the West's status.

In Washington, Senators and members of Congress brought the meme into their debating chambers. William Edwards (D-CA) and Fortney Stark (D-CA) had news items on the Islamic bomb read into the record.[14] The issue even influenced later discussions about Carter's family, as Lester Wolff (D-NY)—in an angry debate on the Libyan business connections of Carter's wayward brother Billy—argued that Libya was 'bankrolling Pakistan's nuclear development program with an eye to acquiring an "Islamic bomb"'.[15]

Pakistani protests did little to curtail speculation. Officials denied that the, now implicitly admitted, nuclear programme was anything other than an enterprise created by and for the Pakistani state. Zia cryptically commented to the BBC: 'It does not mean that Pakistan one day will make a bomb and it will fly it off [sic] in an umbrella to its Arab friends and say here is the bomb, now throw it down the drain. How can that be done?'[16] The media reporting infuriated Pakistani Minister of Foreign Affairs Agha Shahi, who repudiated connections between the Islamic world and the nuclear project, contending that the media commentary represented part of an anti-Pakistani campaign.[17] An editorial followed in the semi-official *Pakistan Times* railing against the Islamic bomb as a "Western created

---

[13] Leo Abse, House of Commons Debate, July 3, 1979, *Hansard Online* hansard. millbanksystems.com/commons/1979/jul/03/tokyo-summit-meeting#S5CV0969P0_19790703_HOC_223 (accessed May10, 2013).

[14] William D. Edwards, 'MCPL Nuclear Alert Series, IV', July 12, 1979, *Congressional Record*, 96th Session (hereafter *CR96*), 18414–18415; Fortney H. Stark, 'MCPL Nuclear Alert Series V', September 6, 1979, *CR96*, 23432–23433; Fortney H. Stark, 'Pakistan's Nuclear Program', October 23, 1979, *CR96*, 29252–29253

[15] Lester L. Wolff, 'Resolution of Inquiry into the Matter of Billy Carter', September 10, 1980, *CR96*, 24958. On 'Billygate', see Salim Yaqub, *Imperfect Strangers: Americans, Arabs, and U.S.-Middle East Relations in the 1970s* (Ithaca: Cornell University Press, 2016), 316–318.

[16] Michael Charlton, Interview with Mohammad Zia ul-Haq, Transcript, June 21, 1979, TNA FCO96/955, 1.

[17] United Kingdom Embassy (hereafter UKE) Islamabad to Foreign and Commonwealth Office (hereafter FCO), 'Pakistan Nuclear', July 2, 1979, TNA FCO96/956, 1.

myth" that was "part of the process of rallying the non-Islamic world against the Islamic people".[18] Despite having used Islamic bomb language prior to 1979, Zia, Shahi, and the Pakistani media were unwilling to recognise that they were the concept's parents.

Throughout the summer and autumn of 1979 the press continued to posit a pan-Islamic nuclear project centred on Pakistan. American journalist Don Oberdorfer buried administration statements on the lack of evidence for an Islamic bomb at the foot of an article on congressional disquiet.[19] Sunanda Datta-Ray, reporting from Calcutta for the *Guardian,* referenced the "intense speculation about the imminence of the Libyan financed 'Islamic bomb'".[20] A subsequent article on British links to the clandestine Pakistani purchasing programme ignored official comment about the meme's speculative nature and once more suggested Libyan financing for the project.[21] In India D.K. Palit and P.K.S. Namboodiri published *Pakistan's Islamic Bomb,* arguing that there was a "Pakistani-Arab ambition to build a nuclear bomb for Islam".[22] The book contributed nothing new to the debate, offering no proof of 'Arab' funding for Pakistan's programme and leaving the project's Islamic nature implicit.[23]

On 9 December *Observer* journalists Colin Smith and Shyam Bhatia contended that 'Dr Khan Stole the Bomb For Islam', following up on mid-year stories alleging British links to the Pakistani nuclear programme.[24] Smith and Bhatia amplified this a week later with an article considering the wider implications of pan-Islamic nuclear proliferation.[25] These pieces inspired Dalyell to

---

[18] UKE Islamabad to FCO, 'Pakistan Nuclear: Agha Shahi's Press Briefing', July 2, 1979, TNA FCO96/956, 2–3.

[19] 'Arms sales to Pakistan Urged to Stave Off A-Bomb There', *WP,* April 6, 1979, A7.

[20] 'Pie in the nuclear sky may save Charan Singh', *TG,* August 19, 1979, 7.

[21] 'Pakistan bomb link denied', *TG,* August 23, 1979, 3.

[22] D.K. Palit and P.K.S. Namboodiri, *Pakistan's Islamic Bomb* (New Delhi: Vikas Press, 1979), v.

[23] UKE Islamabad to FCO, 'Pakistan's Islamic Bomb: An Indian Survey', August 1, 1979, TNA FCO37/2206.

[24] 'How Dr Khan Stole the Bomb for Islam', *The Observer* (hereafter *TO*), December 9, 1979, 11.

[25] 'Atoms for War', *TO,* December 16, 1979, 12.

raise the matter in parliament.[26] After pointing towards a "potential world holocaust" originating in the Arab world or Asia, he outlined his "dream of nightmare proportions", in which Islamic solidarity was the root of nuclear proliferation in an unreliable, irrational Muslim world.[27] Dalyell's speech symbolised how embedded the Islamic bomb idea had become when the media and peripheral political figures discussed the interactions between nuclear technology and the Islamic world.

Public comment on an Islamic bomb—from Oberdorfer's reports, to Smith and Bhatia's exposés, to Dalyell's speeches—tended towards the orientalist and alarmist. As the issue was evolving in 1978–79, Edward Said published the landmark *Orientalism*. Although scholars have debated Said's, at times ahistorical, observations and conclusions ever since, his thoughts on media images of Islam are apposite when considering the Islamic bomb.[28] "Lurking behind all these images," he argues, "is the menace of *jihad*. Consequence: a fear that Muslims (or Arabs) will take over the world."[29] Fear did lie behind public discussion of the Islamic bomb: fear of Middle Eastern nuclear war; and more modest—but still significant—fears of changes in the balance of power between the Muslim world and the West. As sociologist Jonathan Lyons notes, there is a "single, persistent Western discursive formation of violence in Islam that remains largely immune to serious challenge on historical, linguistic, and theological bases".[30] Moreover, non-proliferation and the Islamic bomb feed into the belief that developed nations have a monopoly over the legitimate use and technologies of violence.[31] Cultural anthropologist Hugh Gusterson comments that, for decades,

---

[26] 'MPs to Debate Atom Bomb Revelations', *TO*, December 16, 1979, 1; 'MP Praises Observer', *TO*, December 23, 1979, 14.

[27] Dalyell, House of Commons, December 18, 1979.

[28] Andrew J Rotter usefully dissects Said's work in 'Saidism Without Said: Orientalism and U.S. Diplomatic History', *American Historical Review*, 105:4 (2000), 1207–1210.

[29] Edward Said, *Orientalism* (London: Penguin, 2003), 287. The 1970s also saw the 'Arab terrorist', the perfidious 'oil sheik', and later the 'Islamic revolutionary' becoming more prominent in Western popular culture's demonology. See Yaqub, *Imperfect Strangers*, 183–207.

[30] Jonathan Lyons, *Islam Through Western Eyes: From the Crusades to the War on Terrorism* (New York; Columbia University Press, 2012), 124.

[31] Ibid., 112.

creators of foreign policy and public opinion saw the spread of nuclear weapons to the Islamic world as particularly dangerous.[32] In the case of Pakistan in the 1970s, both Gusterson and Lyons are correct in their assessment of public discourses, but as the documentary evidence demonstrates, the policymakers of the 1970s were far less influenced by fear of pan-Islamic proliferation. As this study shows, the Islamic bomb was—for policymakers—a problem of propaganda rather than an imminent reality.

Reporting on the Islamic bomb also formed part of a continuum of proliferation alarmism extending back to 1945. For decades governmental, scientific, military, and media figures had made dire predictions about the spread of nuclear weapons. At the core of this was the notion of a proliferation cascade, an unstoppable chain reaction of new nuclear states emerging after an 'Nth country' had achieved nuclear capability.[33] The 'cascadologists' feared that this Nth country acquiring nuclear weapons would threaten the national security of other nations, thereby leading to a domino effect. Fears surrounding the Islamic bomb were of a different stripe. This was not a potential cascade based on national security or even the 'prestige' of being a nuclear state. It was a cascade based on the perception that sharing a common (and poorly understood) religion meant that one country achieving nuclear status would lead to an automatic, and willing, spread of that status.

The 'Islamic bomb' in Government

Many works on America, Britain, and the Islamic bomb misunderstand how policymakers comprehended and interpreted Pakistani rhetoric. The Islamic bomb *was* present in discussions of Pakistan's nuclear programme, but understandings were frequently more nuanced than in public discourses. There may indeed have been anti-Muslim, even racist, sentiments present in policymakers' minds, but the available evidence does not show that this influenced policy to any significant degree. There was never any concrete evidence of an Islamic bomb, but the idea prompted deliberation about how to deal with the publicity it created. There were policymakers who, at times, placed a greater

---

[32] Hugh Gusterson, 'Nuclear Weapons and the Other in the Western Imagination', *Cultural Anthropology*, 14:1 (1999), 112.

[33] John Mueller, *Atomic Obsession: Nuclear Alarmism from Hiroshima to Al-Qaeda* (Oxford: Oxford University Press, 2010), 89–91.

emphasis on the possibility of an Islamic bomb. However, both the American and British governments recognised that there was little—if any—hard evidence to back up assertions of an integrated Islamic quest for nuclear weapons.

While the Islamic bomb became embedded in the media coverage of Pakistan, nuclear weapons, and the Middle East, policymakers in Washington and London sought to cut through the speculation, despite the increased pressure that the Islamic issue and the Pakistani purchasing programme put on non-proliferation policy. Islam had been a thread running through British and American assessments of Pakistan's nuclear ambitions since 1974. Eleven days after the Indian detonation FCO analysts suggested that Pakistan may seek firmer links with "Iran and the Arab states", a natural move for Islamabad given Pakistan's troubled history with India and known links with friendly Islamic states.[34] British diplomats in Islamabad opined that there was mounting domestic pressure from harder-line religious elements— such as the Jamaat-e-Islami—to pursue a nuclear option.[35] Furthermore, senior diplomatic officers thought that Pakistan might well seek to put pressure on India through its ties with the Organisation of the Islamic Conference.[36] The most direct connection between a Pakistani nuclear programme and the Muslim world came from an FCO SAD assessment two weeks after the detonation in India. SAD commented that:

> In order to minimise the real cost for Pakistan and also to increase the Arab commitment to him [Bhutto] he would almost certainly seek substantial financial assistance from the Middle East oil producers. Indeed, he might even endeavour to involve them in some kind of nuclear partnership. A development of this kind in the already potentially explosive situation in the Middle East, particularly with its anti-Soviet undertones, would produce its own dangers.[37]

---

[34] Seaward to Wilford, 'Implications of the Indian Nuclear Test', May 29, 1974, TNA FCO66/654, 5.

[35] The JI (literally Islamic Party) was a socially conservative party that campaigned for governance according to Islamic principles.

[36] Imray to Drew, 'Pakistani Reactions to the Indian Nuclear Test', May 29, 1974, TNA FCO66/663, 1.

[37] Chalmers to Male, 'Indian Nuclear Explosion', June 3, 1974, TNA FCO66/654, 5.

In the years after the test, Islam appeared in various contexts. In a meeting with Kissinger in late 1975, India's Ambassador to the United States Triloki Nath Kaul pointed out the rising danger of pan-Islamism. Kissinger's response was illuminating: "We may become allies yet."[38] In 1977 British officials had noted discussion of the reprocessing plant as a potentially pan-Islamic facility, the *Pakistan Times* having asserted Islamabad's willingness to assist Muslim countries in acquiring nuclear technology. However, Robin Fearn at the British Embassy in Islamabad noted, "it is a theme which may be developed as a factor in US hostility to Pakistan's acquisition of the plant", but he also suggested: "Perhaps it need not be taken too seriously."[39] French consular staff in Islamabad thought the same, indicating to their British counterparts that there was no evidence that the Muslim world attached *any* significance to the reprocessing plant as an "Islamic facility" and the plant's entire *raison d'être* was to retain the nuclear weapons option for Pakistan in the hope of gaining parity with India.[40] American intelligence suggested Muslim nations had promised aid for the nuclear programme, Pakistan having apparently convinced the quite disparate Islamic states of Saudi Arabia and Libya that "the Muslim world must no longer suffer the humiliation of being second class citizens in a nuclear age".[41] At a joint meeting in Washington at the end of 1977 American and British delegates agreed that, under Zia, Pakistan was looking towards the Islamic world for financial support.[42]

During 1978 Zia's rhetoric about "reinforcing the power of the Muslim world" and Bhutto's death-cell testimony about the imminence of Pakistani nuclear capability had been debated in multilateral gatherings (for example,

---

[38] Memcon, 'Meeting with Foreign Minister Chavan', October 7, 1975, *Foreign Relations of the United States* 1969–76 (hereafter *FRUS* 69–76), Vol.E8, Doc.214.

[39] Fearn to Carberry, 'Pakistan: Nuclear Reprocessing Plant', June 24, 1977, TNA FCO37/2066, 1–2.

[40] Burdess to Wilmshurst, 'Pakistan Nuclear Affairs–The Reprocessing Plant', November 14, 1977, TNA FCO96/728, 1.

[41] 'Notes on Proliferation—Global Highlights', November 10, 1977, Jimmy Carter Presidential Library (hereafter JCPL), Remote Archives Capture system (hereafter RAC) NLC-28-63-3-16-2.

[42] 'Record of a Meeting Held at the State Department at 11.30 on December 2, 1977', December 6, 1977, TNA FCO96/728, 3.

at the G7 summit in Bonn). However, as Warren Christopher's diplomatic cables in late 1978 demonstrate, those making non-proliferation policy placed no credence in the blustery rhetoric emanating from Islamabad. Despite this, the Pakistani leader's remarks laid the groundwork for the much more wide-ranging public and private debates about the spread of nuclear weapons in the Islamic world that emerged in 1979.

Between the Iranian Revolution and the Islamic bomb becoming a public issue, the idea preyed upon American and British officials. Visiting London, Jack Miklos (Deputy Assistant Secretary of State for the Near East and South Asia) and Paul Kreisberg (of the Bureau of Political-Military Affairs) suggested that Pakistan had "offered to be the supplier of nuclear weapons to the Arab world". For them, this explained perceived Pakistani casualness about the risks of American economic sanctions.[43] US Ambassador to Pakistan Arthur Hummel echoed these beliefs, suggesting that Zia might use future nuclear capability to win financial support from "Arab oil producers".[44] Missives from the British Embassy in Islamabad mirrored this worry, arguing that an American embargo's impact on the Pakistani economy might be minimal "if accounts of Arab backing for Pakistan's nuclear programme are correct". In such a case the Symington Amendment's invocation would move Pakistan closer to the Islamic world, rather than achieving non-proliferation aims.[45] Although Pakistani moves towards closer alliances with the Muslim world were not signifiers of a desire to spread the Islamic bomb, this created a paradoxical point in policymakers' minds; by taking action to head off the Islamic bomb the USA might increase the chances of alienating Pakistan.

For some American and British policymakers public speculation provoked anxiety about the propaganda implications. Not only would a break in relations with a moderate Muslim state like Pakistan adversely affect the similarly restrained Gulf states' sentiments, there was apprehension about the Islamic bomb's impact on Israeli and Indian attitudes towards Pakistan. Cyrus Vance—in agreement with his British counterparts—did

---

[43] 'Summary Record of a Meeting Held by Mr. H.A.H. Cortazzi', March 5, 1979, TNA FCO96/949, 3.

[44] Brzezinski to Carter, 'Daily Report', March 1, 1979, JCPL, RAC NLC-1-9-8-12-0.

[45] UKE Islamabad to FCO, 'Pakistan: Economic Angles', March 5, 1979, TNA FCO96/949.

not fret about the reality of an Islamic bomb, but about the effects of the looming public outcry on India and the Middle East.[46]

India remained a major concern, as New Delhi vacillated over whether or not Pakistan was pursuing an Islamic bomb.[47] The Indians frequently made much of the issue, unsurprisingly given the long history of Indo-Pakistani conflict. However, John Thomson (British High Commissioner to India) offered a more nuanced response suggesting that there might only be ephemeral support for the Pakistani nuclear programme amongst Muslim nations.[48] At the highest level, Callaghan agreed with Indian Prime Minister Morarji Desai that if there *were* evidence of Arab involvement in the Pakistani programme, this would make the situation much more serious, but Callaghan could offer no evidence for the Muslim funding theory.[49] At this stage American and British policymakers and diplomats recognised that the Islamic bomb idea was a provocation, not a reality, prompting Israeli fears of a nuclear threat from the Islamic world, increasing long-standing Indo-Pakistani tension, and potentially pushing India towards full nuclear weaponisation.

In US-UK discussions shared concerns emerged. John Bushell quizzed Hummel, who contended that sharing nuclear technology was the *quid pro quo* for Islamic support for Pakistan's programme. Hummel suspected that supporters included Libya, Saudi Arabia, and perhaps others.[50]

---

[46] 'Pakistan and the Symington Amendment', March 17, 1979, United States National Archives and Records Administration (hereafter NARA) Record Group 59: State Department Central Files (hereafter RG59); Records of Warren Christopher, 1977–1980 (hereafter RWC), Box 56, Pakistan; Vance to Carter, 'Nuclear Problems in the Sub-continent–Status Report', March 19, 1979, JCPL, RAC NLC-15-37-5-11-8.

[47] United Kingdom High Commission (hereafter UKHC) New Delhi to FCO, 'Nuclear Developments', January 30, 1979, TNA FCO96/947; UKHC New Delhi to FCO, 'Nuclear Developments', February 1, 1979, TNA FCO37/2200.

[48] UKHC New Delhi to FCO, 'Pakistan: Nuclear Weapons programme', March 9, 1979, TNA FCO96/948.

[49] FCO to UKHC New Delhi, 'India/Pakistan Nuclear', April 4, 1979, TNA FCO96/951.

[50] USE Islamabad to State, 'Nuclear Aspects of DepSec Visit Discussed with UK and French Ambassadors', March 7, 1979, National Security Archive Electronic Briefing Book (hereafter NSAEBB) 'The United States and Pakistan's Quest for the Bomb' (hereafter USPQB), Doc.26A, 2–3.

Meeting with senior State Department officials in Washington, Anthony Parsons (formerly Britain's Ambassador in Iran) observed that "events in Pakistan were one of the most horrifying developments since 1945", and the possibility that Arab money might be available to Pakistan only added to the danger for the Middle East.[51] Parsons had never shied away from fearful pronouncements about Islamic nuclear capability. "Pakistan is paranoid in its attitude towards India," he wrote, "and I do not at all like the association between the Pakistani nuclear programme and Arab money (not proven but likely) in the present atmosphere prevailing in the Moslem world. It would not be difficult to construct a Nevil Shute type scenario out of all of this."[52] However, the FCO thought it unlikely that Arab countries would knowingly fund the Pakistani nuclear programme, even though many Muslim states might be glad that a co-religionist had achieved nuclear capability.[53]

Although the FCO largely dismissed Parsons' comments, his view—influenced by Middle Eastern unrest and the Iranian Revolution that he had directly experienced—exposed an underlying fear of a violent, irrational Muslim world dragging the planet to the sort of nuclear fate depicted in Shute's bestselling 1957 novel *On The Beach* (and its 1959 film adaptation). Shute's representation of an atomic apocalypse had galvanized readers and reviewers alike and it had become the iconic image of a nuclear end of the world.[54] Parsons' comments—that made the spread of nuclear weapons to the Islamic world the most apocalyptic change since the atomic bomb's creation—could be taken as illustrative of official fears about an Islamic bomb. However, his argument's main thrust was that Pakistan was an unstable, paranoid state with a deeply unsatisfactory government.[55] Parsons' statements, while expressing fear of Muslim nuclear weapons, were more founded in classically Western

---

[51] 'Record of a Discussion in the State Department', March 16, 1979, TNA FCO96/950, 1.

[52] 'India, Pakistan and Nuclear Weapons', March 8, 1979, TNA FCO96/950.

[53] 'Pakistan's Military Nuclear Programme: Pressures and Inducements', March 23, 1979, TNA Records of the Cabinet Office (hereafter CAB) 130/1073.

[54] Daniel Cordle, 'Beyond the apocalypse of closure: nuclear anxiety in the postmodern literature of the United States', in Andrew Hammond (ed.), *Cold War Literature: Writing the Global Conflict* (Abingdon: Routledge, 2006), 75fn7.

[55] 'Record of a Discussion in the State Department', March 16, 1979, TNA FCO96/950, 1.

concerns about 'irrational' and 'unstable' oriental peoples and rulers. The concerns expressed by Parsons did, however, have a genuine basis in fact. Pakistan *did* have a troubled and volatile political history, with extended periods of military rule and martial law.

Gerard Smith also thought that Pakistan was "secretly transgressing norms" of nuclear behaviour adhered to by most states. Smith was also fearful about the prospects for the Middle East, noting that if a "Moslem bomb" was a genuine threat, the Israeli reaction must be considered. Nuclear weapons in the Middle East and South Asia required much wider deliberation if an Islamic bomb were "on the cards".[56] State Department officials thought this aspect posed a serious threat to US national interests in the Middle East and the Persian Gulf, with Pakistan moving "more towards the militant Islamic camp".[57]

Despite the apocalyptic visions of some individuals, at this stage, policy-makers on both sides of the Atlantic saw the Islamic bomb paradigm as provocative, rather than an imminent threat. The most worrying aspects were the potential reactions of Israel and India, the two states who felt most threatened by, respectively, the Islamic world and Pakistan. The consensus was that, although it was a worrying idea, no one could offer any hard evidence that an Islamic bomb was about to become a reality.

While the Islamic bomb remained pure speculation, harder evidence of clandestine Pakistani efforts to build uranium enrichment facilities continued to emerge. From the British perspective, this purchasing pattern required urgent and determined action.[58] The US government had similar fears, acknowledging that enrichment was now the core of Pakistan's bomb effort and could lead to a potential derailment of US-Pakistani relations by triggering the Symington Amendment.[59] Within the Carter administration there was confusion over the formulation and implementation of policy, as non-proliferation came into conflict with human rights and regional security. Furthermore, a non-proliferationist Congress

[56] Ibid., 5.

[57] 'PRC Paper on South Asia', March 23, 1979, USPQB, Doc.32A.

[58] Carter [FCO] to Thomson, 'Pakistan: Nuclear Weapons Intentions', January 5, 1979, TNA FCO96/947.

[59] Memorandum for Brzezinski, January 30, 1979, JCPL, RAC NLC-10-18-2-27-0; UKE Washington to FCO, 'Pakistan Nuclear Developments', January 27, 1979, TNA FCO96/947.

created problems for the administration, and Zia's stonewalling turned promising avenues into blind alleys.

In Whitehall GEN 74 pondered the timescale for Pakistan to achieve nuclear capability. According to experts Pakistan could produce weapons-grade uranium using the stolen centrifuge designs by 1982 at the earliest, although 1984–85 was more likely. The main constraint on Pakistani efforts was the requirement for precision-made bellows for the enrichment cascades.[60] British customs officials had stopped a shipment of bellows from Ireland and were pushing the Irish government to prevent further exports.[61] French investigations discovered that the Pakistanis had contracted the engineering firm Calorstat to provide the same equipment. London and Washington shared anxieties about the purchasing effort and submitted diplomatic representations to the French government.[62] This new evidence resulted in British ministers once again tightening export controls and seeking "pressures and inducements" to dissuade Pakistan.[63]

By March intelligence on Pakistani purchasing was flooding in to Washington and London. The State Department attempted to influence the Swiss government when it became apparent that multiple Swiss firms were supplying components.[64] Bern was reluctant to act, offering vague assurances that they would talk with the companies concerned.[65] America also pressured the FRG when reports suggested that West German companies were involved in supplying Islamabad.[66] In light of this the USA and

---

[60] GEN 74, 'Pakistan: Possible Timing of a Nuclear Explosion', February 27, 1979, TNA CAB130/1073, 1–2. This is the first British document in which Khan is named.

[61] 'Pakistan: Nuclear Weapons Development Programme', February 23, 1979, TNA FCO96/948.

[62] FCO to UKE Paris, 'Pakistan: Nuclear Weapons Development', March 1, 1979, TNA FCO96/948; Macrae to Alston, 'Pakistan: Nuclear Weapons Development', March 8, 1979, TNA FCO96/949.

[63] GEN 74, 'Pakistan's Nuclear Intentions', February 27, 1979, TNA CAB130/1073.

[64] State to USE Bern, 'US-Swiss Discussions on Pakistan Nuclear programs', March 6, 1979, JCPL, RAC NLC-16-115-5-25-5.

[65] UKE Bern to FCO, 'Pakistan Nuclear Weapons Development', March 9, 1979, TNA FCO96/949.

[66] State to USE Bonn, 'Pakistani Nuclear Program', March 6, 1979, JCPL, RAC NLC-16-115-5-26-4.

the UK continued their close cooperation in attempting to shut down the tentacular purchasing networks. For Washington the situation was becoming increasingly fraught as mounting evidence of Pakistani intentions made it impossible to prevent the Symington Amendment's imposition.

The US government faced the contradictory needs of forcing the Pakistanis to abandon their nuclear aspirations, while preserving good relations in a strategically significant region.[67] Here, non-proliferation policy and re-emergent Cold War imperatives collided as détente crumbled. The Iranian Revolution had destroyed the US alliance with Tehran, just as it appeared that the Soviets were involving themselves more deeply in neighbouring Afghanistan. However, Pakistan was not seen simply as a bastion of US regional influence, but also, possibly as a significant part of wider Cold War arms control. The Tackman radar stations in Iran, which had monitored the Soviet missile test range at Tyuratam, had been lost because of the Revolution, and needed to be replaced with new sites. The most geographically obvious location for these was in Pakistan.[68] A harsh non-proliferation policy risked driving Pakistan away from the USA, striking a further blow to US regional influence *and* superpower arms control.

The Carter administration faced a wave of intelligence and publicity that made the Symington Amendment's imposition a virtual certainty. January to March saw numerous efforts aimed at retarding the inevitable by engaging with Zia's government. The State Department viewed the problem as one of regional security and stability; with the situation in Iran a rupture with Pakistan could pose serious regional problems for the USA.[69] The elite, multiagency Policy Review Committee agreed to delay the Symington Amendment's imposition on the grounds of ongoing diplomatic efforts and Pakistan's critical regional importance.[70] Hummel

---

[67] 'Policy Review Committee Meeting', March 9, 1979, JCPL, RAC NLC-132-73-6-5-3, 2.

[68] Rhodri Jeffreys-Jones, *The CIA and American Democracy*, 3rd edition (New Haven: Yale University Press, 2003), 221; Dennis Kux, *Disenchanted Allies: The United States and Pakistan, 1947–2000* (Washington D.C.: Woodrow Wilson Center Press, 2001), 241

[69] Saunders and Pickering to Newsom, 'Mini-PRC Meeting on the Pakistan Nuclear Problem', January 20, 1979, USPQB, Doc.23A, 2.

[70] 'Summary of Conclusions: Mini-PRC on Nuclear Matters', January 22, 1979, JCPL, RAC 24-102-7-4-1, 2.

had confronted Zia with intelligence about the nuclear programme and the Pakistani President had responded angrily, offering inspection rights to any nuclear facility in Pakistan as proof of his peaceful intentions.[71] The administration seized upon this opportunity, as it demonstrated to Congress the correctness of resisting the Symington Amendment's application.[72] In the end, Zia dashed hopes for inspections when—unmoved by warnings about US legislation—he informed Warren Christopher that he would not permit scrutiny of nuclear facilities and refused to rule out a PNE.[73]

The growing publicity attendant upon Pakistani ambitions highlighted the problematic nature of open US challenges to Zia's government. In contrast to the 'softly softly' approach favoured by some State Department officials, Gerard Smith advocated a high-profile, international campaign against the Pakistani bomb, echoing Anthony Parsons' view that the situation posed the "sharpest challenge to the international structure since 1945". Smith argued that when faced with an eroding global consensus against nuclear weapons, "the prospect of 'Moslem' bombs is as likely as a German and Japanese bomb (consider what their jingos would make of these countries remaining 3d class powers.).". [74] The threat in Smith's eyes was not the Islamic bomb, but the impact of Pakistani nuclear attainment on the international scene, leading to a cascade of key non-nuclears deciding to pursue the nuclear option. Smith argued that the current non-proliferation policy towards the sub-continent was too parochial and demanded that the situation be placed in a global context.[75] Realising that massive publicity about the Pakistani programme was inevitable, State Department non-proliferationists,

[71] Kux, *Disenchanted Allies*, 236.

[72] Memorandum, January 30, 1979, JCPL, RAC NLC-128-9-16-7-6, 2; 'Call on Mr P H Moberly by Mr T Pickering', February 26, 1979, TNA FCO37/2200, 1; UKE Washington to FCO, 'Indo-US Relations: Sino/Pakistan and Nuclear Dimensions', February 21, 1979, TNA FCO96/948, 3–4.

[73] Vance to Carter, Memorandum, March 2, 1979, JCPL, RAC NLC-128-14-5-2-7, 2; Kux, *Disenchanted Allies*, 239.

[74] Smith to Christopher, 'Memorandum to the Deputy Secretary', March 27, 1979, NARA, RG59, RWC, Box 56, Pakistan III, 1

[75] Ibid.

such as Pickering, favoured the high-profile "sunshine approach", placing the full glare of publicity on Pakistan, potentially turning international opinion against Islamabad.[76]

Zia's intransigence, the Symington Amendment's looming implementation, and the flow of information on the purchasing project led the State Department to scramble for a new policy. Hummel argued that no unilateral or multilateral pressure by the USA or its allies could head off Pakistan's programme, and only a "bold initiative" would meet Islamabad's security requirements and constrain their nuclear ambitions.[77] Tom Pickering (Assistant Secretary of State for Oceans and International Environmental and Scientific Affairs) and Harold Saunders (Assistant Secretary of State for Near East Affairs) suggested a "bold initiative", an "audacious buy-off" comprised of extensive security assistance—consisting of arms sales and economic aid—aimed at assuaging Pakistani fears.[78] Additionally, they suggested, there could be exploration of an Indo-Pakistani agreement to neither build nor test nuclear weapons if assistance did not result in movement on the nuclear front. Pickering and Saunders also noted that, "The likelihood of an 'Islamic bomb' with its consequences in the Arab-Israeli dispute will increase Congressional concerns over anything we might propose doing for Pakistan."[79] Pickering and Saunders were worried that the Islamic bomb idea would cause increased consternation in a non-proliferationist Congress. However, the proposal never gained traction. Warren Christopher's assistant, Steve Oxman, thought Pickering and Saunders were "dreaming" if they imagined the package would look like anything other than a bribe for Pakistan and if they believed Congress would permit such a package in the face of persuasive evidence of Pakistani nuclear ambitions.[80] The DoD and ACDA also opposed security assistance

---

[76] 'Record of a Discussion in the State Department', March 30, 1979, TNA FCO 96/951, 2–3.

[77] USE Islamabad to State, 'Pakistan's Nuclear Program: Hard Choices', March 5, 1979, NARA, RG59, RWC, Box 56, Pakistan II.

[78] Oxman to Christopher, Note, March 5, 1979, NARA, RG59, RWC, Box 56, Pakistan II.

[79] Saunders and Pickering to Vance, 'A Strategy for Pakistan', March 5, 1979, NARA, RG59, RWC, Box 56, Pakistan II, 2.

[80] Handwritten marginalia, 'A Strategy for Pakistan', March 5, 1979, 7.

for Pakistan, lest it be mistaken for tacit approval of the nuclear programme.[81] Human-rights proponents in the State Department also rejected increased security assistance, protesting Zia's imposition of Islamic sharia law punishments in Pakistan.[82]

The reaction to the Pickering-Saunders proposal highlights the issues the Carter administration faced in dealing with Pakistan, the challenges of balancing non-proliferation, conventional arms control, and security assistance were becoming all too apparent. The proposal did, in one sense, presage a sea change in administration policy towards Pakistan. The suggestion of exploring a non-test/non-production agreement between Pakistan and India would, in time, become a core element of administration strategy. This strategy eventually focused on the non-testing portion and—when it was articulated in early June—moved policy from *prevention* to *mitigation* of Pakistani nuclear capability.

By April events were taking place that typified the tension between the various strands of Carter's foreign policy that plagued his time in office. As Smith had noted at the end of March, the administration needed to think more widely about how Pakistan was affecting the global environment. Pakistani security was certainly a regional issue and non-proliferation was a global issue. Addressing one meant addressing the other, and vice versa. On April 6, 1979—two days after Bhutto's execution—Carter invoked the Symington Amendment, embargoing arms sales and aid to Pakistan because of mounting evidence of a uranium enrichment programme. The aid cut-off created more headlines and antagonised a Pakistani political establishment already dealing with the violent domestic convulsions provoked by Bhutto's execution.[83] The Symington Amendment's implementation angered the Pakistanis. Islamabad was aggrieved that the USA

---

[81] Slocombe to Lake, 'FY-79 Security Assistance Supplemental', March 2, 1979, NARA, RG59, RAL, Box 5; Blechman to Lake, '1979 Security Assistance Supplemental', March 5, 1979, NARA, RG59, RAL, Box 5; Blechman to Gelb, 'Paks Nobiscum?', January 26, 1979, NARA, RG59, RAL, Box 5, 3.

[82] Schneider to Newsom, et al., 'FY '79 Security Assistance Supplemental', March 5, 1979, NARA, RG59, RAL, Box 5, 2.

[83] 'U.S. Aid to Pakistan Cut After Evidence of Atom Arms Plan', *NYT*, April 7, 1979, 1; 'U.S. Cutting Aid to Pakistan in A-Facility Dispute', *WP*, April 7, 1979, A1; 'Pakistan Denies It Plans A-Bomb: Denounces Washington Aid Cutoff', *NYT*, April 9, 1979, A1.

had continued to supply nuclear fuel to the Indian reactor at Tarapur, an issue that was causing difficulties for the Indo-American relationship.[84] There was scant enthusiasm in the Carter administration for the imposition of sanctions, as Pakistan would suffer little real harm from the withdrawal of bilateral economic aid. However, failure to react publicly to the increasingly blatant Pakistani activities would signal open acquiescence to Islamabad's nuclear ambitions and diminish US credibility on non-proliferation issues.

At the same time as the Carter administration was grappling with the issue of a non-proliferation credibility gap, the international media extensively publicised the Khan Affair.[85] As a result of the publicity the FCO in London expected an onslaught against Britain, and readied briefings that defended British actions and said the absolute minimum about the Pakistani programme.[86] The story eventually broke wide open, publicised by the BBC, Swiss television, the *New York Times*, the *Daily Telegraph*, and many other news outlets, which all followed the lead given by ZDF.[87] During the hard-fought British general election campaign of May 1979 the FCO attempted to deflect questions on to the Dutch and aligned Britain with American concerns about the enrichment programme.[88] On May 4, as it became clear Callaghan had lost the election, his government—pushed by Carter—made a final representation to the Pakistanis regarding British anxieties about the enrichment programme.[89] The changes taking place in British politics militated against a serious reception for the démarche. It later became evident

[84] On the Tarapur issue during late 1978 and early 1979, see George Perkovich, *India's Nuclear Bomb: The Impact On Global Proliferation* (Berkeley: University of California Press, 1999), 218–219. On the problems posed by nuclear exports during the 1970s, see J. Samuel Walker, 'Nuclear Power and Nonproliferation: The Controversy Over Nuclear Exports, 1974–1980', *Diplomatic History*, 25:2 (2001), 215–249.

[85] UKE Washington to FCO, 'Pakistan Nuclear', April 7, 1979, TNA FCO96/951, 1.

[86] Lavers to Baxter, 'Pakistan: Nuclear Weapons Programme', April 6, 1979, TNA FCO96/951, 1–2.

[87] UKE Bern to FCO, 'Pakistan Nuclear', May 3, 1979, TNA FCO96/953.

[88] Alston to Whyte, 'Pakistan Nuclear', May 3, 1979, TNA FCO96/953, 2.

[89] FCO to UKE Islamabad, 'Speaking Note', May 3, 1979, TNA FCO37/2203.

that the Pakistanis had discounted the representation as the "dying whim of the outgoing government".[90]

In the midst of the media furore the FCO placed little credence in the idea of a pan-Islamic nuclear capability originating in Pakistan. Egyptian diplomats expressed anxiety to their British counterparts about the swirling rumours regarding Libya, suspicious of Pakistani attempts to position themselves as suffering anti-Islamic discrimination.[91] Despite provocative Indian speculation that Pakistan's sole nuclear desire was to produce an Islamic bomb funded by Arab money, key FCO officials were unanimous in doubting the real or potential existence of an "Arab bomb".[92] The FCO JNU's David Carter described evidence for this as "woefully thin".[93]

Meanwhile, Islamabad responded to the Symington Amendment's imposition by attempting to make the embargo a pan-Islamic issue. Agha Shahi argued that it was "discriminatory, based on false charges, and designed to keep nuclear power out of the hands of Muslim countries".[94] Shahi contended that the restriction was the fault of a "Zionist lobby" and denied that Libya or *any* Islamic nation was funding the nuclear programme.[95] During this tense period State Department guidance for US consular officials emphasised the problem of regional and global proliferation and *not* religious affiliation:

Q: What would be the implications for the Middle East of what has been described as a "Muslim bomb" to balance the Israeli bomb?

A: As you know the Israelis have repeatedly stated in the past that they would not be the first to introduce nuclear weapons in the region. In our

---

[90] FCO to UKE Islamabad, 'Pakistan: Nuclear', June 4, 1979, TNA FCO37/2205.

[91] White to Holloway, 'Pakistan/Afghanistan', April 17, 1979, TNA FCO96/952.

[92] UKHC New Delhi to FCO, 'Pakistan', April 19, 1979, TNA FCO96/952, 1; Alston to White, 'South Asia–Nuclear Issues', April 19, 1979, TNA FCO96/952, 1; Mallaby to Alston, 'South Asia–Nuclear Issues', April 23, 1979, TNA FCO96/953, 1.

[93] Carter to Granger, 'Pakistan', April 20, 1979, TNA FCO37/2203.

[94] White House Situation Room to Brzezinski, 'Additional Information Items', April 10, 1979, JCPL, RAC NLC-1-10-3-25-9, 2.

[95] USE Islamabad to State, 'GOP Reacts to US Aid Decision: More Political Leaders Attack US Move', April 9, 1979, RG59, RWC, Box 57, Pakistan.

view, any proliferation of nuclear weapons anywhere can only have the most serious consequences for world security.[96]

US discussion guidelines for consular officials illustrated that policymakers— despite what individuals might think in private—realised the Islamic bomb was a propaganda problem by stressing that the issue was *not* one of discrimination, emphasising extensive US nuclear cooperation with Islamic nations like Indonesia, Iran, and Turkey.[97]

Over time it became apparent that the evidence for an Islamic bomb *was* limited at best, non-existent at worst. In bilateral US-UK discussions, Pickering highlighted fragmentary Australian indications about Libya and Iraq.[98] He observed that, while Saudi Arabia had been a substantial aid donor to Pakistan, it was doubtful the Saudis explicitly intended to finance the nuclear programme, especially as Riyadh had been making disapproving noises about Pakistani atomic intentions.[99] John Bushell, in his valedictory dispatch from Islamabad, echoed doubts about the willingness of Muslim states to align themselves with the Pakistani nuclear project, and a concurrent unwillingness in Pakistan to ally with more extreme, Iranian-style, political Islam. Bushell argued that "cooperation with Muslim brothers, yes, alliance on the basis of fundamentalist Islam, no thank you".[100] The Ambassador contended that Pakistan was indeed a significant Muslim nation in terms of its population, but other Islamic states might treat the thought of Pakistan as an "arsenal of Islam" with caution.[101] Bushell asked, "In Islamic terms an 'arsenal' Pakistan may be: but now an arsenal in nuclear terms also? With the problems of its politics and policies post-Bhutto, which Arabs can seriously want to become

---

[96] State to USE Islamabad, 'Contingency Press Guidance', April 6, 1979, JCPL, RAC NLC-16-116-2-10-3, 4.

[97] State to USE Amman et al., 'The Pak Nuclear Problem and the Suspension of US Aid', May 23, 1979, JCPL, RAC NLC-16-116-4-10-1, 3.

[98] State to USE Canberra, 'Pakistan's Nuclear Program', April 7, 1979, JCPL, RAC NLC-16-116-2-18-5.

[99] Alston to Fearn, 'Pakistan Nuclear Programme', April 20, 1979, TNA FCO96/952, 3.

[100] Bushell to Owen, 'Pakistan–Valedictory Dispatch', April 26, 1979, TNA FCO96/955, 7

[101] Ibid.

engaged with Pakistan?"[102] Following up, the outgoing Ambassador argued that the further Pakistan went with a nuclear programme, the harder it would be for Zia to give it up, particularly if nuclear technology became an asset in Pakistani relations with the wider Muslim world.[103]

By late spring the US government was making efforts to persuade interested parties that the Islamic bomb was little more than propaganda. The Americans voiced fears of a nuclear arms race on the sub-continent and tried to demolish the Indian belief that the real danger posed by Pakistan lay in an Israeli/Islamic nuclear confrontation.[104] In Washington Pickering faced questions in Congress. He noted that the administration believed Pakistan was aiming for a nuclear weapons capability, but refused to discuss alleged Libyan financing in open session, leading to further media comment about the Gaddafi connection.[105]

In meetings between Vance, Smith, and Agha Shahi, Shahi stressed that an Islamic bomb was pure speculation, highlighting that Pakistan had turned down Saudi offers of finance for the reprocessing plant. Castigating the Islamic bomb as nonsensical was a theme in Shahi's representations throughout 1979.[106] US officials felt that Pakistan was working hard to get the right reaction from the Muslim and developing worlds, in case they had to publicly justify the nuclear programme's military nature.[107] From using an Islamic bomb as a threat, the Pakistanis had moved to belittling the very idea they had helped create. For Peter Constable, writing on June 6, Pakistani nuclear sharing was a

---

[102] Ibid, 7–8.

[103] Bushell to Parsons, 'Pakistan, India, and Nuclear Weapons', April 29, 1979, TNA FCO96/953, 2.

[104] Jimmy Carter, *White House Diary* (New York: Farrar, Straus, and Giroux, 2010), 314; UKE Washington to FCO, 'US/Indian Nuclear Discussions', April 30, 1979, FCO96/953, 1.

[105] 'Testimony, Assistant Secretary Thomas R. Pickering, Subcommittee on Energy, Nuclear Proliferation and Federal Services', May 1, 1979, TNA FCO96/953, 12; 'Panel Told Pakistan Gained A-Weapons Ability By "End Runs"', *WP*, May 2, 1979, A15.

[106] Extract from US telegram, FCO to UKE Islamabad, 'Pakistan Consortium Meeting 5–6 June', June 8, 1979, TNA FCO96/955, 2.

[107] UKE Washington to FCO, 'Agha Shahi's Talks in Washington', May 8, 1979, TNA FCO96/954, 3.

serious prospect, which could be addressed by distributing nuclear technology amongst Pakistan's "Islamic friends", giving the USA "a much
better chance of exerting influence against any GOP move to contribute to
a so-called Islamic bomb".[108] Thus, Constable advocated undercutting
Pakistan's position by offering Muslim states the fruits of civilian nuclear
technology.

Meanwhile, the State Department and the CIA vacillated on the Islamic
proliferation issue. The State Department acknowledged that the Islamic
bomb was the subject of feverish speculation and believed that Pakistan
might have a *material* interest in spreading nuclear technology, but analysts
had no substantive evidence for pan-Islamic nuclear cooperation.[109]
However the CIA suggested that offers of political and financial support
from oil-rich sympathisers in the Islamic world could tempt Pakistan.
Indeed, Saudi Arabia, Libya, or Iraq might have induced Pakistan to share
sensitive nuclear equipment and to propose terms for future nuclear cooperation.[110] The agency later reversed its position on the Iraqi example,
casting doubt on Pakistani willingness to provide nuclear technology or
materials to Baghdad because of Islamic solidarity.[111]

As media speculation mounted, Western European leaders came under
pressure from the nation most fearful of Islamic nuclear capability. Israeli
Prime Minister Menachem Begin contacted British, French, and West
German leaders warning of the danger posed by the Islamic bomb and
demanding action. Writing to Thatcher, Begin described in dire terms the
consequences of a Pakistani nuclear weapon in the hands of Gaddafi, a
weapon that could "become a mortal danger" for the people of Israel.[112]
The attached Israeli briefing drew links between Pakistan and the Arab

---

[108] State to USE New Delhi, 'Non-proliferation in South Asia', June 6, 1979,
Wilson Center History and Public Policy Program Digital Archive (hereafter
WCDA), http://digitalarchive.wilsoncenter.org/document/114198 (accessed
September 9, 2013), 6–7.

[109] Thornton to Brzezinski, 'Minutes–May 23 PRC Meeting on Pakistan and
Subcontinent Issues', May 25, 1979, JCPL, RAC NLC-24-102-10-6-5, 4.

[110] NIO for Nuclear Proliferation to Director of Central Intelligence, 'Monthly
Warning Report—Nuclear Proliferation', July 24, 1979, USPQB, Doc.41, 2.

[111] CIA, 'Interagency Intelligence Memorandum: Iraq's Nuclear Interests,
Programs, and Options', October 1, 1979, JCPL, RAC NLC-6-34-4-10-3, 10.

[112] Begin to Thatcher, Letter, May 17, 1979, TNA FCO96/954.

world, but none of Begin's intelligence was new, and nothing confirmed the existence of pan-Islamic nuclear cooperation.[113]

The FCO's Paul Lever did not subscribe to Begin's assertions, stating:

> While we share their concern, we believe that they may be making over much of Pakistani-Arab links. Although the Pakistanis are getting financial aid from Arab states the limited evidence available to us (and the Americans) does not support the suggestion that there is any plan to produce an "Islamic Bomb" or to produce weapons-usable material in Pakistan for other Islamic countries.[114]

The speaking note prepared for Thatcher's May 23 meeting with Begin reflected this viewpoint, shared throughout the FCO and other departments.[115] Despite the media exposure, JNU Chief Robert Alston emphasised that the conclusion of earlier British intelligence reporting, that there was little evidence of Arab assistance for the Pakistani nuclear programme, still held true.[116] In the face of widening media coverage the FCO advised British consular officials worldwide that there was "virtually no evidence" for "Arab financing" of the Pakistani nuclear programme.[117]

Thatcher addressed the issue via a letter to Begin that she took a personal hand in drafting.[118] She sympathised with Israel's position, but repeated the FCO's analysis, pointing out: "None of the evidence currently available to us suggests there is any arrangement to transfer weapons-useable material from Pakistan to other Islamic states or organisations."[119] She went on to outline

---

[113] 'Pakistani Activity in the Nuclear Field', appended to Begin to Thatcher, May 17, 1979.

[114] Lever to Cartledge, 'Pakistan's Nuclear Programme', May 22, 1979, TNA FCO96/954, 4.

[115] Alston to Moberly, 'Pakistan's Nuclear Programme', May 22, 1979, TNA FCO96/954; 'Briefs for Quadripartite Ministerial Dinner in The Hague on 29 May', May 22, 1979, TNA FCO37/2204.

[116] Alston to Moberly, 'Pakistan's Nuclear Programme', June 5, 1979, TNA FCO96/955, 1. This JIC report remains classified. The conclusions can be apprehended from Alston's statements in his briefing.

[117] 'Pakistan Nuclear', May 3, 1979, TNA FCO96/953, 2; FCO to UKE Washington et al., 'Pakistan Nuclear: Publicity', May 4, 1979, TNA FCO37/2203, 2.

[118] Alston to Moberly, June 5, 1979.

[119] Thatcher to Begin, Letter, June 19, 1979, FCO96/955, 1.

the many steps Britain had taken to thwart the clandestine Pakistani purchasing programme and urged Begin to consider his *own* country's role in preventing Middle Eastern nuclear proliferation.[120]

A month later the new British Foreign Secretary Lord Peter Carrington probed Indian Foreign Minister Shyam Nandan Mishra at the Commonwealth Heads of Government meeting in Lusaka, Zambia. Carrington asked if the Pakistanis were developing an Islamic bomb or a Pakistani bomb. Mishra's aide, Jagat Mehta, could not discern an "integrated Islamic political strategy" behind the bomb programme. Carrington replied that if it did prove to be an Islamic bomb, "it would make the Middle East even more unstable".[121] The Thatcher-Begin letters and the related FCO discussions illustrate the way in which fears created by the Islamic bomb paradigm needed to be addressed. Despite the media coverage—and as the FCO pointed out—there was little evidence pointing to a pan-Islamic nuclear project.

As the British government transitioned from Labour to Conservative there was no reduction in the pace of British and American efforts to shut down Pakistani access to sensitive materials. There was a requirement to persuade key suppliers, such as France, Switzerland, and West Germany, of the need to take more and stronger action on export controls, despite considerable resistance, particularly from the Germans and Swiss, neither of whom were convinced that the danger was as pressing as the UK and the USA insisted.[122] France agreed to watch Calorstat and Helmut Schmidt eventually assured Carter that his government would investigate the improvement of export controls.[123] The Swiss—whose companies were major suppliers to the Pakistanis—were unresponsive, arguing that the NSG guidelines and the NPT stipulations meant there was little they could do at present.[124] The Swiss government—like the FRG—adhered to

---

[120] Ibid., 1–2.

[121] 'Note of a Conversation Between the Foreign and Commonwealth Secretary and Indian Foreign Minister in Lusaka', August 3, 1979, TNA PREM19/155.

[122] Carter to Clark, 'Pakistan Nuclear', May 24, 1979, FCO96/954; 'UK/US Non-proliferation Bilateral, Pakistan Nuclear', May 11, 1979, TNA FCO 96/953; 'Call at FCO by Mr. T. Pickering', May 14, 1979, TNA FCO96/953, 9–10.

[123] UKE Paris to FCO, 'Pakistan Nuclear', May 17, 1979, TNA FCO 96/954, 1; Schmidt to Carter, Letter, May 21, 1979, DDRS, DDRS-272419-i1-4.

[124] Pakenham to Alston, 'US/Swiss Discussion on Argentina/Pakistan', June 8, 1979, TNA FCO96/955, 1.

the existing regulations, even when it became apparent that those regulations were inadequate in the face of new global networks dedicated to the acquisition of sensitive nuclear technologies.

Members of Thatcher's government stressed to senior Pakistanis they were just as concerned about the nuclear issue as Callaghan and Owen had been. Douglas Hurd, Minister of State in the FCO, pressed the matter home, pleading with the Pakistani Ambassador to understand that, although Britain did not want to single out its friends in South Asia, the UK was anxious about the nuclear programme.[125] The new British Ambassador in Islamabad, Oliver Forster, noted that in his first meeting with Agha Shahi, the Foreign Minister had launched into a "bitter tirade" about the Americans and "Jewish lobbies" in the USA that were behind the campaign against Pakistan, contrasting this with what he saw as a more proportional British response. Forster disliked this attempt to drive a wedge between the USA and the UK, but declined to respond to Shahi's provocations. Forster suggested, in a point that Thomson in New Delhi agreed with, that an aggressive "big stick" policy by the UK would provoke the same violent and emotional reaction. Quietly persuasive diplomacy was the way forward.[126]

Quiet diplomacy was a theme that also surfaced in multilateral discussions. Gerard Smith proposed an informal summit to impress upon Western European nuclear supplier nations the situation's urgency. UK delegates would attend but the government expressed reservations about any appearance of developed world countries "ganging up" on Pakistan, Carrington noting that he had serious misgivings about the meeting becoming public knowledge.[127] While the French refused to participate, Carrington instructed his representatives to support the USA, underline the seriousness of Pakistan's ambitions, and emphasise the need for international controls and communication.[128] Robert Gallucci opened the meeting by stressing the US intelligence assessment that the Pakistani centrifuge programme strongly suggested a nuclear weapons project.

---

[125] FCO to UKE Islamabad, 'Pakistan Nuclear', June 7, 1979, TNA FCO96/955.

[126] UKE Islamabad to FCO, 'Pakistan Nuclear', July 23, 1979, TNA FCO96/957; New Delhi to FCO, 'Pakistan Nuclear', July 26, 1979, TNA FCO96/957.

[127] FCO to UKE Vienna, 'Pakistan Nuclear', June 19, 1979, TNA FCO96/955.

[128] UKE Washington to FCO, 'Vienna Discussions on Pakistan', June 25, 1979, TNA FCO96/956; FCO to UKE Vienna, 'Pakistan Nuclear', June 26, 1979, TNA FCO96/955.

The attendees reached the consensus that subtle diplomacy and the more effective application of export controls were the best way forward.[129]

As media attention on Pakistan's so-called Islamic bomb increased with the CBS report, the *8 Days* article, and the publicising of the Khan Affair, public speculation began to affect bilateral relations between both the USA and Pakistan and the UK and Pakistan. Robin Fearn in Islamabad described the Pakistanis as being in a state of "mounting exasperation" over the never-ending revelations about their clandestine nuclear programme.[130] The most damaging story was American journalist Richard Burt's feature on Pakistan in the *New York Times*.[131] As one of three potential solutions to the Pakistani problem, Burt suggested that the USA was planning military strikes against nuclear installations.[132] Pakistani Foreign Secretary Sardar Shah Nawaz protested to Hummel, claiming that the article had been "inspired" by the US government.[133] In the midst of this Shah Nawaz submitted a letter Zia had composed before the Burt article. According to Zia the US Congress had misunderstood the Pakistani nuclear programme, which, making matters worse, was described as a "Muslim atom bomb". Zia offered Carter a "firm assurance that Pakistan's nuclear programme is entirely peaceful in nature and that Pakistan has no intention of acquiring or manufacturing nuclear weapons".[134] This assurance—something that US officials had sought for months—was not as unequivocal as the administration desired and fell short of ruling out nuclear testing or the

---

[129] UKE Vienna to FCO, 'Informal Consultation on Pakistan's Nuclear Weapon Activities', June 27, 1979, TNA FCO96/956, 1, 4–5.

[130] UKE Islamabad to FCO, 'Pakistan Nuclear', July 2, 1979, TNA FCO96/956, 1.

[131] Ronald Reagan appointed Burt, a prominent critic of Carter, Director of the Bureau of Politico-Military Affairs in the State Department.

[132] 'U.S. Will Press Pakistan to halt A-Arms Project', *NYT*, August 12, 1979, 1.

[133] USE Islamabad to State, 'Letter from President Zia-ul-Haq to President Carter', August 14, 1979, JCPL, Records of the National Security Staff, Box 96 (hereafter RNSS96), Pakistan: Presidential Correspondence: 1-12/79. Within the Pakistani foreign service, the foreign secretary was the bureaucratic head of the ministry, while the foreign minister was diplomatic and executive head.

[134] Zia to Carter, Letter, August 9, 1979, JCPL, RNSS96, Pakistan: Presidential Correspondence: 1-12/79, 2, 4.

transfer of materials to other states, but that did not stop Zia repeating similar formulations for the rest of the year.[135]

The Burt article elicited hasty US government repudiations in public and private, and concurrent Pakistani moves to increase defences around the 'peaceful' enrichment facilities at Kahuta.[136] Here was a key instance where media coverage surrounding Pakistan's nuclear aspirations and the notion of an Islamic bomb had a demonstrable and damaging effect on the chances for a diplomatic solution by further alienating Pakistan. State Department spokespeople hurried to deny that the USA was planning to strike at its South Asian ally, while the Pakistani Ministry of Foreign Affairs summoned Hummel for a dressing-down.[137] In conversation with British consular officials Mike Hornblow, of the State Department's Pakistan desk, denied strike plans were in place, but conjectured that Burt might have misunderstood "unknown individuals" who were speculating about Indian or Israeli military action.[138]

American rebuttals failed to prevent a stern rebuke from the Pakistani government, which castigated the Burt article and the CBS report as part of a campaign to incite "Israel, India, and even the Soviet Union to destroy Pakistan's budding nuclear facilities".[139] Such was the Pakistani response that the US Embassy in Islamabad opined there was likely to be lasting damage to bilateral relations and a reduction in the scope for "rational dialogue" on the nuclear issue.[140]

---

[135] Tarnoff to Brzezinski, 'Letter of August 9 from President Zia of Pakistan to President Carter', August 27, 1979, JCPL, RNSS96, Pakistan: Presidential Correspondence: 1-12/79, 1; Tarnoff to Brzezinski, 'Response to Letter of August 9 and September 29 from President Zia of Pakistan to President Carter', October 10, 1979, JCPL, RNSS96, Pakistan: Presidential Correspondence: 1-12/79; Zia to Carter, Letter, September 29, 1979, JCPL, RNSS96, Pakistan: Presidential Correspondence: 1-12/79, 2.

[136] Fabian to Lavers, 'Pakistan Nuclear', July 30, 1979, TNA FCO 96/957, 1; Came to Gould, Untitled, July 30, 1979, TNA FCO37/2206.

[137] Pakistan Ministry of Foreign Affairs, Press Release, August 14, 1979, TNA FCO96/958.

[138] Fortescue to Holloway, 'US/Pakistan', August 14, 1979, TNA FCO96/958.

[139] UKE Islamabad to FCO, 'Pakistan Nuclear', August 16, 1979, TNA FCO96/958, 1–2.

[140] Ibid., 2.

The Pakistani Ambassador to Britain attacked the media speculation as inspired by the US government. The suggestion of an Islamic bomb prompted him to ask "why should Pakistan, which depended a great deal on economic support from its Islamic friends, so exacerbate the Arab/Israel situation as to threaten the continuation of this help"?[141] Zia expanded upon this theme when British parliamentarians visited Pakistan, treating them to a tirade on the "American conspiracy" and the "myth of an Islamic bomb".[142]

Within the US General Advisory Committee on Arms Control and Disarmament, there was speculation that it was the Israelis—supposedly working on a plan named Entebbe 2—who were most motivated to strike at Pakistani nuclear facilities out of fear of Middle Eastern proliferation.[143] ACDA's Charles van Doren noted that Burt had made things a lot harder for the USA. While military options had *not* been under consideration, the categorical denials about the fracas issued by the State Department made it all the more difficult to *ever* consider such an option.[144]

Violent misunderstandings, rooted in misconceptions about US attitudes towards Islam, provoked another damaging episode in US-Pakistani relations. Armed radicals stormed the Al-Masjid al-Haram in Mecca, igniting protests around the globe when rumours circulated that the United States and Israel were behind the defilement of Islam's holiest site, rumours that Carter believed were started by Ayatollah Khomeini.[145] The most serious protest was the burning of the US Embassy in Islamabad. This occurred in the month American hostages

---

[141] Murray to Archer, Untitled, August 15, 1979, TNA FCO96/958, 2.

[142] Forster to White, 'Visit of British Parliamentarians', September 23, 1979, TNA FCO96/959, 2.

[143] General Advisory Committee on Arms Control and Disarmament, 'Friday Morning Session', September 14, 1979, USPQB, Doc.42, 12–13.

[144] Ibid., 15. There is no clear evidence of whether or not the USA did actually formulate plans for military strikes on Pakistan. Corera argues—from uncited "private memos" and post-facto interviews—that the option had been suggested. These memos have not surfaced during archival research. See Gordon Corera, *Shopping for Bombs: Nuclear Proliferation, Global Insecurity, and the Rise and Fall of the A. Q. Khan Network* (London: Hurst and Co., 2006), 28.

[145] Carter, *White House Diary*, 371.

were taken in Tehran and Brzezinski was fearful that the situation was becoming one of America versus Islam.[146] Dreading more, and increasingly deadly, attacks on US facilities, Vance ordered the evacuation of non-essential personnel from sensitive posts throughout the Middle East and South Asia.[147] Zia later privately commented—espousing a view of his people more often associated with Western observers—that the attack was evidence the USA needed a strong leader in Pakistan to control the emotional and volatile Pakistani nation.[148] Thomas Thornton—interviewed in 1995—argued that after the embassy incident US relations with Pakistan were "about as bad as with any country in the world, except perhaps Albania or North Korea".[149] Back in 1979 Thornton had commented to Brzezinski that there was little way he could see to engage in sensible policymaking about Pakistan. A worry—shared by the National Security Adviser—was that any punitive action on nuclear matters could destroy what remained of the relationship with a strategically important ally.[150]

Faced with Islamic opposition to the United States, and a deteriorating situation in South Asia, the Cabinet in London saw the attacks as evidence that "the influence of Islamic extremism" was spreading from Iran to Pakistan.[151] Despite this FCO observers opined that non-proliferation diplomacy, which relied upon generalisations about an Islamic bomb, could prove damaging rather than useful: "It seems dangerous, for instance, to put about suggestions of an 'Islamic bomb'. In general, we think it would be a mistake to make quite sweeping generalisations without backing them up with proposals for action which might be taken to remedy the situation."[152] Here, the FCO displayed a nuanced view,

---

[146] Brzezinski, *Power and Principle: Memoirs of the National Security Adviser, 1977–1981* (London: Weidenfeld and Nicholson, 1983), 484.

[147] Gerges, *America and Political Islam*, 67.

[148] Haqqani, *Pakistan*, 183.

[149] Thomas Thornton, quoted in Kux, *Disenchanted Allies*, 245.

[150] Thornton to Brzezinski, 'Evening Report', November 26, 1979, JCPL, RAC NLC-10-25-4-21-6, 2.

[151] 'Conclusions of a Meeting of the Cabinet Held at 10 Downing Street', November 22, 1979, TNA CAB128/66, 2.

[152] Asia Group Meeting, 'Agenda Item 3(a): Pakistan's Nuclear Weapons Development Programme', December 13, 1979, TNA FCO37/2208, 4.

recognising that lumping the entire Muslim world into a single monolithic group was counterproductive. Likewise, British analysts argued that panicked fearmongering of the kind seen in the media served no purpose without solutions to address the problem's root causes. Finally, like many of their American counterparts, British policymakers saw financial incentives as a greater motivator for Pakistan to proliferate to nations such as Libya or Saudi Arabia than any sense of Islamic solidarity.[153] The Islamic links posited in the *Observer*'s 'Dr Khan' article were, in the FCO's view, so speculative as to be unworthy of comment.[154]

Just as the Islamic bomb became embedded in public discourses, the Soviet invasion of Afghanistan in December 1979 forced rapid bridge-building between the West and the Islamic world. While covert governmental and overt congressional pressure continued on Pakistan during the 1980s, the dominant American foreign policy apparatus—and its British partner—recognised that action against Pakistan must be subservient to anti-Soviet action in Afghanistan. It became the case that it was not enough to repair relations with Pakistan, but better relations between the West and the wider Muslim sphere were needed to foster a coalition against the USSR. When he visited the Middle East in January 1980 Peter Carrington was delighted to find that Saudi Arabia and Pakistan were working together to mobilise Muslim opinion against the Soviets.[155] In such an atmosphere, mention of an Islamic bomb required curtailing lest it create rifts in the emerging alliance between the 'free' and 'Muslim' worlds.

CONCLUSION

For British and American policymakers, understandings of the Islamic bomb were subtler than those held by the media and peripheral political figures. Senior politicians, government officials, and diplomats recognised that faith was a cloak concealing a nationalistic desire for nuclear

---

[153] 'Asia Group: Verbatim Report, December 13; Pakistan Nuclear Development Programme', December 13, 1979, TNA FCO96/961, 2.

[154] 'Additional Advice for the Prime Minister: Nuclear Proliferation (Second *Observer* Article)', December 18, 1979, TNA FCO96/961, 2.

[155] Richard J. Aldrich and Rory Cormac, *The Black Door: Spies, Secret Intelligence, and British Prime Ministers* (London: William Collins, 2016), 359.

capability. They correctly assessed that Pakistan had an economic and political stake in portraying itself as the custodian of Muslim nuclear power but found no evidence to suggest plans for wider, pan-Islamic proliferation.

For policymakers on both sides of the Atlantic, media coverage of an Islamic bomb transformed public perceptions of Pakistan's nuclear ambitions from a regional, South Asian matter to a multiregional problem affecting the Middle East and beyond. Although policymakers assessed that proliferation because of faith was not especially likely, the public perception of a potential Islamic nuclear capability that this created needed to be combated. The US press guidance in April, Thatcher's June response to Begin, and the diplomatic fallout from the Burt article demonstrate that this media-generated perception forced reactions from the foreign policy establishments in Washington and London. Mention of the Islamic bomb—regardless of its flimsy reality—required a response in the febrile, fragile atmosphere of the late 1970s. There was a scintilla of doubt about trans-Islamic proliferation. The what-ifs surrounding the Islamic bomb, and the interconnected Khan Affair exposed niggling fears—founded in long-standing tropes about the Islamic world's violence and irrationality—that if Muslim states did acquire nuclear weapons, the threat was much more serious than other forms of proliferation. The Islamic bomb—the merest mention of which could provoke anxiety and outrage—represented a collision of culture, geopolitics, and international security.

While the Islamic bomb paradigm prompted the American and British governments to react to the propaganda problem it created, the intense public discussion about supposed pan-Islamic nuclear ambitions did little to alter the actual policies or strategies used against Pakistan. What did dramatically alter those policies was the intransigence shown by Zia and his subordinates. As the next chapter demonstrates, it was this resistance to any form of accommodation that would see US and UK policy undergo significant change in 1979.

CHAPTER 7

"Dead End" The Failure of Political Solutions to Pakistan's Nuclear Ambitions, January 1979 to December 1979

Parallel to the Islamic bomb becoming a significant public issue, Washington and London continued to seek political solutions to the problem of Islamabad's nuclear ambitions. While still making efforts to inhibit Pakistan's clandestine enrichment programme, and thus increase the time it would take to attain nuclear capability, American and British non-proliferation policy towards Zia's government underwent a significant change during 1979. By the summer Jimmy Carter's administration and Margaret Thatcher's new government had realised that the Pakistanis were *not* going to abandon their atomic aspirations. Policy therefore changed from attempting to *prevent* Islamabad's acquisition of nuclear capability to *mitigating* its worst effects. This was not because of fear of pan-Islamic proliferation, but because of the recalcitrance and stubbornness of Zia and his government.

Core to this mitigation strategy was seeking Pakistani assurances not to test a nuclear device. Carter and his advisers attempted to maintain pressure on Zia by resisting demands for arms shipments and aid, but by mid-1979 had resigned themselves to an eventual Pakistani nuclear capability. Peter Constable—Deputy Assistant Secretary for Near Eastern and South Asian Affairs at the State Department—typified this attitude when he argued that the US government had reached a "dead end" and recommended a strategy based on asking Zia not

© The Author(s) 2017
M.M. Craig, *America, Britain and Pakistan's Nuclear Weapons Programme, 1974–1980*, Security, Conflict and Cooperation in the Contemporary World, DOI 10.1007/978-3-319-51880-0_7

to test.[1] The cut-off of aid under the Symington Amendment in April 1979 marked the maintenance of pressure but, beneath the sanctions, little hope remained that Islamabad would ever abandon the bomb.

Although British policymakers spent time considering possible pressures and inducements for Pakistan, they were far less wedded to directly pressuring Zia than their American counterparts. This created occasional friction between the allies, as Washington believed London lacked a serious commitment to non-proliferation goals. Initial British solutions revolved around a Sino-Indian-Pakistani regional security treaty. When this proved unworkable the UK suggested a global solution in the form of a universal declaration on nuclear trade and non-proliferation. Despite London's zeal for a political endeavour believed to complement and enhance the NPT, there was little enthusiasm in the United States and Europe. Regardless of the different approaches taken, by the end of 1979 American and British solutions to the Pakistani nuclear problem had coalesced around policies of mitigation, rather than prevention. However, even the strategy of mitigation would be compromised by events at the year's end. The Soviet invasion of Afghanistan meant that Islamabad would soon be offered weapons and money, despite the accelerating nuclear programme.

Searching for Solutions

Consultations between the USA and the UK during early 1979 highlighted the confusion within foreign policy circles about the right way to approach the Pakistani nuclear problem. In mid-January Thomas Pickering met with British representatives in London and Washington, making it clear that although the State Department wanted to pressure Pakistan, excessive publicity would make the congressional situation even more difficult.[2] Just after Pickering left London, the NSC's Tom Thornton arrived without tangible plans for Pakistan. Thornton did not

---

[1] Department of State (hereafter State) to United States Embassy (hereafter USE) New Delhi, 'Non-proliferation in South Asia', June 6, 1979, Wilson Center History and Public Policy Program Digital Archive (hereafter WCDA), http://digitalarc hive.wilsoncenter.org/document/114198 (accessed September 9, 2013), 1–2.

[2] Moberly to Alston, 'Pakistan Nuclear', January 16, 1979, The National Archives of the United Kingdom (hereafter TNA) Records of the Foreign and Commonwealth Office (hereafter FCO) 96/947.

favour military support for Zia, preferring dialogue between Pakistan, India, and China to promote a stable South Asia. The FCO SAD's Kelvin White advocated trying to ease Pakistani fears about India by getting the Indians to approve safeguards and inspections. Thornton agreed, but had no idea how to achieve this.[3] A week later Jack Miklos arrived with more opinions, supporting enhanced US financial support for Pakistan.[4] Thus, the FCO found itself having to divine American intentions. British officials concluded that Pickering wanted to hit the Pakistanis hard, Thornton wanted to leave them alone for a while, and Miklos wanted to cosset them.[5] In the face of this confusion Whitehall drove forward with plans for regional, then universal, political solutions.

Bilateral, multilateral, and regional nuclear treaties had been a part of discussions on the South Asian nuclear situation ever since 1974. The Pakistanis had proposed a South Asian Nuclear Weapon Free Zone (SANWFZ) in the United Nations General Assembly (UNGA) every year since the Indian test. This proposal—based on the expectation it would put international pressure on India—had never gained traction. More importantly, India consistently rejected such proposals. Pakistan had also frequently expressed a willingness to sign the NPT as soon as India did. As Naeem Salik argues, this "provided a convenient shelter for Pakistan to hide behind when subjected to international pressure".[6] The downside of this tactic was that Pakistani policy was hostage to Indian policy decisions and actions.[7] It was within the context of American indecision, sub-continental recalcitrance, and the realisation that 'sticks and carrots' were not enough that the UK attempted to forge an agreement to arrest an Indo-Pakistani nuclear arms race.[8]

[3] 'Summary Record of a Meeting Held in the Foreign and Commonwealth Office Between Mr D F Murray and Mr Tom Thornton', January 16, 1979, TNA FCO96/947, 2–3.

[4] 'Summary record of a Call on Mr Donald Murray by Mr Jack Miklos', January 22, 1979, TNA FCO96/947, 2.

[5] White to Weston, 'Pakistan: Nuclear', January 25, 1979, TNA FCO96/947, 1.

[6] Naim Salik, *The Genesis of South Asian Nuclear Deterrence: Pakistan's Perspective* (Oxford: Oxford University Press, 2009), 153–154.

[7] Ibid., 154.

[8] Fearn to Lavers, 'Pakistan: Economic Angles', March 5, 1979, TNA FCO96/949, 1.

British thinking on halting that arms race underwent a speedy evolution during the spring and early summer of 1979. From proposals for a modified SANWFZ, there was a rapid transformation when ACDD Chief Christopher Mallaby proposed an Indo-Pakistani regional security treaty (RST).[9] Mallaby wanted to reshape the old SANWFZ idea, changing the name, geographical scope, and political concessions that both sides would have to make.[10] Senior officials thought the proposal worth considering if China could be included to assuage Indian fears. Moberly and Cortazzi informed the US Embassy about British plans for the RST proposal. US consular staff responded favourably and agreed that UK leadership in the matter might elicit better results from South Asia and China.[11] French officials indicated that the RST—should Britain get it moving—was something they would support.[12] It was at this point that John Thomson outlined what he saw as the political complexities militating *against* a regional solution and sketched the plan that he devoted considerable energy to advocating: a global, "universal declaration" on nuclear trade and non-proliferation.[13] FCO arms control experts were sceptical about Thomson's proposal, while accepting the difficulties of implementing the RST. ACDD suggested various alternatives, from building on the ongoing (and ultimately fruitless) Comprehensive Test Ban (CTB) negotiations to a massive "son of NPT".[14] A wrinkle appeared in these discussions when David Owen rebuked his staff for discussing matters with the USA without

[9] Bushell to Moberly, 'South Asian Nuclear Weapons Free Zone (SANWFZ)', February 21, 1979, TNA FCO96/948; Mallaby to Moberly, 'India, Pakistan and Nuclear Weapons', March 8, 1979, TNA FCO96/948.

[10] Ibid.

[11] Cortazzi, 'India, Pakistan, and Nuclear Weapons', March 9, 1979, TNA FCO96/949, 2.

[12] 'Record of a Conversation Between Mr H A H Cortazzi and M. Jean Noiville: South Asian Nuclear Matters', March 21, 1979, TNA FCO96/952; James to Cortazzi, 'South Asian Nuclear Security Treaty', April 12, 1979, TNA FCO 96/952, 1.

[13] Thomson to Holloway, 'Regional Security Threat', March 15, 1979, TNA FCO96/950, 4.

[14] Burns to Parsons, 'Pakistan Nuclear', Undated, TNA FCO96/950, 6–9.

first appraising ministers of the proposals, and for largely ignoring Thomson's suggestions.[15]

By mid-June Margaret Thatcher had ushered Callaghan from office. This change coincided with—but did not create—a shift in policy towards Pakistan's nuclear aspirations. The Symington Amendment's imposition, and the accumulating publicity about the Khan Affair and the propaganda dilemma posed by the Islamic bomb, demanded new and wide-ranging thinking on policy.[16] Thomson, who by this time had ensured his universal declaration became the dominant political means of dealing with the sub-continental nuclear question, effected the strategy change.[17] This change had not been smooth. The JNU and ACDD continued to advocate the RST, despite Indian hostility to such a proposal and Owen's scepticism in the weeks before he left office.[18] The delay in formulating a new policy stemmed from the American desire to try out their approaches before employing "new thinking".[19]

Washington's approach was subtly different from London's and illustrates the Carter administration's gradual turn from prevention to mitigation. While the British strategy prior to the adoption of Thomson's plan focused on an Indo-Pakistani *security* treaty, the USA tended towards a treaty linked to both states neither producing nor using nuclear weapons. There was agreement between Pickering and State Department human-rights advocates that it might be worthwhile pursuing a solution to the Indo-Pakistani

---

[15] Alston to Moberly, 'Pakistan's Nuclear Ambitions', March 22, 1979, TNA FCO96/950, 1.

[16] Cortazzi to Rose, 'Pakistan and China: Nuclear Questions', April 12, 1979, TNA FCO96/950, 1.

[17] Moberly to Bondi, 'Non-proliferation After INFCE', June 11, 1979, TNA FCO96/955, 1.

[18] Alston to Moberly, 'Non-proliferation Policy', June 1, 1979, TNA FCO37/2205; 'India/Pakistan Nuclear: Record of a Meeting', June 5, 1979, TNA FCO96/955, 1; United Kingdom High Commission (hereafter UKHC) New Delhi to Foreign and Commonwealth Office (hereafter FCO), 'India/Pakistan Nuclear', April 12, 1979, TNA FCO96/952; Wall to Mallaby, 'Pakistan and Nuclear Weapons', April 2, 1979, TNA FCO96/951, 1.

[19] FCO to United Kingdom Embassy (hereafter UKE) Islamabad, 'Pakistan, India, and Nuclear Weapons', April 6, 1979, TNA FCO96/951; FCO to 10 Downing Street, 'Pakistan Nuclear Programme', April 30, 1979, TNA FCO96/953.

problem through a bilateral nuclear accord.[20] The NSC thought such an accord, focusing on a joint non-production/non-use agreement between India and Pakistan, worthy of implementation.[21] This proposal—although still advocating an active non-proliferation stance through the quest for promises not to produce nuclear weapons—was a small step on the road to mitigation. Such a suggestion implicitly accepted an eventual Pakistani *ability* to produce nuclear weapons alongside the perceived pre-existing Indian capability. Vance argued that the US-Pakistan security relationship and the Carter administration's non-proliferation policy were often in conflict. According to Vance, the most notable facet of this conflict was the situation in Afghanistan and the threat it posed to Pakistani and US interests in South Asia.[22]

Zia was relieved that US ties with Pakistan were not entirely focused on the nuclear issue, and from this came Hummel's observation that Zia might be persuaded to agree a "freeze" on the enrichment programme, allowing the resumption of aid and giving time for the construction of broader, more lasting non-proliferation agreements.[23] Hummel's suggestion came on the back of his assessment a few weeks earlier that there was almost nothing that could be done by the USA and its allies to halt Pakistan's programme.[24] Although it was another long shot, the State Department alerted the FCO to Hummel's freeze initiative in order to use classified British intelligence on the enrichment programme in discussions

[20] Schneider to Christopher, 'Pakistan', March 9, 1979, United States National Archives and Records Administration (hereafter NARA), Department of State, Central Files, Record Group 59 (hereafter RG59), Records of Warren Christopher (hereafter RWC), Box 56, 1.

[21] Memorandum for Brzezinski, 'While You Were Away', March 13, 1979, Jimmy Carter Presidential Library (hereafter JCPL), Remote Archives Capture system (hereafter RAC) NLC-10-19-1-19-9, 1.

[22] Raphael to Christopher et al., 'Policy Toward Pakistan', April 9, 1979, NARA, RG59, RWC, Box 57, Pakistan, 1–2.

[23] Brzezinski to Carter, 'Daily Report', April 11, 1979, JCPL, RAC NLC 1-10-4-1-4, 1-2; Christopher to Carter, Memorandum, April 16, 1979, JCPL, RAC 128-14-6-12-5, 2.

[24] USE Islamabad to State, March 5, 1979, 1.

with the Pakistanis.[25] For Pickering it was vital to pin Islamabad down to a deal before bringing New Delhi into any bilateral accord.[26] However, the short-lived freeze proposal died in the face of Zia's obduracy, supporting Hummel's assertion that little could be done about Islamabad's nuclear ambitions.[27]

After the freeze proposal's failure, it became apparent that—once again—nothing short of massive US support to alleviate Pakistan's security fears would have any influence on nuclear ambitions. Shah Nawaz and Munir Ahmad Khan admitted as much to Hummel and Robert Gallucci in Islamabad on April 25, while maintaining the claim of a peaceful nuclear programme.[28] Agha Shahi and Zia were of a similar mind when they met with Vance and Hummel respectively in early May, both Pakistani statesmen reiterated the long-standing offer to sign a bilateral or regional Indo-Pakistani non-proliferation agreement.[29] As usual, Indian reluctance to sign a regional agreement that did not include communist China undermined this offer.[30] Gerard Smith was exasperated by Vance's agreement with Shahi to explore congressional attitudes towards arms sales, believing that this would confuse key allies over the American stance on the Pakistani problem.[31] Furthermore, US diplomats in Islamabad argued that opportunities for further bilateral USA-Pakistan progress were limited.[32] Because of these colliding factors the NSC suggested a complex approach that combined a range of actions and strategies. The approach

[25] State to USE London, 'Pakistan's Nuclear Problem', April 13, 1979, JCPL, RAC NLC-16-116-2-40-0, 1-2.

[26] Alston to Fearn, 'Pakistan Nuclear Programme', April 20, 1979, TNA FCO96/952, 2-3.

[27] Thornton to Brzezinski, 'Evening Report', April 23, 1979, JCPL, RAC 24-100-4-22-6, 1.

[28] USE Islamabad to State, 'Luncheon Discussion of Nuclear Issues', April 26, 1979, NARA, RG59, RWC, Box 57, Pakistan, 3.

[29] Tarnoff to Brzezinski, 'Issues Paper: U.S. Nuclear Strategy Towards India and Pakistan', May 16, 1979, JCPL, RAC NLC-132-75-5-9-8, 1.

[30] Ibid., 2.

[31] Saunders and Pickering to Newsom, 'Consultations with Congress on Arms Sales to Pakistan', May 12, 1979, NARA, RG59, RWC, Box 57, Pakistan.

[32] UKE Islamabad to FCO, 'Pakistan Nuclear', May 13, 1979, TNA FCO96/953, 1.

brought together Pickering's and Saunders' Indo-Pakistani non-use/non-production proposal, a potential resumption of aid, a coordinated diplomatic campaign aimed at key Pakistani allies, continued efforts to frustrate the clandestine programme, and enhanced talks with India.[33] Vance, in conversation with Peter Carrington, clung to the hope that he could persuade India and Pakistan to agree to a regional pact.[34]

Within the elite Policy Review Committee, the conflict between ardent non-proliferationists such as Smith and the regionally focused State Department bureaux was clear. Smith wanted to set up an international group to assess the Pakistani problem, something that Christopher felt should be on an informal basis lest it appear that the developed world was ganging up on Pakistan.[35] Smith argued that it was not simply a matter of dealing with Pakistani perceptions, but with *global* perceptions about America's commitment to non-proliferation.[36] A suggestion previously put forward by Paul Kreisberg of the Policy Planning Staff (PPS) was for an independent, well-respected figure to act as a mediator in South Asia.[37] The suggestion received tacit approval, but only if there were reasonable prospects for an Indo-Pakistani non-development/non-use agreement. Preparatory to this, Ambassador Robert Goheen in New Delhi should have exploratory discussions with the Indians. CIA Chief Admiral Stansfield Turner injected a note of alarm into proceedings when he drew attention to intelligence reports that suggested Pakistan might carry out a nuclear test before November, well in advance of US and UK estimates of when it might have a nuclear explosive capability.[38] Rumours of an imminent Pakistani test in part drove US diplomacy in the coming months. Avoiding a test would form the mitigation strategy's cornerstone. Were Pakistan to explode a nuclear device it would damage American

[33] Tarnoff to Brzezinski, May 16, 1979, 3.

[34] 'Record of a Discussion Between the Secretary of State and Mr Cyrus Vance at the FCO', May 21, 1979, TNA FCO37/2204, 2.

[35] 'PRC on Pakistan and Subcontinent Matters', May 23, 1979, JCPL, RAC NLC-132-75-5-2-5, 2.

[36] Ibid., 10.

[37] Kreisberg to Christopher, 'A Mediator for the South Asian Nuclear Problem', May 22, 1979, NARA, RG59, Records of Anthony Lake (hereafter RAL), Box 5.

[38] 'PRC on Pakistan and Subcontinent Matters', May 23, 1979, 4.

credibility by showing the world that the non-proliferation policy had failed. Persuading Islamabad to remain in a state of nuclear ambiguity would give the appearance of non-proliferation success, while allowing regional security concerns—such as the developing situation in Afghanistan and the tensions between India and Pakistan—to be addressed.

There was precious little hard intelligence to support these rumours about a test, and their appearance in the media threaded their way through discussions throughout late May and beyond. Rumours connected to debates within the Carter administration led Peter Constable to elucidate the change from prevention to mitigation. Robert Goheen received instructions to approach Desai with the bilateral plan and potential third-party mediation.[39] The embattled Indian Prime Minister would not countenance a non-development/non-use agreement. Desai reminded Goheen that India had already made a non-development pledge and, if Zia did the same, that would be as good as any formal arrangement. More dramatically, Desai stated that if Pakistan tested a bomb—or indicated it was about to test one—India would act at once "to smash it".[40] In light of this confrontational response Goheen eschewed mentioning third-party mediation plans. Rumours of an impending Pakistani test reached the media because of comments by none other than Indian Foreign Minister Vajpayee during his visit to the USA in late April.[41] The American attempts to gain an Indo-Pakistani agreement bore little fruit because of intransigence on both sides, media speculation about the Pakistan bomb project, and a lack of focus on the administration's part.

The correspondence with Goheen bracketed Constable's gloomy June 6 telegram in which he fretted about the Islamic bomb. Constable—backed by Vance—argued that the USA had "come to a dead end in our bilateral and multilateral efforts to prevent the spread of nuclear weapons technology to the nations of South Asia".[42] The only solution was a new strategy

---

[39] State to USE New Delhi, 'Nuclear Dialogue with India', June 2, 1979, National Security Archive Electronic Briefing Book (hereafter NSAEBB) 'The United States and Pakistan's Quest for the Bomb' (hereafter USPQB), Doc.35A, 2–3.

[40] USE New Delhi to State, 'India and the Pakistan Nuclear Problem', June 7, 1979, USPQB, Doc.35B, 1–2.

[41] George Perkovich, *India's Nuclear Bomb: The Impact On Global Proliferation* (Berkeley: University of California Press, 1999), 218.

[42] State to USE New Delhi, June 6, 1979, 1.

that—rather than seeking to prevent the spread of nuclear technology to South Asia—sought Pakistani assurances they would not detonate a PNE, a formal assurance of India's commitment not to develop nuclear weapons further, and Chinese assurances against deploying nuclear force in a hypothetical Sino-Indian conflict.[43] While the Carter administration had resigned itself to the fact that Pakistan would develop some form of nuclear capability, the State Department sought to prevent testing of that new-found capability. At the heart of this desire to avoid public displays of atomic attainment was a fear that the global non-proliferation regime could not bear the weight of India's and Pakistan's ascent to nuclear capability. Constable therefore argued that:

> If we fail to act decisively, we will also jeopardise our global non-proliferation strategy, which could collapse under the weight of two additional nuclear weapons states. By treating South Asia as a "special case", we may have a better chance to head off nuclear weapons development here, and also to preserve in the rest of the world those elements of our non-proliferation strategy which are working[.][44]

Here was the clearest expression yet from a senior foreign policy figure in the Carter administration that the US government had resigned itself to the fact that Pakistan could not be prevented from gaining nuclear capability, even if that capability was couched in the disingenuous terms of a PNE.

Although policy had shifted subtly, Carter was unwilling to make Pakistan an explicit exception to policy. In response to Constable's memo Gerard Smith contended that acquiescing to Islamabad acquiring unsafeguarded nuclear facilities would be a mistake that would "drain most of the consistency out of your [Carter's] nuclear policy". In annotating the memo Carter noted his agreement with Smith.[45] Because of the impasse, and the threat posed to global non-proliferation *and* political relationships in the sub-continent, Vance tasked Smith with creating a committee to find a solution.[46]

[43] Ibid., 2.

[44] Ibid., 3.

[45] Smith to Carter, 'Nonproliferation in South Asia', June 8, 1979, USPQB, Doc.36.

[46] Brzezinski to Vance, 'The South Asian Nuclear Problem', June 19, 1979, USPQB, Doc.37.

The June debates over strategy towards Pakistan represented a significant change in policy, little remarked upon in the existing literature. No longer was the Carter administration attempting to halt proliferation in South Asia. Faced by Zia's stonewalling of every attempt to resolve the situation in the non-proliferation regime's favour, policy had hit a dead end. Despite Carter's agreement with Smith that a policy exception would be untenable, the decision to seek assurances over testing demonstrates that prevention had failed and mitigation was the goal. Persuading the Pakistanis not to test a device—PNE or otherwise—would preserve the public appearance of working towards non-proliferation and head off a further deterioration in Indo-Pakistani relations.[47] This would save American face and maintain credibility on non-proliferation. While this change was taking place in Washington a similar change was taking place in Britain, as the Thatcher government made efforts to gain international backing for its universal declaration.

UNIVERSAL SOLUTIONS?

By July—and despite internal debates over wording—the British universal proposal was ready for submission to key foreign capitals. The 'Draft Declaration on Nuclear Trade and Non-proliferation' was just three double-spaced pages long, and laid out nuclear energy's importance, the need to avoid proliferation, the necessity for international consensus and cooperation, for technical assistance to developing countries, and for arrangements to be made regarding nuclear technology transfers.[48] The declaration's stipulations were, in essence, an international suppliers and purchasers agreement coupled with a restatement of the NPT's Article IV.[49] London, in adopting the declaration as the main plank of

---

[47] In her 2014 study Or Rabinowitz argues that the Carter administration discarded the idea of a non-testing agreement. However, her analysis is hampered by not having accessed the critical materials available at the Carter Library. See Or Rabinowitz, *Bargaining on Nuclear Tests: Washington and Its Cold War Deals* (Oxford: Oxford University Press, 2014), 138.

[48] Moberly to Bondi, 'Non-proliferation After INFCE', June 11, 1979, TNA FCO96/955, 3–5.

[49] Article IV outlined the rights of parties to the treaty to research, produce, and use nuclear energy for peaceful purposes without discrimination.

their non-proliferation policy, tacitly accepted that Pakistan was going to gain nuclear capability, but still wanted to delay the advent of a Pakistani bomb as long as possible. The declaration did not make any effort to undermine the Pakistani programme, but offered another layer of proliferation controls to try and delay eventual acquisition and to mitigate the effects of further proliferation *from* Pakistan.

Before the declaration was made ready for circulation there were the first high-level non-proliferation meetings between the USA and the UK since Britain's change of government. It quickly became clear that avoiding a Pakistani nuclear test was core to future strategies. Douglas Hurd met with Smith and pointed out that, while the pressure being exerted by Western governments on Pakistan was feeble, the USA and UK should be working with the non-aligned nations on the grounds that a Pakistani nuclear test would have damaging implications for international atomic trade and thus harm nuclear prospects for the developing world.[50] While Smith stated without much optimism that Vance had tasked him with producing an interagency study aimed at solving the sub-continental nuclear problems, Thomson extolled his declaration's virtues, pointing out that a global solution avoided the appearance of bullying any one nation.[51] Smith left London with a copy of the draft and the promise of American feedback.[52]

It became apparent that the much more global British policy conflicted with the American regional focus. The FCO opinion was that the declaration was a "positive way forward, rather than a negative pressurising policy".[53] While the Smith study was still ongoing Michael Pakenham characterised American strategy as returning to twin bilateral strands. The USA would search for ways to solve the Tarapur problem with India, while dealing with Pakistan by reversion to "sticks and carrots".[54] Much of this

[50] Alston to Moberly, 'Call by Ambassador Smith on Mr Hurd', June 29, 1979, TNA FCO96/956.

[51] 'Call at FCO by Ambassador Smith', June 29, 1979, TNA FCO96/957, 1–3.

[52] Carter to Cromartie, 'Draft Declaration on Nuclear Trade and Non-proliferation', July 2, 1979, TNA FCO37/2206, 1.

[53] White to Forster, 'Pakistan: Relations With HMG', July 27, 1979, TNA FCO96/957, 2.

[54] Pakenham to Alston, Memorandum, August 3, 1979, TNA FCO96/958, 1.

revolved around either withholding or offering the supply of advanced military equipment and potentially giving *more* publicity to the clandestine Pakistani programme - the "sunshine approach". Regarding the universal declaration there was little sympathy in Washington for the proposal, where officials preferred to postpone further institutional relationships between nuclear purchasers and suppliers until the 1980s.[55] The Americans harboured reservations about the declaration, although Pickering had admitted that they had no coherent alternatives.[56] Patrick Moberly observed that the State Department felt that the declaration would end up as a "lowest common denominator approach", becoming so diluted as to have no meaningful power. Likewise, he suggested that the British approach sat awkwardly with American emphasis on supplier controls and congressional legislation, such as the Symington and Glenn Amendments and the Nuclear Non-proliferation Act of 1978 (although these were not directly cited). The French had also raised objections relating to the ongoing INFCE process. Senior negotiators in the French government observed that any universal declaration on nuclear trade could be seen by the developing world as unwisely pre-empting the results of INFCE.[57]

During August Hugh Cortazzi visited Washington to drum up American support for the declaration, briefed to make it clear that the Thatcher government did not favour "bullying" Pakistan and preferred a political solution that did not target individual countries.[58] In discussions with Jane Coon (Deputy Assistant Secretary for the Sub-continent) who had considerable experience of Pakistan and non-proliferation, Cortazzi emphasised the limited scope for the development of relations with Pakistan when that country was entering an economic and political crisis. Coon was sympathetic, and noted that she shared Britain's more restrained assessment of when Pakistan might be able to test a nuclear weapon, agreeing that sober, quiet diplomacy was a better alternative to the "sunshine approach". Still, Coon reiterated American objections to the declaration. From her

---

[55] Ibid., 2, 4.

[56] Alston to Thomson, 'Nonproliferation', August 24, 1979, TNA FCO37/2207, 1; Moberly to Thomson, 'Non-proliferation', August 10, 1979, TNA FCO37/2207, 1.

[57] Ibid., 1–2.

[58] Carter to Cortazzi, 'Briefs for Call at State Department, 9 August', August 7, 1979, TNA FCO96/958, 2.

perspective it would mean acquiescing to sensitive nuclear facilities in Pakistan when there was no guarantee that safeguards were adequate.[59] Coon's remarks on acquiescing to nuclear facilities did not reflect the fundamental change at the higher levels of American non-proliferation policymaking. Cortazzi departed, having failed to change minds. As for other states, the Canadians and French were sitting on the fence, while the Australians and Germans were unenthusiastic about the concept, notwithstanding brief initial interest.[60]

Despite vigorously promoting the universal declaration, the British government eventually conceded it was never going to gain traction. However, with the Smith Study stalled the USA still did not have an alternative. Vance was trying to paper over the widening cracks in his department, the main fault line being between ardent non-proliferationists, such as Smith and Pickering who at a minimum wanted safeguards applied to Pakistani facilities, and the 'regionalists' from the various geographical bureaux, who would be content with assurances of good behaviour.[61] Zia had made vague noises about assurances in letters to Carter over the preceding months and during his September visit to New York, but these were far short of the more formal guarantees that even the regionalists desired.[62] PPS Chief Anthony Lake outlined the available options to deter proliferation while maintaining good relations with Pakistan. In most cases, these options—such as the provision of advanced F-16 fighters—conflicted with other administration policies on conventional arms restraint and there had been little consideration of what to do if *all* approaches failed. Furthermore, he noted deep divisions on all the potential avenues for progress.[63]

---

[59] Weston to Cortazzi, 'Pakistan', August 10, 1979, TNA FCO96/958.

[60] Alston to Thomson, August 24, 1979, 1.

[61] Henderson to FCO, 'Smith Study on Nuclear South Asia', October 6, 1979, TNA FCO96/959.

[62] Zia to Carter, Letter, September 29, 1979, JCPL, Records of the National Security Staff, Box 96 (hereafter RNSS96), Pakistan: Presidential Correspondence: 1-12/79, 2; Tarnoff to Brzezinski, 'Response to Letters of August 9 and September 29 from President Zia of Pakistan to President Carter', October 10, 1979, JCPL, RNSS96, Pakistan: Presidential Correspondence: 1-12/79.

[63] Lake to Vance, 'The Pakistan Strategy and Future Choices', September 8, 1979, Wilson Center Digital Archive (hereafter WCDA), http//digitalarchive.wilson-center.org/document/114217 (accessed September 9, 2013), 4–7.

There was hope for progress when Agha Shahi visited Washington in October. The nuclear issue dominated the meeting, even though Zia had again cancelled elections and was proceeding with his Islamisation process, something that conflicted with Carter's human-rights agenda.[64] Intelligence and media rumours of an imminent Pakistani nuclear test added urgency to the meeting.[65] From the fractious discussions came assurances that Pakistan would not develop nuclear weapons nor transfer sensitive nuclear technology, promises that aligned with what Britain was attempting to achieve with its universal declaration. Assurances not to develop nuclear weapons did not, however, preclude the development of a PNE along the lines of India's 1974 explosion. Shahi was far less forthcoming on the most public statement of nuclear attainment, testing, although he assured Vance that Pakistan would not have the capability to test for at least six months.[66] State Department and NSC reports to the FCO highlighted the fault lines in the administration. Tom Pickering observed that Shahi discussed the nuclear project in terms of national pride and prestige, but the US side in the talks continually emphasised the impediment to good relations that the "pointless" bomb programme created. Furthermore, deterioration in the regional situation would not automatically cause a change in US attitudes, while a nuclear test would prompt the relationship's "fundamental reconsideration".[67] Thus, Pickering had tacitly conceded that it was admission of nuclear capability (PNE or otherwise) through testing—and not the capability itself—that was now the administration's prime concern. From the NSC side, Thomas Thornton was furious with the State Department's line in the discussions. According to Thornton there was agreement between Carter, Brzezinski, and Secretary of Defense

[64] Dennis Kux, *Disenchanted Allies: The United States and Pakistan, 1947–2000* (Washington D.C.: Woodrow Wilson Center Press, 2001), 240–241.

[65] 'U.S. Aides Say Pakistan Reported to Be Building an A-Bomb Site', *New York Times* (hereafter *NYT*), August 17, 1979, A6; CIA, 'Alert Memorandum on Pakistani Plan for an Early Nuclear Test', October 10, 1979, USPQB, Doc.43, 1.

[66] Vance to Carter, Memorandum, October 17, 1979, JCPL, RAC NLC-128-14-12-11-9.

[67] UKE Washington to FCO, 'Nuclear Pakistan and the U.S.', October 19, 1979, TNA FCO37/2209, 1–2.

Harold Brown that, if the Afghan situation deteriorated, the USA should offer aid to Pakistan. Thornton concurred with British assessments of a Pakistani test's imminence (which stated that the earliest Pakistan could test was in two to three years) and argued that "unduly alarmist" appraisals of when Pakistan might explode a device were being used by Vance's clique to pursue a strategy of pressuring Pakistan, regardless of the impact on other policies.[68]

While the Shahi meeting's outcomes were relayed to Whitehall, parallel discussions were taking place aimed at healing a rift in the transatlantic non-proliferation alliance. In London, on October 12, a quadripartite meeting had taken place between American, British, French, and West German representatives. In the run-up to the conference, British analysts saw the Pakistani problem as acute. It would damage international non-proliferation effort, increase the chance of war with India, and a Pakistani test would bring a "Muslim bomb" that much closer. Here again, 'cascadology' reared its head, in conjunction with the old trope that a lack of action in one instance of potential proliferation would irreparably damage the entire edifice. For the British the only answer was through positive diplomatic solutions.[69] There was a consensus on all these points at the meeting and a shared willingness to seek new and imaginative diplomatic solutions.[70] After the fact, the Americans were depressed by what they perceived as British unwillingness to do anything about Pakistan. Thornton placed Britain just ahead of the recalcitrant FRG in terms of a desire to address the problem.[71] The FCO was anxious to avoid this grim view of Britain's attitude becoming common currency in Washington, even though British officials saw the USA as having few, if any, constructive plans. Carrington eventually sent a stern telegram outlining the many British efforts to frustrate

[68] On British test estimates see FCO to UKE Washington, 'Nuclear Pakistan', October 19, 1979, TNA FCO96/960; on Thornton's views see UKE Washington to FCO, 'Nuclear Pakistan and the U.S.', (Thornton telegram) October 19, 1979, TNA FCO37/2209, 2.

[69] 'Speaking Note for Consultations on Pakistani Nuclear', October 11, 1979, TNA FCO96/959, 1–2.

[70] FCO to UKE Islamabad, 'Pakistan Nuclear', October 15, 1979, TNA FCO96/960, 1–2.

[71] Robinson to Moberly, 'Nuclear Pakistan', October 22, 1979, TNA FCO96/960.

Pakistani nuclear ambitions and the considerable efforts they had undertaken to achieve a political solution.[72]

When Gerard Smith toured European capitals in early November the FCO was at pains to point out how seriously Britain took the situation. The prospect of a Pakistani test loomed over everything, especially as Zia had refused to rule one out.[73] British diplomats argued there was little that could be done if the Pakistanis were determined to test, but after they had "proved their virility" with a test there could be opportunities for progress.[74] From New Delhi Thomson—now Britain's High Commissioner in India—averred that testing was likely to provoke a violent Indian response and again advocated the lapsed universal declaration.[75] The FCO concluded that British objectives must be to discourage testing, maintain tight controls on sensitive supplies, and continue diplomatic contacts with Pakistan.[76] This was, in the baldest terms, acceptance from Britain's foreign policy community that prevention had failed and that mitigation was now the only option. Moberly located Pakistan at the centre of a nexus of proliferation anxieties; a Pakistani test could open the door to a proliferation cascade, perhaps even reinvigorating the old question of West German nuclear weapons. American and British non-proliferation policy—typified by the NPT—was frequently predicated on keeping nuclear weapons out of West Germany's hands.[77] The British reaction to a test would affect credibility and the ability to exert influence in all future cases. Moberly was disappointed that a tougher, more

[72] FCO to UKE Washington, 'Nuclear Pakistan', October 25, 1979, TNA FCO96/960.

[73] 'Zia Says he Has Not Ruled Out An Atom Bomb Test In Pakistan', *NYT*, October 28, 1979, 9.

[74] UKE Islamabad to FCO, 'Nuclear Pakistan', October 26, 1979, TNA FCO96/960, 1.

[75] UKHC New Delhi to FCO, 'Nuclear Proliferation', October 30, 1979, TNA FCO96/960, 2.

[76] Anon., 'India/Pakistan and Non-proliferation', October 20, 1979, TNA FCO96/960, 8–9.

[77] Shane J. Maddock, *Nuclear Apartheid: The Quest for American Atomic Supremacy from World War II to the Present* (Chapel Hill: University of North Carolina Press, 2010), 125–126, 247–248.

combative line against Pakistan was not recommended.[78] Cortazzi agreed that a strong line was needed, but felt that Britain carried little weight in Pakistani eyes and was therefore limited in influence.[79] When Smith, Pickering, and van Doren arrived in London, Cortazzi was at pains to point out that the UK was not complacent, despite what the US delegation might think. He suggested that the universal declaration was a still a viable idea, as it addressed the problems presented by Pakistan *and* India.[80] Smith not only responded by highlighting how important the CTB negotiations were in showing the developing world that the nuclear weapon states were taking vertical proliferation seriously, but he also espoused scepticism about the universal solution's viability because of its linkages to the international nuclear trade.[81]

Smith offered a gloomy assessment of his sojourn in Europe upon his return to Washington. West Germany lacked commitment to non-proliferation, the Netherlands was supportive, France stressed it was the only nation other than the USA to take concrete action against Pakistan, and Britain doubted that sticks and carrots would have any effect. The UK and FRG had also cited the Chinese point of view on the entire issue.[82] Chinese officials had defended the right of states to acquire nuclear weapons and advised against a tough line on Pakistan due to its importance in the South Asian anti-Soviet structure.[83] Despite this, Chinese diplomats had frequently advised their American counterparts that they disapproved of Pakistan's "unwise" weapons programme. A later CIA estimate challenged this position, suggesting China was aiding the

[78] Moberly, 'The Sub-continent: Nuclear Issues', October 30, 1979, TNA FCO96/960, 1–2.

[79] Cortazzi, 'The Sub-continent: Nuclear Issues', October 30, 1979, TNA FCO96/960.

[80] 'Anglo-US Talks, FCO, London', November 1, 1979, TNA FCO37/2208, 2.

[81] Ibid., 4; FCO to Washington, 'Pakistan Nuclear', November 5, 1979, TNA FCO96/960, 1.

[82] Smith to Vance, 'Consultation in Europe on Pakistan', November 15, 1979, USPQB, Doc.45, 2.

[83] Alston to White, 'Visit of Chairman Hua to Paris: Pakistan Nuclear', October 25, 1979, TNA FCO96/960; FCO to Peking, 'Premier Hua's Visit', November 6, 1979, TNA FCO96/960, 4.

Pakistani programme as part of its anti-Soviet posture.[84] Smith saw the PRC as "preaching the need to bolster Pakistan as a barrier to Soviet adventurism in the region" and argued there was "little enthusiasm in Europe to emulate our position with Pakistan".[85]

The British still attempted diplomacy to encourage change in Pakistan, despite American doubts. There were fraught meetings in London between Shah Nawaz, Minister for Foreign & Commonwealth Affairs Peter Blaker, and Hugh Cortazzi. Shah Nawaz found suggestions of Libyan funding for the nuclear programme "utter nonsense" and harangued his hosts about all the efforts that Pakistan had made to defuse the situation.[86] Cortazzi pointed out that Zia's cancellation of elections and the imposition of Islamic punishments were seriously affecting Pakistan's image in British eyes, an image that could be further damaged should Pakistan carry out a nuclear test.[87] Kelvin White of SAD thought Shah Nawaz's comments were "flannel" but noted that, over lunch, Shah Nawaz had stated that Pakistan should enter a state of nuclear ambiguity, just like Israel.[88] This was exactly what both Britain and America, in their own ways, sought from the Pakistani leadership. Cortazzi then visited Pakistan and met with Zia and Shahi. Despite once more outlining British concerns and clarifying that Britain had no desire to deprive the developing world of access to nuclear power, beyond expressions of goodwill and mutual respect, little was gained by Cortazzi's sojourn.[89]

By this time the State Department was searching for any solutions to the Pakistan problem. Aside from continuing with diplomatic efforts and attempts at reassuring Pakistan about its security, a proposal was

---

[84] CIA, 'A Review of the Evidence of Chinese Involvement in Pakistan's Nuclear Weapons Program', December 7, 1979, JCPL, RAC NLC-15-37-6-3-6.

[85] Smith to Vance, 'Consultation in Europe', November 15, 1979, 1.

[86] , 'Summary Record of a Call on Mr Peter Blaker by Shah Nawaz', November 12, 1979, TNA FCO96/961, 1–2.

[87] Ibid., 6.

[88] White to Masefield, 'Pakistan: Talks With Shah Nawaz', November 16, 1979, TNA FCO96/961, 1–2.

[89] UKE Islamabad to FCO, 'Pakistan Nuclear', December 3, 1979, TNA FCO96/961, 1–2; UKE Islamabad to FCO, 'Anglo/Pakistan Relations', December 5, 1979, TNA FCO96/961, 1–3.

made to offer Zia funding for a civilian nuclear energy programme, the reason he and Bhutto had claimed for the reprocessing and enrichment projects in the first place.[90] This was Washington's last throw of the dice. The sacking of the US Embassy in Islamabad did yet more damage to the US-Pakistan relationship, and the hostage crisis in Tehran occupied the administration's attention.[91] In the end the Soviet invasion of Afghanistan forced the Carter administration to make concessions on the supply of arms to Pakistan, the sticks and carrots on which so much faith had been placed.

Conclusion

The political efforts to combat Pakistani nuclear aspirations underwent dramatic, but little commented upon, change during 1979. For the ardently non-proliferationist Carter administration, the gradual switch from prevention to mitigation was a significant alteration in policy. This switch in focus came about because of Pakistani intransigence and the unwillingness of Zia and his subordinates to make any concessions on the nuclear front. The request for Pakistan not to test a nuclear device was at its crux. Such a demand would maintain non-proliferation credibility by removing the potential for a very public embarrassment. In the final analysis it was not the cultural factor of religion that exerted the greatest influence, but ideas of credibility and face-saving. Factional disagreements within the US foreign policy establishment and an almost overwhelming faith in the ability of bribery or coercion to solve proliferation problems also hampered the search for solutions. The move to mitigation did not, however, go unchallenged. Ardent non-proliferationists on both sides of the Atlantic sought to maintain a more stringent non-proliferation policy. British efforts to provide a global political answer foundered because of reticence amongst the nuclear weapon states and nuclear supplier states—particularly America—to consider yet another nuclear agreement at a time when SALT II, the CTB, and—in particular—INFCE were all being

[90] Saunders, Pickering, and Lake to Vance, 'November 14 PRC Meeting on South Asian Nuclear Issues', November 10, 1979, USPQB, Doc.44, 7.

[91] UKE Washington to FCO, 'Nuclear South Asia', December 13, 1979, TNA FCO96/961.

discussed. As 1979 ended the Soviet intervention in Afghanistan char
the terrain even more. In addition to the tacit acceptance of Pakistani
nuclear ambitions contained within the attempts to gain assurances of no
testing, the invasion meant that Pakistan would be offered arms and aid,
even while pursuing its nuclear weapons programme.

CHAPTER 8

"Peanuts" The Cold War, Khan, Islam, and the Death of Non-proliferation, January 1980 to January 1981

On January 18, 1980, Muhammad Zia ul-Haq stood before American journalists and disparaged Jimmy Carter's offer of military and economic assistance for Pakistan, describing a proposed $400 million aid package as "peanuts".[1] Carter's abandonment of the embargo put in place by the Symington Amendment's April 1979 implementation stemmed from the USSR's intervention in Afghanistan. US policy towards Pakistan had changed profoundly during 1979, from attempting to prevent Islamabad from gaining nuclear capability to endeavouring to mitigate the effects of an inevitable attainment. Now Carter's policies altered again, casting conventional arms restraint aside in an attempt to maintain the appearance of an active non-proliferation stance in South Asia and to bolster Pakistan in the face of Soviet adventurism. American *and* British policy morphed during 1980 from one of mitigation coupled with an embargo on arms sales and aid, to one of mitigation, with arms sales and aid becoming de-linked from nuclear policy. After the invasion of Afghanistan,

---

[1] United States Embassy (hereafter USE) Islamabad to Department of State (hereafter State), 'Zia's Remarks to US Newsmen on US Aid Offer, Bilateral Aid, Nuclear Issue', January 18, 1980, Digital National Security Archive (hereafter DNSA), NP01720, 2.

© The Author(s) 2017
M.M. Craig, *America, Britain and Pakistan's Nuclear Weapons Programme, 1974–1980*, Security, Conflict and Cooperation in the Contemporary World, DOI 10.1007/978-3-319-51880-0_8

Carter's policies on human rights, conventional arms restraint, and non-proliferation vanished.[2]

Vital to this story were American, British, and Pakistani attempts to maintain credibility. For the Western states, this meant an appearance of non-proliferation activity. The core means for realising which were requests to Pakistan not to undertake nuclear testing. Testing would be the most public demonstration of nuclear attainment and, if Pakistan *were* to test, the impact would throw the entire trilateral relationship into doubt. In the face of Soviet activities in Afghanistan, not Washington, London, nor Islamabad could afford for this to occur.

Media attention focusing on the nuclear programme also caused frustration, with stories about the Khan Affair, the 'Islamic bomb', and the clandestine purchasing project threaded through media coverage. This coverage—particularly that related to the Khan Affair—shone an uncomfortable light on efforts to impose global export controls. The reluctance and recalcitrance of key European nuclear technology supplier states were also part of a complex and often frustrating situation for Washington and London. However, the clandestine purchasing programme was one area where the UK and US governments did continue to make genuine efforts to retard the Pakistani nuclear programme through export controls and international diplomacy.

Islam and Khan, Again

Clandestine procurement, the Khan Affair, the Islamic bomb, and the media attention paid to these connected issues created serious problems for the US and UK governments during 1980. Margaret Thatcher's government in particular came under pressure from parliamentarians and journalists, both of whom kept the Pakistani nuclear programme in the public eye. The propaganda predicament this created, while the US-UK alliance attempted to fashion a Muslim bulwark against Soviet expansionism, was a thorn in the side of administrations juggling the challenges of non-proliferation and the Soviet threat.

From mid-1979 onwards British officials had been aghast that the Dutch had neither informed the UK of their misgivings about Khan nor

---

[2] Scott Kaufman, *Plans Unraveled: The Foreign Policy of the Carter Administration* (DeKalb: Northern Illinois University Press, 2008), 231.

terminated his employment when he came under suspicion in 1975.[3] The FCO opined that initial Dutch explanations were inadequate and demanded a clearer account.[4] There was dissatisfaction about Dutch handling of the matter, their initial report characterised as "unconvincing and difficult to understand".[5]

By January 1980 London was frustrated by Dutch delays in publishing their official Khan Affair report. Thatcher, Peter Carrington, Douglas Hurd, and other ministers were being bombarded with parliamentary questions on the affair, exports, and the Islamic connection. Consequently, the FCO pressed their Dutch counterparts to release the report in the hope of gaining respite from the ceaseless parliamentary attentions.[6] British diplomats in The Hague reported that Dutch contacts regarded their own government as evasive, worried about the potential for embarrassment, and plagued by infighting between the Foreign Ministry and the Ministry of Economic Affairs.[7] Gijs van Aardenne, Minister of Economic Affairs, was portrayed as deliberately delaying release, provoking Secretary of State for Energy David Howell to write and ask pointedly for a publication date.[8] British irritation remained up to the report's release on February 29. In preparation, the FCO briefed embassies to prepare for extensive media coverage, stressing British innocence in the affair, concern about the laxity of Dutch

[3] Carter (FCO), Handwritten Note, July 6, 1979, The National Archives of the United Kingdom (hereafter TNA) Records of the Foreign and Commonwealth Office (hereafter FCO) 96/957.

[4] FCO to United Kingdom Embassy (hereafter UKE) The Hague, 'Theft of Secrets from Almelo: "The Khan Affair"', July 23, 1979, TNA FCO37/2206; UKE The Hague to FCO, 'Theft of Secrets from Almelo: "The Khan Affair"', July 31, 1979, TNA FCO96/957.

[5] Carter to Thorp, Untitled, August 20, 1979, TNA FCO96/959.

[6] FCO to UKE The Hague, 'URENCO: Khan Affair', January 4, 1980, TNA FCO96/1077.

[7] UKE The Hague to FCO, 'Khan Affair', January 7, 1980, TNA FCO96/1077, 1; Hervey to Alston, 'Khan Affair', January 16, 1980, TNA FCO96/1077.

[8] UKE The Hague to FCO, 'URENCO: Khan Affair', February 1, 1980, TNA FCO96/1077; Howell to van Aardenne, Letter, February 11, 1980, TNA FCO96/1107.

security, emphasising export controls, and the extent of diplomatic representations to Pakistan.[9]

Thatcher's government remained under pressure from opposition parties in parliament. The Liberal Lord Avebury, the Labour Lord Wynne-Jones, and the usual 'awkward squad' of anti-nuclear MPs like Frank Allaun and Tam Dalyell all probed Pakistan's programme. Dalyell personified parliamentary pressure on the Thatcher government. His barrage of questions covered the Khan Affair, covert purchasing, and the Islamic bomb. On January 17 the Prime Minister responded to Dalyell's probing on the Khan report, assuring him that the government was taking steps to prevent the affair's repetition.[10] Shortly thereafter, Dalyell reiterated his nightmare of an Islamic bomb:

> If I have nightmares, they are about a Pakistani bomb or a Libyan bomb. We are now told that the Iraqis are doing nuclear weapons for some years, should also be about an Iraqi bomb. Those nations might use a nuclear bomb. It is for that reason that I go on and on, at Prime Minister's Question Time, about the Khan incident.[11]

The bombardment continued as Dalyell put over forty questions to the government during January and February alluding to the "development of a Pakistani or Islamic nuclear weapon".[12] His question on "Arab links with Pakistan's nuclear weapons programme" received a non-committal

[9] FCO to United Kingdom High Commission (hereafter UKHC) New Delhi, 'Khan Report', February 29, 1980, TNA FCO96/107.

[10] Tam Dalyell, Written Answers (Commons), 'Nuclear Security', January 17, 1980, *Hansard Online*, http://hansard.millbanksystems.com/written_answers/1980/jan/17/nuclear-security#S5CV0976P2_19800117_CWA_24 (accessed October 30, 2013).

[11] Tam Dalyell, Speech (Commons), 'Nuclear Weapons', January 22, 1980, *Hansard Online*, http://hansard.millbanksystems.com/commons/1980/jan/24/nuclear-weapons#column_726 (accessed October 30, 2013).

[12] 'Pakistanis rejected aid to protect nuclear programme', *The Times* (hereafter TT), March 8, 1980, 5; Tam Dalyell, Speech (Commons), 'Nuclear Security', January 31, 1980, *Hansard Online*, http://hansard.millbanksystems.com/commons/1980/jan/31/nuclear-security-1#S5CV0977P0_19800131_HOC_285 (accessed October 30, 2013).

answer from the FCO that such associations were speculative, rendering official comment impossible. In the background the FCO underscored the persistence—especially in connection with Libya—of such rumours, but emphasised a lack of conclusive evidence for such links.[13] Frank Allaun supported Dalyell. Allaun had, in 1978, publicised the inverter issue, and used that instance of Pakistani "deceit" to interrogate Thatcher on the wisdom of arming Pakistan to resist a Soviet incursion. As had been the case from 1974 onwards, the larger question was of whether or not the proliferation question could exist in a compartmentalised form, hived off from other matters such as conventional arms supply. Thatcher explicitly compartmentalised the issues, contending that the government had sought Islamabad's assurances on nuclear technology transfers. Thatcher argued that Pakistan was now in the "front line" of a revived Cold War, making the issue of arms sales a separate matter.[14]

Dalyell's relentlessness frustrated a government trying to balance support for Pakistan with attempts to delay Islamabad's eventual nuclear capability. Senior ministers tasked Norman Lamont, DoE Parliamentary Under Secretary, with briefing Dalyell in the hope of persuading the MP to moderate his inquisition. Lamont told Carrington that he had informed Dalyell how troubled the government was about the Khan Affair, but were constrained by the situation's delicacy and international implications. Lamont suggested that his *tête-à-tête* with Dalyell might have moderated the MP's campaign.[15] This was a forlorn hope. FCO arms control experts concluded that the state of affairs had become absurd, each answer supplying Dalyell with ammunition for his next salvo.[16] The MP persisted, following up on Libyan-Pakistani nuclear connections, connections that the government maintained were speculative or confidential. Prime

[13] 'Draft Supplementary PQ (Mr Dalyell)', January 25, 1980, TNA FCO96/1103.

[14] Frank Allaun, Question (Commons), 'Nuclear Security', January 29, 1980, *Hansard Online*, http://hansard.millbanksystems.com/commons/1980/jan/29/nuclear-security#S5CV0977P0_19800129_HOC_158 (accessed October 30, 2013).

[15] Lamont to Carrington, 'The Khan Affair: Tam Dalyell's Parliamentary Questions, Record of a Meeting with Tam Dalyell', January 28, 1980, TNA FCO96/1077, 1–2.

[16] Reeve to Moberly, 'PQ by Mr Tam Dalyell MP: Non-proliferation Treaty', March 7, 1980, TNA FCO37/2370.

Ministerial briefings still argued there was no "corroborative evidence" for such allegations.[17] Furthermore, the FCO reasoned that, to develop nuclear weapons, Libya must abrogate its NPT commitments *and* a recently concluded IAEA safeguards agreement.[18] Dalyell's relentless interrogation led to the suggestion of *another* confidential briefing, this time with the FCO's Douglas Hurd.[19] Throughout the year, and far into the decade, Dalyell kept up the pressure on Thatcher's government regarding Pakistan.

By the time the FCO commented on the "absurdity" of Dalyell's relentless questioning, there had been extensive discussions within the government, and between Britain and America, emphasising that attempts to hobble the Pakistani purchasing programme could not cease. Michael Pakenham, at the British Embassy in Washington, observed that further revision of US and UK export controls might be required, now that the Pakistanis realised efforts were being made to impede their access to critical materials.[20] The FCO was also concerned that, because America and Britain were making public moves to bolster Pakistan, some key nuclear technology suppliers "may relax on the Pakistan nuclear issue, not only in their diplomatic contacts with the Pakistanis but also in the field of export controls". The FCO regarded the American position as critical, arguing that many other states took their cue from the USA.[21] On January 18 the State Department's Thomas Pickering observed to Robert Alston of the JNU that, while no further pressure would be put on Pakistan regarding the nuclear issue and aid offers were not contingent upon a change in Islamabad's nuclear policy, the Carter administration had made it clear that non-proliferation credibility remained relevant. Furthermore, the US government would continue pursuit of tough export

---

[17] Tam Dalyell, Written Answers (Commons), 'Nuclear Proliferation', March 13, *Hansard Online*, http://hansard.millbanksystems.com/written_answers/1980/mar/13/nuclear-proliferation (accessed October 30, 2013); 'For the Prime Minister: Parliamentary Question (Background)', Undated, TNA FCO37/2370, 1.

[18] Ibid., 2.

[19] Alston to Moberly, 'Mr Tam Dalyell MP: Questions on Non-proliferation', March 18, 1980, TNA FCO37/2370.

[20] Pakenham to Alston, 'Nuclear Exports: UK Controls', January 11, 1980, TNA FCO96/1103, 2.

[21] FCO to UKE Washington, 'Pakistan Nuclear', January 16, 1980, TNA FCO96/1103.

controls and remained committed to encouraging compliance by other states.[22] Communications to US embassies at the end of January clarified this stance, underscoring the continuing need for US-UK diplomatic cooperation to frustrate the enrichment programme.[23]

Mounting evidence of Pakistani purchasing efforts and the lack of controls in key supplier states confirmed the need for sustained attention on export restrictions. The geographical diversity of a Pakistani purchasing programme that tapped into competitive international markets compounded the difficulty and complexity of managing export controls. As more information appeared it became clearer that the Pakistani network was well organised and global in reach. The West German attitude remained awkward, the FRG government was unwilling to take the action on export controls that the UK and the USA desired.[24] British diplomats kept their German contacts under pressure, but the outlook was gloomy. By October the British Embassy in Bonn contended that the Germans had "given up" on the question of exports, were obstructive, and disinclined to put proper regulations in place.[25]

The USA and UK saw Turkey as a worrying case because of alleged Islamic ties, Turkish disillusionment with the West, Turko-Pakistani military relations, and potential Turkish interest in acquiring nuclear technology for its own use. Like Britain, Turkey was in dire economic straits and, according to one British official, would "sell her soul to the devil" for foreign trade.[26] Less florid analyses argued that the Turks probably saw Pakistani nuclear aspirations as destabilising and—although the Turkish government relied

[22] Alston to Moberly, 'Non-proliferation: India and Pakistan: US Views', January 18, 1980, TNA FCO96/1103.

[23] State to USE Bern et al., 'US Non-proliferation Policy and Renewed Assistance to Pakistan', January 30, 1980, National Security Archive Electronic Briefing Book (hereafter NSAEBB) 'The United States and Pakistan's Quest for the Bomb' (hereafter USPQB), Doc.36.

[24] UKE Bonn to FCO, 'Nuclear Exports: Export Controls', January 8, 1980, TNA FCO96/1103.

[25] UKE Bonn to FCO, 'Export Control Regulations', October 29, 1980, TNA FCO96/1106; UKE Bonn to FCO, untitled, November 21, 1980, TNA FCO96/1107.

[26] Rawlinson to Martin, 'Turkey: Possible Sale of Nuclear Related Equipment to Pakistan', March 5, 1980, TNA FCO37/2370.

on Islamic fundamentalist support in parliament—it was doubtful if that influenced Prime Minister Süleyman Demirel's administration in this case.[27] The USA made a series of representations to Ankara. The Turks were sympathetic to the problem but, as with many other nations, including the FRG, felt they could do little because of inadequate export controls and a lack of justification for action.[28]

In northern Europe the Norwegians were more amenable to British and American approaches. Newspaper stories about firms supplying materials to Pakistan had embarrassed the Norwegian parliament.[29] The USA and the UK expressed concern and, in return, gained guarantees from Oslo about tightening controls and curtailing the export of sensitive items to South Asia.[30]

Thatcher's government remained anxious about British industry's role in supplying materials for Pakistan's clandestine programme. The inter-departmental Official Group on the Control of the Export of Special Materials and Equipment identified several shipments to Pakistan and argued that the implications of a sub-continental nuclear arms race had become even more serious. Export controls had a key role to play in winning time for a political solution.[31] Parliamentary and public interest in the issue made it even more important that orders—such as those placed with (and declined by) Emerson Electrical Industrial Controls (EEIC)— be prevented from reaching their destination.[32]

Switzerland was a particularly intractable supplier state and their case typified the challenge of controlling sensitive exports in a global marketplace. The diplomatic efforts of 1979 had borne little fruit, and Swiss firms such as CORA and VAT remained—according to American and British

[27] Ibid.

[28] 'Pakistan Nuclear: Summary of Discussion with US Representatives on 23 April', April 29, 1980, TNA FCO37/2370, 5; Alston to Roberts, 'US/Turkish Discussions on Inverters for Pakistan', June 25, 1980, TNA FCO96/1105.

[29] Jones to Alston, 'Pakistan Nuclear', April 24, 1980, FCO96/1105.

[30] Alston to Jones, 'Pakistan Nuclear', April 28, 1980, TNA FCO96/1105; Jones to Alston, 'Pakistan Nuclear', May 12, 1980, TNA FCO96/1105.

[31] Alston to Press, 'Official Group on the Control of the Export of Special Materials and Equipment', January 23, 1980, TNA FCO37/2370.

[32] Carter to Gittelson, 'Inverters for Pakistan', February 26, 1980, TNA FCO96/1104.

intelligence—principal suppliers to the Pakistani programme. In US-UK meetings Robert Gallucci and Robert Alston noted the difficulties posed by Switzerland and clarified the need for close cooperation to combat the activities of Swiss firms.[33] Preparing for a meeting between Carrington and his Swiss opposite number, Pierre Aubert, Alston opined, "The Swiss are the least cooperative of all suppliers of nuclear equipment in exercising restraint in the interests of non-proliferation. They try to hide behind the letter of the minimum conditions agreed by the Nuclear Suppliers Group."[34] In Washington NSC staff commented that the Swiss record on Pakistan was "just appalling" and that all American protestations had gone unheeded.[35] Bern, despite démarches from Washington and London, remained reluctant to intervene. NSC Executive Secretary Peter Tarnoff characterised Switzerland as ignoring exports not explicitly identifiable as nuclear. Even Warren Christopher's approach to Swiss Ambassador Raymond Probst on the latter's departure from the position failed to achieve results.[36]

Swiss reluctance to target its 'leaky' firms fed the media furore surrounding the Pakistani programme. The *Washington Post's* reporting on the matter—frequently using the 'Islamic bomb' as a framing device—alleged that the Carter administration had threatened to curb nuclear cooperation with Switzerland unless Bern improved export controls.[37] Christopher argued that this was untrue and that the State Department had only delayed Swiss requests for the transfer of US-origin fuel to the UK and Italy for

[33] 'Pakistan Nuclear: Summary of Discussion with US representatives on 23 April', April 29, 1980, TNA FCO37/2370, 1–2.

[34] Alston to Moberly, 'Visit of Swiss Foreign Minister: Non-proliferation', April 24, 1980, TNA FCO37/2370, 2.

[35] Global Issues to Brzezinski, 'Evening Report', July 1, 1980, Jimmy Carter Presidential Library (hereafter JCPL), Remote Archives Capture system (hereafter RAC) NLC-28-56-1-14-4.

[36] Tarnoff to Brzezinski, 'Swiss Assistance to Pakistan's Nuclear program', August 12, 1980, JCPL, RAC NLC-28-55-8-12-0, 1–2.

[37] 'Swiss Send Nuclear Aid to Pakistan', *Washington Post* (hereafter *WP*), September 21, A1; 'U.S. Says Evidence Shows Pakistan Planning A-bomb', *WP*, September 21, 1980, A20; 'U.S., Swiss At Impasse on A-Policy', *WP*, September 22, A1.

reprocessing until the Alpine state took action on exports.[38] Despite media coverage and diplomatic pressure, the Swiss frustrated efforts to bring them into the export control fold.[39] By the Carter administration's last few weeks, the situation remained unresolved, despite a more constructive tone from Bern and a turnabout in US attitudes that—concurrent with the waning of enthusiasm for non-proliferation—saw denying approval for the transfer of Swiss fuel as an impediment to gaining Swiss cooperation.[40]

In Britain, the question of firms being disadvantaged by a commitment to strict controls had been debated since the inverter issue arose. In late 1979 the outcry over the Khan Affair had rekindled these debates. In light of the Khan revelations Hurd argued for tighter controls, pointing out that, although Britain could not do this alone, strong action would send a clear signal to other supplier nations.[41] The DoT's Cecil Parkinson accepted the situation's seriousness, but stressed the importance of foreign trade to Britain and argued that if other countries (such as Switzerland) were not enforcing controls, why should the UK damage its economy by rigorously screening exports?[42] From this brief debate, export controls *were* tightened to place further restrictions on items that could potentially be used in a nuclear weapons programme.[43] The same debate took place in mid-1980, prompted by Swiss dissembling about their exports. Parkinson harked back to his 1979 comments and, although

---

[38] Christopher to Carter, Memorandum, September 23, 1980, JCPL, RAC NLC 128-15-9-12-1, 2.

[39] Brzezinski to Carter, 'Information Items', October 31, 1980, JCPL, RAC NLC-1-17-4-29-7, 2; Alston to Roberts, 'Centrifuge Technology: Switzerland and Pakistan', November 7, 1980, TNA FCO96/1107; State to USE Bonn, 'FRG Nuclear Exports to Pakistan', November 30, 1980, JCPL, RAC NLC-16-121-3-26-9, 2.

[40] Muskie to Carter, 'Transition Issues', November 10, 1980, JCPL, RAC NLC-15-61-5-24-7, 8; Muskie to Carter, 'US-Swiss Nuclear Cooperation Issue', December 29, 1980, JCPL, RAC NLC-128-16-19-2, 2.

[41] Hurd to Parkinson, 'Control of Nuclear Exports', August 14, 1979, TNA FCO96/958, 1–2.

[42] Parkinson to Hurd, 'Control of Nuclear Exports', August 28, 1979, TNA FCO96/959, 1.

[43] Hurd to Parkinson, 'Control of Nuclear Exports', September 26, 1979, TNA FCO96/959.

he disagreed with a strong British stance on exports while competitors were profiting from the Pakistani programme, he reluctantly agreed to further tightening of controls.[44]

While Washington and London attempted to restrict Pakistani access to sensitive materials, the professed need to gain the trust and support of Muslim states in combating the Soviets in Afghanistan butted up against continued media deployment of the meme of an Islamic bomb.[45] Throughout 1980 print and broadcast media imbued the Pakistani nuclear programme with transnational religious overtones. Just as in 1979, when the issue became a major story, the media used the Islamic bomb—often in a lazy or sensationalist manner—as shorthand for an expected proliferation cascade with dire consequences for the Middle East. The persistence of this meme created an uncomfortable tension between Pakistani nuclear aspirations, the Afghanistan crisis, and efforts to build a Muslim anti-Soviet alliance.

From December 1979 stories had circulated about the delivery of uranium 'yellowcake' from the Islamic African nation of Niger to Libya and then to Pakistan, with French involvement.[46] The possibility that these Islamic states had received raw uranium supplies outside of international controls disturbed FCO officials.[47] Rumours of violent uranium convoy hijackings and diversion to Libya and Pakistan precipitated hurried denials from Nigerien-French mining concerns SOMAIR and COMINAK.[48]

[44] Parkinson to Hurd, Letter, July 21, 1980, TNA FCO37/2371.

[45] The British government also unsuccessfully attempted to negotiate Soviet withdrawal under the 'Afghanistan neutrality' plan. Such efforts involved extensive contacts with Pakistan and members of the Islamic Conference. See Gabriella Grasselli, *British and American Responses to the Soviet Invasion of Afghanistan* (Aldershot: Dartmouth, 1996), 90–94.

[46] 'French Uranium for Libya', *The Sunday Times*, December 2, 1979, E28. 'Yellowcake' is a semi-processed form of uranium used in the preparation of fuel for nuclear reactors. A later 'uranium from Africa' story was a component of the drumbeat for war in the run-up to the 2003 invasion of Iraq. See Gabrielle Hecht, 'The Power of Nuclear Things', *Technology and Culture*, 51:1 (2010), 1–30 and *Being Nuclear: Africans and the Global Uranium Trade* (Cambridge: The MIT Press, 2012).

[47] Seton to Carter, Handwritten Memo, January 9, 1980, TNA FCO96/1103.

[48] 'Translation of a Telex from Head of Public Relations of CEA to Mr Chadwick', January 4, 1980, TNA FCO96/1103.

The reporting on this matter once again argued that a Pakistani bomb could be a Libyan-funded "Islamic bomb".[49] In the United States the CIA maintained its belief in the possibility of an Islamic bomb. In an assessment of Pakistani connections in the Middle East the agency again suggested that Libya and Pakistan were cooperating on nuclear weapons technology.[50] The same organisation later contended that Pakistani attainment of 'the bomb' might well axiomatically imply Libya nuclear capability.[51] The CIA's position is at least in part explicable by reference to the agency's position at the end of the 1970s. Embarrassed by the revelations of 1975 and under pressure to reform, the foregrounding of new 'threats to national security' would serve as a countermeasure. In the case of Iran, the general feeling within the US government was that intelligence had failed to anticipate the rise and triumph of the clerics. Thus, a fixation on the Muslim threat was overcompensation for this oversight.

In March it became clear to British politicians and officials that a major media investigation into the alleged Islamic bomb was underway. EEIC again approached the British government with disturbing news. BBC journalists, researching a documentary on Pakistani nuclear ambitions, had approached the company.[52] Management had denied the BBC permission to film at the factory, but that did not stop production going ahead, with the programme appearing in June. In the intervening months stories about Pakistani nuclear aspirations were framed within a pan-Islamic context. Discussion of Libya seemed to require a mention of alleged Libyan-Pakistani cooperation.[53] Journalists couched their

---

[49] 'French deny direct sale of uranium to Pakistan', The *Guardian* (hereafter *TG*), January 5, 1980, 5.

[50] CIA (National Foreign Assessment Center), 'Pakistan: The Middle East Connection', February, 1980, CIA-FOIA www.foia.cia.gov/sites/default/files/document_conversions/89801/DOC_0000631184.pdf (accessed October 30, 2013), 5.

[51] CIA (National Foreign Assessment Center), 'Developments in the Libyan Nuclear Program', May, 1980, JCPL, RAC NLC-6-48-4-23-4, 11.

[52] Gittelson to Parkin, 'BBC Television: Pakistan', March 20, 1980, TNA FCO96/1105.

[53] 'President's positions fall to rebel forces in Chad', *TG*, April 5, 1980, 5.

commentary in the now familiar language of an Islamic proliferation cascade. Jack Anderson wrote:

> When Pakistan does get its nuclear bomb, the world will enter a new and more dangerous era. A shaky dictatorship like Gen. Zia ul-Haq's, armed with a nuclear arsenal is frightening enough. What makes the situation far worse is that Pakistan will likely share its nuclear know-how with even less responsible Arab nations, like the fanatic Muammar Qaddafi's Libya, which is a protector of terrorists and an implacable foe of Israel.[54]

Echoing Dalyell's December 1979 address, Anderson's piece illustrates how entrenched the belief that Pakistan would share its nuclear technology with its co-religionists had become. As Rodney Jones argues, Islamic bomb coverage implicitly (and sometimes explicitly) contains within it "worst-case scenarios about threats to the security and perhaps survival of Israel".[55]

On June 16 the BBC aired 'Project 706: The Islamic Bomb' in the popular *Panorama* current affairs strand. The documentary underscored alleged Pakistani-Libyan connections, uranium from Niger, the complicity of British, German, Italian, and Swiss industry, and the threat of 'Islamic' proliferation. The opening monologue of reporter Philip Tibenham drew upon Islamic bomb speculation, informing viewers that:

> This convoy grinding across the empty Sahara is carrying what could be the raw material for the world's first nuclear war. The trucks are heading for a dusty desert air strip with a cargo of uranium yellow cake. It's been mined in the Islamic state of Niger. It'll be flown on to Islamic Libya; then on to Islamic Pakistan. Tonight, *Panorama* reports exclusively on payments of millions of pounds by Libya's Colonel Gaddafi to finance Pakistan's efforts to build the 'Islamic bomb'.[56]

---

[54] 'Pakistan Near Entry Into Atomic Club', *WP*, April 11, 1980, B9.

[55] Rodney W. Jones, *Nuclear Proliferation: Islam, the Bomb, and South Asia* (Beverly Hills: Sage, 1981), 44.

[56] Transcript, 'Panorama Contents Monday 17th June 1980 5340/5525 2042-2100 BBC-1; Project 706: The Islamic Bomb', BBC Written Archives Centre, File T67/13/1, Panorama-Project706-The Islamic Bomb, 1. The programme was also aired by PBS in the United States.

The JNU assessed the documentary as a broadly accurate account of Pakistani efforts, but speculative and inaccurate in detail. Analysts contended that by far the most important allegation was the Libyan/'Islamic' bomb, but that there was *still* no substantive evidence of Libyan financing *or* Pakistani agreement to proliferate. Anonymous allegations were "sensational" but carried "little conviction" (although DoE staff were more convinced of the interviewee's sincerity).[57] On-the-record government responses stressed Libyan adherence to the NPT and Pakistani non-proliferation assurances. For Thatcher's government the most embarrassing element was the programme's transmission during a visit to London by Agha Shahi, which provoked anger amongst the Pakistani delegation.[58] Government analysts argued that, overall, European industrial firms—including EEIC—came off far worse than governments and Dutch laxity received no coverage at all. They argued that the documentary offered an "extremely shallow" assessment of the links between civilian and military nuclear technology. Crucially, British experts did not see the film as harmful to the government or detrimental to international anti-proliferation efforts.[59] In the face of media reporting on the documentary—particularly in the *Guardian*—the Thatcher government stressed Pakistani assurances and a lack of evidence for an Islamic connection.[60]

As the media continued to draw attention to the Pakistani nuclear programme's supposed religious aspects, journalists drew other states into the fold. The *Daily Express* cited Islamic—but secular—Ba'athist Iraq as the next nation likely to "go nuclear" in a lurid story about murders, smugglers, and Parisian chambermaids. As part of a report on the killing of Egyptian physicist Yehya al-Meshad by persons unknown (allegedly because of his role in Saddam Hussein's nuclear programme), the newspaper warned its readers that "Arab states have lost all hope of winning a conventional war against Israel. So a terrifying premium has been placed on the alternative—a nuclear bomb. Pakistan, funded by Libya's Colonel

[57] Roberts to Acland, 'Panorama Documentary on Pakistan Nuclear Programme', June 17, 1980, TNA FCO37/2370, 1.

[58] Lavers to Roberts, 'Pakistan Nuclear: Panorama Programme, 16 June', June 18, 1980, TNA FCO37/2370.

[59] Fullerton to Manley, 'Panorama Programme on Pakistan Nuclear Bomb', June 18, 1980, TNA FCO96/1105, 1–2.

[60] FCO to UKE Islamabad, Untitled, June 18, 1980, TNA FCO37/2371, 2.

Gaddafi, already has the know-how."[61] The *Guardian's* Eric Silver, reporting from Jerusalem, observed, "no one here doubts the danger from an Arab or Islamic bomb".[62] In an editorial also criticising the Israeli nuclear stance, the normally sober *Times* asked if, by 1985, might there not be nuclear weapons in the hands of "fanatical Islamic revolutionaries" or could the Libyan-Pakistani Islamic bomb have eventuated?[63] This last piece did not go unchallenged. Syed Aziz Pasha, General Secretary of the Union of Muslim Organisations of UK and Eire, castigated the *Times* for causing "anger and distress", not only to British, but also to all Muslims. Aziz contended that the, so-called, Islamic bomb did not exist and argued that no other nuclear programme had ever been named for the originating state's religious affiliations. The paper rather primly added: "The phrase 'Islamic bomb', which has passed into common usage, did not originate with *The Times*."[64]

The protest by Aziz did not stop *The Times* publishing another piece on Pakistan that connected nuclear ambitions and Islam. Going back to Bhutto's death cell testimony, the article concluded: "Pakistan's nuclear activity—and the implications of an Islamic bomb, if ever such a thing should exist—threaten to usher in a new age of uncertainty."[65] Veteran journalist James Cameron, even while satirising attitudes towards nuclear weapons, asserted that Islam was "the originator of spreading God's word by the sword" and that he would not "especially like to be sent into eternity by a General Zia finger on the button, let alone a Khomeini finger".[66] Even in satire, the media portrayed Islam as violent and the Islamic bomb as leading to an almost inevitable Middle Eastern apocalypse.

---

[61] Jeffrey Richelson, *Spying on the Bomb: American Nuclear Intelligence from Nazi Germany to Iran and North Korea* (New York: W.W. Norton & Co., 2006), 321; 'The chambermaid who stumbled on a secret A-war', The *Daily Express*, July 8, 1980, 9. Saddam tended to place his nuclear ambitions within a pan-Arab, not pan-Islamic, context, a distinction that many Western observers failed to entirely appreciate.

[62] 'French nuclear sale angers Israel', *TG*, July 18, 1980, 6.

[63] 'Is There An Islamic Bomb?', *TT*, July 22, 1980, 13.

[64] Syed Aziz Pasha, Letter 'Nuclear bomb', *TT*, July 29, 1980, 13.

[65] 'National pride could push General Zia to Islamic bomb', *TT*, August 13, 1980, 6.

[66] 'Gods of war', *TG*, August 19, 1980, 8.

The American press also weighed in, mixing together the Khan Affair, Islam, Swiss obduracy, and Libyan anti-Americanism. Revelations surrounding 'Billygate', the Libyan business connections of Carter's wayward brother, then under investigation by the Justice Department, did not help the situation.[67] In a *Washington Post* article that liberally referenced the *Panorama* documentary, an anonymous US official made a veiled swipe at the Europeans, commenting, "Some countries were lax and bureaucratically inept...but some others knew what was happening and allowed it to go ahead for political or commercial reasons."[68] By the year's end the same newspaper reported alleged fissures in the Islamic bomb project, with Libya portrayed as frustrated by a lack of Pakistani progress.[69] The issue that dominated the American media when it came to Islam was not the potential for a Muslim nuclear weapon. The Iranian hostage crisis occupied far more column inches and hours of broadcasting. The encounter with an unrealised nuclear weapon remained less significant than the very real encounter with the new form of political Islam represented by Iran.

In public at least, the idea that a Pakistani nuclear bomb was an Islamic bomb had—by 1980—become so embedded in the media that it went almost totally unchallenged. The Libyan connection, Bhutto and Zia's inflammatory rhetoric, and the perceived certainty of a pan-Islamic proliferation cascade became accepted as fact. Rebuttals by Muslim leaders and official governmental comment that such assertions were speculative at best went unheeded. Unlike in 1979, a remarkable aspect of this is the *lack* of comment on the issue within the available official documents, the declassified record showing little in the way of high-level internal discussion. What is unsurprising is the disparity in coverage in Britain and America. In Britain the topic was far more pervasive. British partnership in URENCO, the Khan Affair's public prominence, and the European focus of Pakistani procurement explain the British media's fascination with an Islamic bomb. In America the issue was subsumed beneath the Iranian hostage crisis and Afghanistan. The

---

[67] Burton I. Kaufman and Scott Kaufman, *The Presidency of James Earl Carter, 2nd edition, revised* (Lawrence: University Press of Kansas, 2006), 228–230.

[68] 'U.S. says evidence shows Pakistan planning A-bomb', WP, September 21, 1980, A20.

[69] 'Deterring a Qaddafi Bomb', WP, December 23, 1980, A15.

Islamic bomb—while terrifying—was a yet-to-be-attained capability. In Iran, by contrast, what was portrayed as a violent, irrational, radical Islamic state directly threatened American citizens.

NON-PROLIFERATION AND AFGHANISTAN

As Soviet troops entered Kabul, and détente finally died, Carter determined to make the Kremlin's actions as costly as possible. He noted in his diary, "I sent messages to our allies, key non-aligned leaders, plus all the Muslim countries—urging them to speak out strongly against the Soviet action."[70] This Soviet act of "brutality" posed a direct threat to Persian Gulf oil fields and to Afghanistan's neighbour, Pakistan.[71] Soviet intervention fed perceptions that—from Angola to the Horn of Africa and to Vietnam—the forces of communism were once more on the march.[72] Pakistan—a pivotal state in the region since the Iranian revolution—received immediate consideration in the invasion's aftermath, the need to bolster Pakistan overriding other regional concerns.[73] The relationship with Britain remained vital to building a consensus amongst America's European allies. For Thatcher's government Afghanistan presented the first major test of their commitment to the transatlantic foreign policy relationship.[74] Thatcher supported Carter when it came to the need for

[70] Jimmy Carter, *White House Diary* (New York: Farrar, Straus, and Giroux, 2010), 382.

[71] Carter, *Keeping Faith*, 471–472; Jimmy Carter, 'Address to the Nation on the Soviet Invasion of Afghanistan', January 4, 1980, PPPJC, www.presidency.ucsb.edu/ws/index.php?pid=32911&st=&st1=#axzz2j6opB200 (accessed October 29, 2013).

[72] Richard Vinen, 'Thatcherism and the Cold War', in Ben Jackson and Robert Saunders (eds.), *Making Thatcher's Britain* (Cambridge: Cambridge University Press, 2012), 199.

[73] Husain Haqqani, *Pakistan: Between Mosque and Military* (Washington D.C.: Carnegie Endowment for International Peace, 2005), 184; Zbigniew Brzezinski, *Power and Principle: Memoirs of the National Security Adviser, 1977–1981* (London: Weidenfeld & Nicholson, 1983), 429.

[74] Daniel James Lahey, 'The Thatcher Government's Response to the Soviet Invasion of Afghanistan, 1979–1980', *Cold War History*, 13:1 (2013), 21.

resolute action.[75] And, as Carter indicated, the Islamic world was key. During 1978 and 1979, the Islamic bomb had featured in discussions about how to deal with the Pakistani nuclear programme, but had failed to influence actual policy. The Afghanistan crisis demonstrated the need for the West to use the bonds of Islam to build a vigorous Muslim anti-Soviet alliance. The assembling of an Islamic coalition to resist the Soviets was founded in a belief that this would cause the Islamic world's sympathies to realign towards Washington.[76] Unlike the image created in public, within government Islam was now a positive force for good in the reinvigorated Cold War.

The need to reinforce and reassure Pakistan gave rise to another change in US government policy. Despite the mid-1979 move to mitigating, rather than preventing, eventual Pakistani nuclear capability, the Carter administration had been determined not to offer the Pakistanis arms and aid. Mitigation did not mean that efforts to delay Pakistani nuclear acquisition had ended, and the issues of non-proliferation and the supply of conventional arms were still intimately linked, with export controls and international diplomacy used to hinder the programme. With the emergent crisis in Afghanistan Carter's policy mutated again, moving from one of mitigation with punishment (in the form of withholding arms and aid), to one of simple mitigation, where arms supplies were de-linked from the nuclear issue. Major challenges remained, as Carter's previous zeal for non-proliferation—and Congress' concurrent enthusiasm—created difficulties for attempts to bolster Pakistan. Change was not the US government's sole prerogative. The shift from prevention to mitigation had also been apparent in British policy during 1979. This change remained in 1980, with Thatcher's government following the American lead. There were voices within both governments still calling for a strong non-proliferation policy. Non-proliferation advocates, like Gerald Smith in Washington and Robert Alston in London, made the case for the continued pursuit of active non-proliferation policies in the

---

[75] Carter to Thatcher, transcript of telephone conversation, December 28, 1979, TMSS, THCR3/1/4 (Personal Message T180A/79T), www.margaretthatcher. org/document/112219 (accessed October 29, 2013), 3–4.

[76] Andrew Preston, *Sword of the Spirit, Shield of Faith: Religion in American War and Diplomacy* (New York: Anchor Books, 2012), 578.

case of Pakistan, but their entreaties were largely ignored by more senior figures committed to mitigation.[77]

For America, Britain, and Pakistan, appearances were vital and it was crucial that—in public at least—perceptions did not arise that a sub-continental non-proliferation policy was being abandoned. There were persistent demands to Zia's government for assurances that they would not conduct the most public and visible statement of nuclear attainment, a nuclear test. Thus, Pakistan would enter a state of 'nuclear ambiguity' similar to Israel. Concurrently, Islamabad did not wish to appear as allying too closely with the West in order to maintain standing in the Muslim and non-aligned worlds. Thus, Zia rejected Carter administration offers of aid and military equipment.

From the moment the Soviets intervened in Afghanistan the Pakistani nuclear programme influenced thinking on how to bolster and aid a nation that became the anti-Soviet effort's lynchpin. As the NSC's Thomas Thornton recalled, the American position on Pakistan changed over-night.[78] The embargo on military sales underwent a swift reassessment in the face of a drastically altered situation in South Asia. Even though the Carter administration had acquiesced to eventual Pakistani nuclear capability, Cyrus Vance still contended that the nuclear programme precluded significant economic aid and credit for military purchases. Harold Brown suggested distinguishing between the Pakistani programme as was, and a Pakistani programme involving nuclear testing.[79] This was a key point. A Pakistani test would dramatically demonstrate the failure of the non-proliferation policy. Exerting pressure on Zia not to test at least preserved the outward appearance of a vigorous non-proliferation policy. However, certain legislation blocked the way to full-blown aid for Pakistan: the Symington Amendment.

---

[77] The second NPT Review Conference took place in Geneva during August to September 1980. It was a largely inconclusive affair that failed to produce substantive outcomes. Some of this was down to the considerable friction between the United States and the Soviet Union over Afghanistan, while there was also considerable dissatisfaction amongst the non-nuclear weapon states about the lack of superpower progress towards genuine nuclear disarmament.

[78] Dennis Kux, *Estranged Democracies: India and the United States, 1941–1991* (Washington D.C.: National Defense University Press, 1993), 245.

[79] PRC Meeting, 'Southwest Asia', December 27, 1979, JCPL, RAC NLC-24-102-1-13-7, 5.

The Amendment impeded acceptance of what Vance, articulating the policy of acquiescence, observed as the "facts of nuclear life" regarding Pakistan.[80] Given that the administration had, by mid-1979, abandoned efforts to prevent Islamabad's nuclear acquisition, the "facts of life" were that, at some point, Pakistan would become at least a tacit nuclear weapon state. Several options were presented to Carter, including a presidential waiver, exploring Saudi financing for Pakistani arms purchases, or seeking a one-time congressional appropriation for military purchases. A waiver was unlikely because of overwhelming intelligence pointing to a Pakistani nuclear weapons programme.[81] Regardless of means, from the outset it was imperative that Pakistan be brought into the fold to construct a strong anti-Soviet front in Afghanistan. In Islamabad Arthur Hummel was convinced that if the Amendment could not be bypassed the ability to aid Pakistan would be muted. The Ambassador recommended that the administration deal urgently with the threat to Pakistan, and rationalised the abandonment of efforts to prevent Pakistani nuclear acquisition by stating that the nuclear programme's slow pace permitted the sidelining of non-proliferation matters for the time being.[82]

In the face of a renewed Soviet threat, non-proliferationist Democratic members of Congress, such as Jonathan Bingham (D-NY), Frank Church (D-ID), and Clement Zablocki (D-WI), supported special authorisation for aid to Pakistan, overriding the US legislation.[83] Thus, the most ardently non-proliferationist members of Congress supported the mitigation policy of de-linking arms sales and the nuclear issue. The Symington Amendment notwithstanding, the administration deprioritised the nuclear programme. As Brown admitted to his Chinese hosts in Beijing at the end of January, the administration had moved from a policy of prevention to one of mitigation. "Our big problem with Pakistan was their attempts to get a nuclear program," stated the Secretary of Defense. "Although we still object to their doing so, we will now set that aside for

---

[80] UKE Washington to FCO, 'Afghanistan', December 28, 1979, TNA FCO96/961.

[81] Saunders to Vance, 'NSC Discussion of Support for Pakistan', January 1, 1980, DNSA, NP01707, 2–4.

[82] Brzezinski to Carter, 'Daily Report', January 3, 1980, JCPL, RAC NLC-1-13-7-5-4.

[83] UKE Washington to FCO, 'Nuclear South Asia', January 4, 1980, TNA FCO37/2370.

the time being and concentrate on strengthening Pakistan against potential Soviet action."[84]

With the USSR having invaded an Islamic state, Islam and independent
nationalism became significant issues for the Carter administration in mobilising regional resistance to Soviet adventurism.[85] Gaining Muslim support
for US objectives meant demonstrating that America was not solely guided
by oil and Israel, a demonstration that policymakers hoped would win over
states distrustful of America because of lingering anti-colonialism, pan-
Arabism, and emergent Islamic ideology.[86] Zia also hoped that by sponsor-
ing an "Afghan jihad" he could gain prestige in the Islamic world.[87]

There was a raft of suggestions regarding how to achieve US goals in
relation to Pakistan, from a resurrection of A-7 sales, to debt relief, food
aid, refugee assistance, and oil supplies.[88] These suggestions were tied
together into a $400 million deal offered to Agha Shahi when he visited
Washington in mid-January.[89] Brzezinski had wanted a much more
substantial package for Pakistan, and was frustrated when budgetary
restraints, lingering non-proliferation worries, and human-rights concerns
militated against a larger offer.[90] Seeking a commitment from Pakistan to
not test a nuclear weapon was vital to the proposed deal.[91] Vance
informed the Pakistani foreign minister that the nuclear issue could be
set aside for the time being, but was at pains to point out that a test would

---

[84] Brown to Smith, 'Extract of Memorandum of Conversation', January 31, 1980,
NSAEBB 'New Documents Spotlight Reagan-era Tensions Over Pakistani Nuclear
Problem' (hereafter RTPP), www2.gwu.edu/~nsarchiv/nukevault/ebb377
(accessed October 29, 2013), Doc.33.

[85] 'Agenda: SCC Meeting, January 14, 1980', January 11, 1980, JCPL, RAC
NLC-17-18-28-4-0, 1.

[86] Ibid., 4.

[87] Haqqani, *Between Mosque and Military*, 185.

[88] Turner to Carter, 'Memorandum for the President', January 10, 1980, JCPL,
Records of the Office of the National Security Advisor (Carter Administration)
1977–1981, Box 59, Pakistan 1/80, 1; USE Islamabad to State, 'Economic
Assistance Package for Pakistan', January 10, 1980, DNSA, NP01713, 1.

[89] State to USE Islamabad, 'US-Pakistan Talks: Economic Assistance and Debt',
January 15, 1980, DNSA, NP01716, 2.

[90] Brzezinski, *Power and Principle*, 448.

[91] 'Agenda: SCC Meeting', January 11, 1980, 5.

drastically alter the US-Pakistan relationship. In response Shahi made a vague commitment that Pakistan would "do nothing to embarrass" the United States.[92] Thomas Pickering, during his visit to London, had made it clear to his British counterparts that the Carter administration still took the appearance of proliferation seriously. The US representative emphasised to his hosts that while further pressure would not be put on Pakistan regarding the nuclear issue and aid offers were not contingent upon changes in Pakistani nuclear policy, the US government believed it had made it clear that proliferation remained relevant. Pickering argued that a nuclear test would radically alter US-Pakistani relations, making it almost impossible to obtain congressional approval of military aid.[93] However, Islamabad did not welcome with open arms the $400 million package put to Shahi in Washington. Zia scoffed at the offer, stating it was "peanuts".[94] Zia's disparagement represented not just a blow for administration credibility over Pakistan, but a very public strike at the heart of non-proliferation policy.

The UK, recalling the heady days of Macmillan and Kennedy, was a key US ally in the reinvigorated Cold War. US officials highlighted to the British government that the administration's nuclear policy towards Pakistan had been "put on ice" and that nuclear questions should not prevent swift, decisive action on Afghanistan.[95] The FCO was happy to follow the US lead and saw little that Britain could do for Zia, but did not want to let the question of a more "forthcoming" Western policy towards Pakistan "run into the sands". The UK, Carrington's FCO suggested, was in a position to encourage other NATO nations to support US action. At all levels there was an awareness of Indian concerns about both the Pakistani nuclear programme and Western re-arming of their neighbour, but the FCO believed that action could not be totally constrained by the traditional Indo-Pakistani rivalry.[96]

[92] UKE Washington to FCO, 'US/Pakistan Talks in Washington: 12 January', January 15, 1980, TNA FCO96/1103.

[93] Alston to Moberly, January 18, 1980.

[94] USE Islamabad to State, January 18, 1980, 2.

[95] UKE Washington to FCO, 'Nuclear South Asia', January 4, 1980, TNA FCO96/1103.

[96] FCO to UKE Islamabad, 'Afghanistan: Support for Pakistan', January 1, 1980, TNA FCO96/1103.

Senior British figures, from the Foreign Secretary down, allied them-selves with the American approach. Carrington's high-profile visit to Islamabad in mid-January was not simply to show solidarity with Pakistan but pressed home British views on Zia's nuclear aspirations. Patrick Moberly, one of Carrington's top advisers, reaffirmed the centrality of the no testing paradigm by contending that "an explosion would endanger Pakistan's security by alarming her neighbours, alerting her enemies and scaring off her friends".[97] Moberly also thought it too early to discern how the Soviet intervention in Afghanistan, and the Western reaction to it, influenced Zia on the nuclear issue; he might be more disposed to listen to reason, but equally he might see Afghanistan as giving him leeway to ignore diplomatic appeals.[98] In Islamabad Carrington enjoyed extensive discussions with Zia and Agha Shahi. When covering the nuclear issue Shahi assured Carrington that a Pakistani test would not take place for at least six months and that the "next government" would make any decision for such an action, and reiterated Pakistan's commitment not to transfer nuclear technology nor to manufacture nuclear weapons.[99]

Carrington's post-journey report to Thatcher failed to mention the nuclear issue, illustrating the extent to which it had been deprioritised since mid-1979. The Foreign Secretary did highlight the leading role that Saudi Arabia and Pakistan were taking in mobilising Muslim opinion against the USSR. The Pakistanis, Carrington argued, were quite justified in being affronted at the meagreness of the US offer when their existing military equipment was so out-dated and ill-suited for purpose.[100] In the face of this the Foreign Secretary contended that the UK should encourage Carter to meet Pakistan's military needs up to a level that gave India "no justifiable reasons for concern".[101] The Pakistanis saw Britain's role as a conduit for their own interests. While visiting Islamabad Carrington had taken it upon himself to act as a messenger between Pakistan and India

[97] Moberly to Cortazzi, 'Pakistan: Nuclear Programme', January 11, 1980, TNA FCO37/2370.

[98] Ibid.

[99] UKE Islamabad to FCO, 'Secretary of State's Discussions with Agha Shahi', January 15, 1980, TNA FCO96/1103, 3.

[100] Carrington to Thatcher, 'Afghanistan', January 19, 1980, TNA FCO96/1104, 2.

[101] Ibid., 4.

and the Western states, something that Zia valued.[102] British diplomats
believed they were expected to encourage Saudi Arabia to fund Pakistani
defence purchases, lead Western opinion, and urge the USA to do
more.[103] The British relationship with the Saudis was, however, severely
compromised by the Islamic world's reaction to the 'Death of a Princess'
affair.[104] During multilateral talks in London at the end of January British
officials insisted that the nuclear issue should not be "pushed into the
shadows" because of a Pakistani test's potentially regionally disastrous
consequences. The US side argued that they had made the dire consequences of a test clear.[105] However, British policy in these early days
evolved into a 'wait and see' stance, holding back on major commitments
to Pakistan until Brzezinski and Christopher had visited Islamabad for
talks, and Zia had hosted a major gathering of Islamic states.[106] More
widely, as Daniel Lahey points out, British commitments on Afghanistan
were diplomatic, rather than economic, in nature. Thatcher could ill-afford
anti-Soviet trade restrictions that damaged the fragile British economy.[107]
Thus, initial British policy towards Pakistan during the early days of the
Afghanistan crisis was founded on diplomatic and moral support. Britain

---

[102] 'Summary Record of a Meeting Between the Foreign and Commonwealth
Secretary and President Zia', January 15, 1980, TNA FCO96/1104, 3–5.

[103] UKE Islamabad to FCO, 'Follow-up to Your Visit', January 22, 1980, TNA
FCO96/1103, 3.

[104] *Death of a Princess* was a lightly fictionalised drama-documentary based on the
real-life execution of a young Saudi noblewoman for adultery. It, as Alan
Rosenthal argues, investigated the "social pressures, ideals, and strains of modern
Arab society". Saudi Arabia—and the wider Islamic world—reacted furiously to
the US-UK co-production. The British Ambassador was expelled from Riyadh,
commercial contracts were cancelled, and even overflights by the supersonic
Concorde were banned. See Alan Rosenthal, 'The Politics of Passion: An
Interview with Anthony Thomas', *Journal of Film and Video*, 49:1–2 (1997), 95.

[105] State to USE London, 'London Meets on Assistance for Pakistan in Light of
Soviet Invasion of Afghanistan', January 30, 1980, JCPL, RAC NLC-16-120-3-
40-4.

[106] UKE Islamabad to FCO, 'Arms Sales to Pakistan', January 30, 1980, TNA
FCO96/1104.

[107] Lahey, 'The Thatcher Government's Response to the Soviet Invasion of
Afghanistan', 27–29.

adopted a 'light touch' approach, emphasising a quiet diplomacy that supported US efforts and the campaign against the clandestine purchasing programme.

The Christopher-Brzezinski mission to Islamabad and Riyadh in February left the US officials feeling optimistic about the future of US-Pakistani relations, despite continued Pakistani intransigence.[108] American refusals to extend security commitments to cover an attack by India concerned the Pakistanis, but experienced South Asia hands Thomas Thornton and Jane Coon remained puzzled by Zia's overall attitude.[109] Overtures to Zia had the side effect of upsetting and angering India. In January 1980 Indira Gandhi had returned to power and—prompted by Carrington's efforts—while denouncing Soviet adventurism, was even more anxious about US efforts to turn Pakistan "into an arsenal".[110] Indian disquiet moved Carter to approve the shipment of nuclear fuel for the reactors at Tarapur, overturning a decision by the Nuclear Regulatory Commission (NRC.)[111] In trying to balance Pakistan, India, Afghanistan, and national security, Carter only hastened the demise of his flagship non-proliferation policy.

On leaving Pakistan Brzezinski was, according to one British official, euphoric.[112] The National Security Adviser commented to Sir Robert Wade-Gery, British Deputy Cabinet Secretary, that the "material for a successful Southwest Asia policy is there for us to work with".[113] For Brzezinski, a "successful Southwest Asia policy" was one that continued to deprioritise Pakistani nuclear ambitions while emphasising the communist threat. Reporting back to colleagues in Washington, Brzezinski and Christopher observed that, during talks with the Pakistanis, there had

---

[108] Brzezinski, *Power and Principle*, 448–449.

[109] Dennis Kux, *Disenchanted Allies: The United States and Pakistan, 1947–2000* (Washington D.C.: Woodrow Wilson Center Press, 2001), 251.

[110] Sir Donald Maitland, transcript of interview by Malcolm McBain, December 11, 1997, British Diplomatic Oral History Programme (hereafter BDOHP), www. chu.cam.ac.uk/archives/collections/BDOHP/Maitland.pdf (accessed November 2, 2013), 36; Kaufman, *Plans Unraveled*, 218.

[111] Kaufman, *Plans Unraveled*, 218–219.

[112] Armstrong to Palliser, 'South West Asia', February 5, 1980, TNA Records of the Prime Minister's Office (hereafter PREM) 19/136.

[113] Brzezinski to Wade-Gery, teleletter, February 4, 1980, TNA PREM19/136.

been some indications that a nuclear test might be imminent, but these were judged as purely a negotiating tactic, and Shahi denied there would be a test, in an echo of his earlier remarks to Carrington about the testing issue. Brzezinski argued that the administration should only present Congress with modifications to the 1959 US-Pakistan security agreement and a waiver of the Symington Amendment as part of specific aid proposals, rather than as independent actions. Christopher suggested that Pakistan preferred an "Islamic option", whereby it should count on a US security umbrella but rely on Muslim states for direct cooperation and support.[114]

When reporting to Carter Brzezinski stated that the Pakistanis realised the USA stood four square behind the 1959 agreement and, as a result, there had been no need to increase the $400 million package or offer a formal security treaty. On broader geopolitical matters he contended that mobilising Muslim support required increased efforts on the Palestinian situation and military aid to Saudi Arabia.[115] News from Beijing that Pakistan might imminently conduct a nuclear test put the 'no testing' question in alarming perspective. The US Ambassador to China reported that, although the Chinese had discouraged Pakistani nuclear development, a test *was* imminent.[116] As it transpired, the Ambassador was mistaken, but there was momentary alarm that Islamabad had *already* attained nuclear capability.[117]

Within the Carter administration there were voices demanding a less relaxed non-proliferation policy. Ardent non-proliferationist Gerald Smith took a dim view of the damage US overtures to Zia were doing to global anti-proliferation efforts. However, by this point Smith's views had been smothered by the mitigationist policy adopted by more influential figures in the administration. Smith recounted that, despite continued demands, Zia had still not ruled out a nuclear test. Smith also recognised that

[114] 'Special Coordination Committee Meeting', February 6, 1980, JCPL, RAC NLC-128-10-7-11-0, 2.

[115] Brzezinski to Carter, 'Summary Report and Recommendations: Pakistan/ Saudi Arabia', February 6, 1980, Declassified Documents Reference System, DDRS-264074-i1-3, 1-2.

[116] UKE Peking to FCO, 'Pakistan Nuclear', February 14, 1980, TNA FCO96/ 1104.

[117] Pakenham to Carter, 'Possible Nuclear Test by Pakistan', February 20, 1980, TNA FCO96/1104.

overtures to Pakistan were damaging to Carter's wider non-proliferation efforts. The Ambassador saw risks in courting Pakistan; what, Smith wondered, would happen if Zia used US guns to protect the enrichment plant at Kahuta? How should the USA react if Pakistan tested a device after a Soviet invasion of their territory, just as America was about to come to their aid? Conversely, Smith subscribed to the West German view that military supplies to Pakistan probably gave leverage over their nuclear decision-making and there was general agreement that there must be no decrease in efforts to control exports to Pakistan.[118]

Despite the American and British governments stressing that non-proliferation remained important, it became clear that in Washington the policy of mitigating Pakistani acquisition of nuclear capability was increasingly de-linked from the issue of arms sales. As it became clearer that non-proliferation objectives were being divorced from the broader anti-Soviet policy, the NSC articulated the policy of acquiescence to Pakistani nuclear ambitions, stating that although elements of US support should be founded on a request to Pakistan for the strongest non-proliferation statement possible, aid should not be made conditional on the content and nature of such a statement.[119] As the NSC—and Brzezinski in particular—gained pre-eminence in foreign policy and the perception built that the Soviets were on the march in the developing world, hawkish attitudes towards the USSR overrode the humanitarian elements of Carter's policies.

As the US and UK governments pursued the policy of mitigation, primarily through seeking 'no test' assurances, it became apparent the Pakistanis also put a great deal of store on the wider perception of their own actions. Given the significance that Western media attached to the Pakistani nuclear programme, a consensus in Washington and London emerged that the nuclear issue was *not* a major factor in the Pakistani refusal to accept the American aid package after Agha Shahi formally and publicly rejected the offer on March 5. Instead, Shahi emphasised reliance on Pakistan's growing friendships with China, the Islamic world, and

---

[118] Pakenham to Alston, 'Nuclear Pakistan', February 7, 1980, TNA FCO37/2370.

[119] 'Issues Paper for SCC, February 29, 1980', Undated, JCPL, Zbigniew Brzezinski Collection (hereafter ZBC), Geographic File, Southwest Asia/Persian Gulf, Box 15, 3.

non-aligned states, arguing that a closer relationship with the USA could damage these increasingly vital connections.[120] However, speculation remained that a key reason for refusal was the nuclear issue. Although Shah Nawaz had indicated that the Pakistani government thought that nuclear strings *were* attached to the aid offer, Carter surmised that Zia had concluded the aid package's value was outweighed by the risks of allying too closely with the USA. Thus, Carter concluded, the main financing for aid should come from Islamic states.[121] Thatcher agreed and noted that the UK would quietly continue to encourage closer relations between Pakistan and the wider Islamic world.[122]

The State Department and the FCO agreed that the nuclear issue was not a major factor governing Pakistani decision-making. The area where Robert Gallucci felt nuclear matters may have had influence was over overt US statements on nuclear testing. Gallucci argued that the mantra that a test would "drastically change the relationship" had made Zia doubt the depth of Carter's commitment to Pakistani security. Gallucci continued that if the administration reached an arrangement with Zia (and he stressed that this might never actually happen), it was doubtful it would provide additional leverage in the nuclear area. At best it would allow for continued dialogue.[123] The British Embassy in Islamabad concurred with Gallucci's assessment. Fresh indications had come to light that cast doubt on existing assessments of the timescale for a Pakistani test, although the dates the media had been speculating about were regarded as implausible.[124]

[120] USE Islamabad to State, 'Agha Shahi Publicly Rejects Proposed US Assistance Package', March 6, 1980, DNSA, NP01749.
[121] UKE Islamabad to FCO, 'Pakistan Foreign Policy', March 9, 1980, TNA FCO96/1105; Carter to Thatcher, Letter, March 13, 1980, TMSS, THCR3/1/7 f63, www.margaretthatcher.org/document/112691 (accessed October 30, 2013), 1–2.
[122] Thatcher to Carter, Letter, April 8, 1980, TMSS, THCR3/1/8 f24 [T78/80], www.margaretthatcher.org/document/112705 (accessed October 30, 2013), 1.
[123] UKE Washington to FCO, 'Nuclear Pakistan', March 18, 1980, TNA FCO96/1104, 1.
[124] Fabian to Alston, 'Pakistan Nuclear', March 24, 1980, TNA FCO96/1104.

Alston and the JNU found themselves in a similar position to Gerald Smith in Washington. Alston's anti-proliferation stance was now super-seded by a US-UK position that sought the appearance of non-proliferation through 'no testing' assurances, but that, in reality, had deprioritised the goal of getting Pakistan to abandon its nuclear ambitions. Carrington had insisted that, while Afghanistan was the most pressing issue, Pakistani nuclear activities were viewed no less seriously than before December 1979. Export controls were vital because, if Western governments dropped their guard, the Pakistanis might take it as a sign that a lower priority had been given to non-proliferation objectives.[125] Alston, disturbed that the Afghanistan crisis had not improved prospects for resolving the nuclear dilemma, offered a new policy. In light of Shahi's rejection of Washington's offer, perhaps a proposal could be formulated to boost Pakistani confidence in the face of the Soviet and Indian threats *and* address the nuclear quandary? Thus, Alston and the JNU suggested offering Pakistan a package combining enrichment, reprocessing, and power services that was intended to halt Zia's moves towards weapons capability and re-establish the pre-eminence of the non-proliferation policy.[126] JNU's counterparts in SAD argued that the proposal was founded on false premises, contending that Islamabad's rejection of US aid was nothing to do with its size and scope, and everything to do with appearances. SAD contended that by being *seen* to reject the aid offer, Pakistan increased its standing with Islamic and NAM countries.[127]

A trilateral discussion between the JNU, Oliver Forster, and John Thomson debated Alston's desire to move back towards a preventative non-proliferation policy. Thompson had been the driving force behind 1979's abortive universal declaration on nuclear trade and exports, the failed policy that codified British acquiescence to eventual Pakistan nuclear capability. Afghanistan, Alston argued, had not changed the basic British attitude that the nuclear dilemma was extremely serious, even though American commitment was weakening. The inducements that Alston out-lined were, he contended, formulated to allow Zia to abandon his nuclear

---

[125] FCO to UKE The Hague, 'Pakistan Nuclear', February 18, 1980, TNA FCO37/2370.

[126] Alston to Whyte, 'Pakistan Nuclear', March 18, 1980, TNA FCO37/2370.

[127] Whyte to Alston, Handwritten Notes Appended to 'Pakistan Nuclear', March 18, 1980, TNA FCO37/2370, 2.

aspirations without losing face.[128] In response Thomson emphasised that the package should be presented in such a way as not to upset the Indians. Perhaps a similar offer could be made to them?[129] Forster—while supporting the basic idea—suggested that issues of prestige and national security made Zia unlikely to change his nuclear position, arguing that the invasion made it the wrong time for such an approach and Indian involvement could exacerbate anti-Western feeling in Pakistan.[130] SAD echoed Forster's opinion, agreeing that it was entirely the wrong time to pressure Pakistan. One optimistic note was that Zia had so far failed to extract the economic support he expected from other Islamic states. This might offer an area for negotiation, although observers thought it unlikely that Zia would accept any deal with nuclear strings attached.[131]

The opportunity to explore Pakistani attitudes came in mid-June when Agha Shahi visited London. Shahi expounded upon the Pakistani-hosted Islamic Conference's success in creating a united front against the USSR and expressed satisfaction with the extent of Western support. However, too much Western involvement, Shahi contended, would spoil Pakistan's non-aligned credentials and remove the *appearance* of independent action.[132] The Pakistani Foreign Minister submitted a list of military equipment he wished to obtain from Britain and Carrington promised to attend to it as quickly as possible. Thus Britain's most senior diplomat, by accepting the Pakistani request for military aid, acquiesced to the policy of mitigating the nuclear programme and de-linking arms and non-proliferation. The only comment on nuclear issues was when the Pakistani Ambassador stated that the BBC's 'Muslim bomb' documentary had been "inaccurate and unhelpful".[133] The government was reticent to involve itself and suggested that the Pakistani Embassy contact the BBC directly.[134] In conversation with Thatcher, Shahi raised many of the same

---

[128] Alston to Forster, 'Pakistan Nuclear', April 18, 1980, TNA FCO37/2370.

[129] Thomson to Alston, 'Pakistan Nuclear', May 6, 1980, TNA FCO37/2370.

[130] Forster to Alston, 'Pakistan Nuclear', May 26, 1980, TNA FCO37/2370.

[131] Archer to Alston, 'India and Pakistan Nuclear', June 2, 1980, TNA FCO37/2370.

[132] 'Call by Pakistan Foreign Minister', June 17, 1980, TNA PREM19/320, 1–3.

[133] Ibid., 3

[134] Ibid., 8.

points but—bringing up a subject that Whitehall had thought long buried—expressed worry about India's new fleet of British-made Jaguar strike aircraft.[135] The lack of reflection on nuclear matters in these discussions, and those conducted by Douglas Hurd in September, demonstrate how much events in Afghanistan had pushed the issue aside. Although attempts to retard the Pakistani programme continued through the means of export controls and international diplomacy, just as the FCO had suggested, direct pressure on the Zia government effectively ceased.[136]

Cessation of pressure on Pakistan did not halt British government consideration of non-proliferation. Following multilateral discussions at the June NATO summit in Ankara, Carrington tasked the JNU with drafting a ministerial briefing on non-proliferation. The JNU took the opportunity to try and resist the mitigationist policy by offering a gloomy, alarmist view of the terrain that drew on prevailing Islamic bomb ideas. The report opened bleakly; the price for stopping a proliferation cascade was eternal vigilance as, "every act of proliferation increases pressure on others to follow suit".[137] As far as Middle Eastern proliferation was concerned, this issue was hived off from the challenge of Pakistan. In the Middle East the peace process was the best guarantee of non-proliferation. In Pakistan the best guarantee was the resolution of Indo-Pakistani tension.[138] Despite this compartmentalisation the JNU speculated about the Libya-Pakistani connection, but again no solid evidence was provided. The report concluded by offering the same solutions that had been discussed *ad infinitum*: export controls, international consensus, access to reprocessing facilities, and safeguards.[139] Alston's subsequent paper was similarly alarmist about a proliferation cascade rooted in Pakistan. Stopping Pakistan was considerably more urgent than action against India, especially as it could prevent the introduction of nuclear weapons to the Middle East.[140]

---

[135] Ibid.

[136] FCO to UKE Islamabad, 'Afghanistan', September 9, 1980, TNA PREM19/387.

[137] Alston to Moberly, 'Non-proliferation Issues and Prospects', July 23, 1980, TNA FCO37/2371, 1.

[138] Ibid., 2.

[139] Ibid., 7.

[140] Alston to Coles, 'South Asia: Nuclear Proliferation Issues', July 31, 1980, TNA FCO96/1106, 1.

Despite Alston's and the JNU's efforts, overt non-proliferation did not make it back on to the agenda. Although Britain remained a driving force in the field of export controls, Thatcher's government followed the US lead when it came to refraining from explicit pressure on Pakistan. Zia's obduracy, US lethargy, and the exigencies of a renewed Cold War had all militated against strong action against Islamabad's nuclear aspirations. British and American analysts were correct in one particular respect; the chances of persuading Zia to change direction and eschew nuclear capability were close to zero.

RE-CALIBRATION

The quest for a new approach to the nuclear dilemma was not confined to London. A US survey on global attitudes towards non-proliferation revealed that a majority of non-communist nations considered America an unreliable nuclear partner and, with an overemphasis on supply and the fuel cycle (typified by INFCE and the anti-reprocessing stance), had failed to adequately focus on potential bomb-makers.[141] Concurrent with the British JNU-SAD discussions, the elite PRC attempted to thrash out a revised Pakistan policy. Vance—in one of his last contributions to the debate before resigning from office—stated that the non-proliferation objectives laid out in 1977 still stood, but that there were questions over the methods used to achieve the desired results.[142] Smith then outlined three options: to continue on the present course; to accept a universal, non-discriminatory code on nuclear trade, as advocated by the UK in 1979; or to promote a regime that required safeguards on all new export agreements, indefinite deferral of reprocessing, international plutonium storage, multinational control of sensitive facilities, and improved international cooperation when dealing with 'problem countries'. The DoD, Joint Chiefs of Staff, ACDA, and the State Department all supported option three. At the same meeting the NSC opposed any leniency towards India. The appearance of preferential treatment over fuel supplies to

---

[141] 'March/April 1980 Non-proliferation Survey', April 11, 1980, JCPL, RAC NLC-34-65-14-1-7.

[142] The final straw was the failed attempt to rescue the Iranian hostages, but Vance had bitterly contested other major foreign policy issues with Brzezinski.

Tarapur would badly affect the US position with Pakistan and undercut global non-proliferation efforts.[143]

The NSC—in the process of establishing hegemony over foreign policy decision-making as part of the tussle between Brzezinski and Vance—was dubious about Smith being assigned leadership. As far as the NSC was concerned this would produce a one sided, non-proliferationist report for the President.[144] The NSC, rationalising the South Asian proliferation predicament in terms of the prevailing mitigationist viewpoint, argued that proliferation was a long-term challenge that was subordinate to more immediate concerns.[145] All this came at a time when the proliferation problem was deepening. The CIA argued that the Pakistanis believed (quite perceptively, as it turned out) that the USA was resigned to their having nuclear weapons capability, while India was seen as being determined to move forward with continued PNE testing. The latter thus further increased Pakistani resolve to obtain atomic weapons.[146]

As these discussions were taking place, moves were afoot to reassess the non-proliferation policy in the INFCE meeting's wake. Smith had canvassed the opinions of key congressional figures and received a lengthy response from Bingham and Zablocki. Both politicians feared the effects of a Middle Eastern, Islamic proliferation cascade that would have knock-on effects in South Asia. They argued:

If Pakistan acquires nuclear weapons, India might in turn increase the level of her own weapons work or perhaps try militarily to halt the Pakistani program. And so too might the Soviet Union, which could in any event be looking for a pretext for adventure in Pakistan. The acquisition of weapons by Libya would drastically increase the pressure on our

---

[143] 'Policy Review Committee Meeting', April 9, 1980, JCPL, RAC NLC-15-108-2-3-0, 2-4.

[144] Oplinger to Brzezinski, 'Last Week's PRC on Non-proliferation', April 14, 1980, JCPL, RAC NLC-15-108-2-3-0.

[145] 'Persian Gulf–South West Asia Region', April 29, 1980, JCPL, ZBC, Geographic File, Southwest Asia/Persian Gulf, Box 15, 3.

[146] Special Assistant for Nuclear Proliferation Intelligence to DCI, 'Warning Report—Nuclear Proliferation', April 30, 1980, USPQB, Doc.47, 2-3.

friends the Egyptians and the Sudanese as well as creating other possibilities too numerous even to formulate. One could go on in this fashion almost indefinitely.[147]

Amongst the worry Bingham and Zablocki had a point; the impact of Pakistani acquisition of nuclear weapons was unknown. Which way would India jump? What would the USSR do? Would the weapons end up in Libyan hands? Carter was also subject to perceptive accusations from congressional supporters of non-proliferation that, in light of Pakistan and the nuclear shipments to India, he had abandoned his anti-nuclear policy.[148]

The Carter administration sought to recalibrate the non-proliferation policy in light of a changed global situation, a reinvigorated Cold War, congressional pressure, and allied attitudes. Particularly in the case of Pakistan, this remained one of the administration's preoccupations in the last few months of Carter's time in office. The NSC saw difficulties ahead, especially in the area of relations between key allies in Europe and Asia. Developing world countries were concerned about energy security, and major nuclear states—Britain and France—were pushing ahead with extensive reprocessing programmes. Set against this backdrop was an abject lack of progress on major non-proliferation challenges, such as those involving Pakistan and South Africa.[149] In fact, the recommendations offered for approval fell into the 'more of the same' category: deferment of reprocessing, strict supplier controls, and full-scope safeguards.[150] Even after all the experiences since 1977 little new was suggested. In the White House, Gus Speth—Carter's anti-nuclear environmental affairs adviser—suggested a stronger stance on non-proliferation and a delay in the review of policy until after the November election.[151] Speth's suggestions were—like the

[147] Bingham and Zablocki to Smith, 'Observations on State Department Draft Post-INFCE Planning Paper', May 9, 1980, NARA, RG59, Records of Edmund Muskie (hereafter REM), Box 3, 3–4.

[148] Atwood to Muskie, 'The Congressional Agenda: Issues and Strategies', May 16, 1980, DNSA, NP01781, 9–10.

[149] PRC Presidential Decision Paper, 'Nonproliferation Planning Assumptions', May 12, 1980, JCPL, RAC NLC-34-65-14-2-6, 2

[150] Ibid., 5–6.

[151] Speth to Muskie, 'Attached Non-proliferation Policy Memorandum', June 4, 1980, NARA, RG59, REM, Box 3.

departed Cyrus Vance's—founded in a belief that Carter's 1977 non-proliferation policies were still relevant. Speth offered a detailed policy platform that mandated strict adherence to the original principles propounded by Carter and layered on further domestic and international restrictions.[152] Like Smith, Alston, and other non-proliferation advocates, Speth found his entreaties ignored.

In the midst of debates over non-proliferation, the State Department assessed Pakistani objectives as getting the Soviets out of Afghanistan as quickly as possible, minimising the USSR's incentive to put pressure on Pakistan, and preventing Soviet and Indian interests from coalescing.[153] Pakistani attitudes were not the only thing considered when addressing the nuclear problem. Connections with the wider Islamic world were vital. While the Islamic Conference continued to condemn Soviet aggression, it criticised the USA for not doing enough to promote peace in the Middle East.[154] Resolving the Palestinian situation was seen as vital in bolstering Islamic support for the USA in Afghanistan. The US Embassy in Singapore argued that a lack of American cultural connection to the "extended Middle East", coupled to the rise of political Islam—that incorporated social radicalism and anti-superpower "xenophobia"—was a major new factor.[155]

Even without the impact of wider political and cultural factors, the chances of persuading Zia to veer away from the nuclear option, or for greater legal flexibility on nuclear matters, were bleak. Yet again, the Symington Amendment provided a seemingly insurmountable obstacle. The NSC asked if there was any way Carter could use executive power to bypass the restrictions? Furthermore, the Tarapur situation had agitated non-proliferationists in Congress, making a legislative exception for

[152] Speth to Muskie, 'Implementation of the President's Non-proliferation Policy', June 4, 1980, NARA, RG59, REM, Box 3.

[153] State Department Bureau of Intelligence and Research, 'Pakistan faces a Dangerous and Puzzling World', May 20, 1980, DNSA, AF00947, 1.

[154] Muskie to Carter, 'Memorandum for the President', May 22, 1980, JCPL, RAC NLC-128-15-5-11-6, 2-3.

[155] USE Singapore to the White House, 'The Extended Middle East', July 24, 1980, JCPL, ZBC, Subject File Box 43, Weekly Reports [to the President], 121–135 [12/79–4/80].

Pakistan unlikely.[156] Vance's replacement, Edmund Muskie, contended that bypassing the Symington Amendment would show greater US support for Pakistan. Harking back to Vietnam, Muskie likened it to "the Tonkin Gulf Resolution of the Middle East".[157] David Newsom noted that the USA could help the Pakistanis with military equipment, if the Saudis footed the bill, but on the nuclear front the administration should expect no help from Islamabad.[158] There were no immediate plans to increase the aid offer to Zia, such were the Symington Amendment's restrictions. The administration was happy to sell weapons, but could not legally provide the funds to the cash-strapped Pakistanis. Saudi Arabia was, as always, a potential source of funding, but despite Pakistani efforts to leverage Islamic brotherhood, Saudi money for guns had not yet been forthcoming.[159] Here were prime examples of two key factors. Firstly, prior US enthusiasm for nonproliferation was now severely hampering efforts to constructively aid the nation that had actually *provoked* much of the concern about proliferation. Secondly, the NSC and State Department requests that Carter use executive power to override the Symington Amendment demonstrate how the policy of acquiescence and mitigation—originating in mid-1979 and then further influenced by the Afghanistan situation—had become the mainstream of the administration's attitude towards Pakistan's nuclear ambitions.

By autumn, with the US Presidential election campaign in full swing, prospects for a breakthrough with Pakistan had not improved. Christopher noted that there was virtually no chance of getting a waiver to the Symington Amendment. The disagreements between the administration, the NRC, and Congress over Tarapur had poisoned that well. Christopher also pointed out how much impact media coverage of Pakistan's nuclear project had been having on congressional willingness to authorise military aid. Stories about Swiss supplies and Pakistani purchasing had, and were continuing to have, a negative impact on the prospects for gaining

---

[156] Thornton to Brzezinski, 'PRC On Pakistan, July 17', July 16, 1980, JCPL, RAC NLC-24-103-1-2-8, 2.

[157] 'Minutes–PRC Meeting on Pakistan', July 17, 1980, JCPL, RAC NLC-24-103-1-5-5, 6.

[158] Ibid., 7.

[159] 'Minutes–PRC Meeting on Pakistan', July 22, 1980, JCPL, RAC NLC-33-10-27-1-4, 2, 7.

congressional agreement to a waiver to the Symington Amendment. Only a month before the election, and despite plans to sell the advanced F-16 fighter aircraft and other military materials to Pakistan, he stated, "nothing can be done until next spring".[160] Next spring, something *would* be done, but not by Carter and his team. On October 3 Zia visited the White House. As Dennis Kux notes, by this point Zia had concluded that Reagan would win the election, so saw no need to take the initiative on security assistance. When, in the meeting's final minutes, Carter offered Zia the F-16, the Pakistani President casually noted that as Carter was undoubtedly busy with his election campaign, such matters could wait.[161] Despite positivity in the White House and a remarkable comeback in opinion polls in the week before polling day, Ronald Reagan proved the more adept campaigner and won a landslide victory. On November 4 Carter became the first incumbent since Herbert Hoover in 1932 to fail to win a second term.[162] Foreign policy had played a significant role in ensuring Carter's election in 1976; in 1980 foreign policy was a primary cause of his downfall.

CONCLUSION

For America and Britain 1980 was a low-water mark in efforts to forestall Pakistani nuclear capability. The Carter and Thatcher governments were committed to rigorous controls on nuclear-related exports and expended considerable political capital on efforts to gain international acceptance of their standpoint. However, the global nature of Islamabad's nuclear purchasing programme, and the obduracy of fellow supplier states such as West Germany and Switzerland, diminished the crusade's effectiveness. Diplomatic efforts to persuade Zia to rein in his ambitions were, if anything, even less successful. The crisis in Afghanistan, Pakistani recalcitrance, and the need for all sides to keep up appearances precluded success. Carter, who began with such high hopes for his non-proliferation policy, found himself hoisted by a petard of his own making. The UK and the USA discovered that, faced by a state determined to acquire nuclear weapons, diplomacy and aid carried little weight. This was especially true

---

[160] 'PRC: Pakistan', September 29, 1980, JCPL, RAC NLC-33-10-31-3-7, 2-3.
[161] Kux, *Disenchanted Allies*, 254.
[162] Kaufman and Kaufman, *Presidency of James Earl Carter*, 235.

after the invasion of Afghanistan, when Pakistan was in the front line of a new Cold War. In public the Pakistani programme was continually linked to fear of an Islamic bomb and the spread of nuclear weapons to the Middle East. If nothing else, this was the true public legacy of Bhutto's and Zia's nuclear ambitions. In international diplomacy, once more, it was the issues of credibility and face that played the more significant role in non-proliferation discussions.

CHAPTER 9

Conclusions

The story of American and British involvement in Pakistan's nuclear programme did not end with Jimmy Carter's departure from the White House. Like Carter, Ronald Reagan placed Pakistan at the heart of his anti-Soviet effort in Afghanistan. The major change was a willingness—bolstered by Zia's assurances over production, proliferation, and, most importantly, testing—to offer substantial military and economic aid. The Pakistani leader did not turn down Reagan's 1981 offer of a $3.6 billion package of arms, money, and food as "peanuts". Under Reagan, US-Pakistani cooperation over Afghanistan ballooned into a sizeable operation supporting the Afghan mujahedeen against the Soviets. Margaret Thatcher wholeheartedly supported Reagan's policy, but continued to monitor the clandestine Pakistani nuclear programme. British officials remained active in trying to combat the covert enrichment project, although by the mid-1980s the project was well on the way to success.

Discomfort over Islamabad's nuclear ambitions did not entirely dissipate. Congressmen, senators, and members of parliament expressed unease and anger over the Pakistani nuclear situation. Senators Alan Cranston (D-CA) and John Glenn (D-OH), and Congressman Stephen Solarz (D-NY) were three of the most prominent figures to raise the issue.[1] Solarz went so far as to

---

[1] Dennis Kux, *Disenchanted Allies: The United States and Pakistan, 1947–2000* (Washington D.C.: Woodrow Wilson Center Press, 2001), 275–279.

© The Author(s) 2017

M.M. Craig, *America, Britain and Pakistan's Nuclear Weapons Programme, 1974–1980*, Security, Conflict and Cooperation in the Contemporary World, DOI 10.1007/978-3-319-51880-0_9

have an amendment to the Foreign Assistance Act passed (the 'Solarz Amendment'), barring aid to any country exploding a nuclear device. Thus, the 'no test' paradigm of acquiescence to Pakistani nuclear ambitions became US law.[2] In the UK the indefatigable Tam Dalyell continued to press successive British governments on the issue for the rest of his distinguished parliamentary career.

Individuals within the Reagan administration also remained anxious about Zia's aspirations. In 1982 Secretary of State George Schulz warned that Zia was likely to break his promises and proliferate to "unstable Arab countries".[3] Four years later ACDA Director Kenneth Adelman argued that Zia was continually lying regarding enrichment, and by 1987 senior State Department officials warned that Zia was approaching a "threshold which he cannot cross without blatantly violating his pledge not to embarrass [i.e. not test] the President".[4]

In the late 1990s and early 2000s Western relationships with the Pakistani nuclear programme came under renewed scrutiny because of the 'Test War' between Islamabad and New Delhi and the exposure of the A.Q. Khan network. Pakistan's efforts to attain a uranium enrichment capability by clandestine means were ultimately successful. In 1998 Islamabad carried out its first nuclear tests and joined the ranks of admitted nuclear weapon states. Khan, the man who allowed this to happen by stealing centrifuge plans in 1975, went on to take increasing advantage of global trade, selling centrifuge designs to Iran, Libya, and North Korea, with the tacit permission of the Pakistani state.

This book has given a detailed analysis of American and British responses to the Pakistani nuclear weapons programme in the 1970s. Despite problems created by British commercial interests, London was Washington's staunchest global anti-proliferation partner. When compared to the often

[2] Ibid., 260.

[3] Schulz to Reagan, 'How Do We Make Use of the Zia Visit to Protect Our Strategic Interests in the Face of Pakistan's Nuclear Weapons Activities', November 26, 1982, Nation Security Archive Electronic Briefing Book 'New Documents Spotlight Reagan-era Tensions Over Pakistani Nuclear Problem' (hereafter RTPP), www2. gwu.edu/~nsarchiv/nukevault/ebb377 (accessed October 29, 2013), Doc.16.

[4] Adelman to Poindexter, 'Pakistan's Nuclear Weapons Program and US Security Assurances', June 16, 1986, RTPP, Doc.20; McGoldrick to Negroponte, 'Pakistan', April 9, 1987, RTPP, Doc.23.

acrimonious US relationships with France and West Germany, the US-UK 'non-proliferation special relationship' was functional and productive. The case of Pakistan also demonstrates that the American and British non-proliferation policy was characterised by a remarkable resistance to partisan politicisation. Across three American presidencies and three British prime ministers, Pakistan's nuclear ambitions in the main failed to become an issue utilised for party-political ends. Apart from the laxity of the Nixon years and the early Ford period, Washington and London took non-proliferation seriously and the issue was one that transcended political boundaries. In the case of the 1976 US presidential election it was precisely because of the seriousness of the issue and the casualness with which it had been treated by Nixon, Ford, and Kissinger that proliferation became a political issue.

Gerald Ford's and Henry Kissinger's acquiescence to Pakistani demands for arms while receiving nothing in return demonstrated that regional security concerns overrode non-proliferation concerns, despite signs that Pakistan was pursuing a military atomic programme. Cognisant of Pakistan's stubborn determination to attain nuclear capability, pressured by a Congress awakening to the dangers of proliferation, and provoked by Carter's foregrounding of the problem, the Ford administration switched from broad tolerance to an active engagement in non-proliferation policy.

British investigations revealed the clandestine Pakistani purchasing networks feeding into a covert enrichment programme. In 1976–77, Callaghan's government did not lack the will to tackle this issue, but the incompleteness of existing proliferation controls, a dubious legal case for preventing exports, the recalcitrance of nuclear supplier states, the complexity of Pakistan's network, and a desire not to lose credibility through public embarrassment hampered effective action. Evidence that emerged in 1978 strengthened the case that Pakistan was pursuing a clandestine enrichment programme as part of its nuclear quest. Here the non-proliferation special relationship became fully functional, when the British government—supported and aided by the United States—commenced a campaign conducted in the diplomatic shadows to marshal a united front amongst nuclear supplier states to prevent Pakistani acquisition of sensitive technologies.

In order to challenge this programme there was a division of labour in the non-proliferation relationship. The United States took the lead in diplomatically combating the French reprocessing plant deal, the overt element of the Pakistani programme. In turn, Britain led the campaign against the covert elements of Islamabad's nuclear ambitions—the enrichment project and its associated purchasing activities. First Britain, and then America, took the clandestine Pakistani uranium enrichment programme seriously.

Although created by and for a state, Pakistan's clandestine weapons project and Khan's proliferation network used an international array of non-state actors—money-hungry corporations, shadowy intermediaries, shady arms dealers, amoral industrialists, and duplicitous scientists—to achieve their aims. An examination of the Pakistani case reveals not just the international, state-to-state elements of non-proliferation, but also the transnational elements. The Pakistani programme would not have emerged and developed the way it did prior to the 1970s. And it was the globalised, transnational elements of the Pakistani programme that made it so difficult to combat.

The complementary but different approaches of the USA and the UK tell us much about the relative standing of each state. The Americans were able to put significant pressure on France in a very public way because of their status as the foremost Western power—even in a post-Vietnam, post-Watergate era of diminished respect for the USA. The fact that Britain took the route of subtle diplomatic action confirms the decline of Britain to a middle-ranking power. In many ways this lack of genuine global standing gave Britain greater freedom to act and allowed it the luxury of being able to ignore the requirements of non-proliferation when necessary, freed of the responsibilities associated with American superpower status. However, the period also demonstrates a lack of influence where it mattered: in Islamabad.

Britain and America were forced to balance relations with other world powers in pursuit of their goal. The greatest success for non-proliferation diplomacy was the campaign against the French reprocessing plant. Although diplomatic pressure was a component of this, the French decision that non-proliferation mattered more than the reprocessing plant contract, and the concurrent realisation that they could make more money selling nuclear services when compared with selling the technology itself, was not entirely influenced by allied pressure. Additionally, Carter's decision to exempt Britain, France, and other Western European states from his reprocessing ban allowed France to maintain the domestic industry required to sell these reprocessing services. This diplomatic 'victory' was not without its problems, as when Valery Giscard d'Estaing's government demonstrated an unwillingness to publicly proclaim its decision or to break the contract with Pakistan.

The final French cancellation of the reprocessing plant contract gave the US government a chance to have meaningful discussions with Pakistan over issues of aid and arms sales that had been hampered by Islamabad's determination to acquire the reprocessing plant. However, in attempting to balance non-proliferation and conventional arms control policies, Carter only succeeded in pushing Pakistan away from the United States. There was tension between Carter's multifarious policy aims; the desire to achieve non-proliferation goals conflicted with other global policy imperatives, such as regional security and human-rights commitments.

Zia's unwillingness to abandon the nuclear programme prompted a policy change that had previously gone unnoticed. During 1979 the Carter administration and Thatcher's government shifted from a commitment to the *prevention* of proliferation in South Asia, to a policy of the *mitigation* of proliferation, centred on the quest for assurances from both Pakistan and India not to undertake nuclear tests. America and Britain together strove to maintain the appearance of pursuing non-proliferation through requests for Pakistan not to undertake nuclear testing, the most public demonstration of nuclear attainment. If Pakistan *were* to test, it would throw the entire trilateral relationship into doubt by publicly humiliating the US and UK governments when they had invested significant political capital in anti-proliferation activities, and thus reduce the ability to take anti-proliferation action on the global stage. This change illustrates the difficulties of pursuing non-proliferation when faced with a determined and skilful state apparatus and the role of credibility and saving face in influencing non-proliferation policy. Media stories about the Khan Affair, the Islamic bomb, and the clandestine purchasing project heaped greater pressure on Washington and London to take stronger action against Pakistan. This again emphasises the significance of recalcitrance on the part of key supplier states in the non-proliferation arena. Despite the problem created by supplier state reluctance, the clandestine purchasing programme was the most promising avenue for non-proliferation success and the one area where the UK and US governments did continue to make genuine efforts to retard the Pakistani nuclear programme through export controls and international diplomacy.

Despite the broad harmony of the US-UK 'non-proliferation special relationship', the partnership was not trouble free. Britain's economic situation and the impact this had on arms sales and the nuclear industry caused problems. Throughout the period, British efforts to sell the nominally nuclear-capable Jaguar strike aircraft to India were a persistent problem for sub-continental non-proliferation policies. In pursuing this sale the Wilson and Callaghan governments compartmentalised arms sales and nuclear issues and indicated that, although India had joined the 'nuclear club', the consequences were slight. Even though Callaghan's government was committed to non-proliferation—David Owen being particularly dedicated—commercial considerations frequently trumped non-proliferation objectives. Britain's economic situation in the 1970s was the main concern. Without massive external deals the teetering British aerospace industry would collapse. With Sweden out of the picture the only other competitor to the Jaguar was the French Mirage. British officials could not countenance withdrawing from the bidding process and gifting the French a lucrative contract. The consequence of this was frustration in Washington at British intransigence over Jaguar.

Policymakers in the Carter administration repeatedly attempted to persuade London and New Delhi to abandon the Jaguar negotiations. The USA had seen some success in pressuring Sweden to drop out of the tendering process and attempted to persuade their British allies to do the same. While Jaguar would not represent a significant increase in Indian capability, it was the symbolism of Pakistan's rivals being equipped with modern weapons that could potentially act as nuclear delivery systems (while the United States was using similar weapon systems as 'carrots and sticks' in its dealings with Islamabad) that was the underlying problem. The British government—faced with the potential loss of a lucrative arms deal—wilfully ignored the non-proliferation complications. The ways in which Britain and America were using arms sales were quite different. For Britain, arms sales were an economic concern divorced from non-proliferation action. The sheer commitment to the Jaguar deal demonstrates that, in the face of an economic crisis, the British government was quite willing to subordinate non-proliferation to economic necessity. For the executive branch in the United States, arms sales were never a commercial concern; rather, they were a bribe or a bludgeon used to try to persuade Islamabad to abandon the nuclear programme.

Britain's domestic nuclear industry also complicated non-proliferation diplomacy, with the go-ahead to build the THORP weakening British non-proliferation standing in Pakistani eyes. While the FCO had resisted pressure from the Departments of Trade and Industry to take a softer line on export controls in favour of a stronger anti-proliferation stance, there was no such resistance in the face of the economic realities of the THORP. Stronger export controls might well cause the loss of a few million pounds worth of business, but the potential loss of the revenue generated by the THORP was several orders of magnitude greater. Similar to the Jaguar deal, non-proliferation concerns had to be subservient to the commercial interests of the state when faced with the potential loss of a massive revenue-generating facility. Jimmy Carter was willing to abandon commercial considerations in favour of non-proliferation, not so the Wilson and Callaghan governments. When billions of pounds worth of military aircraft or reprocessing services were at risk, economic self-interest trumped proliferation concerns. With this in mind, the United States was in many ways the more responsible global power and Britain merely confirmed its status as a regional power, happy to act when its own interests remained unthreatened.

It is intriguing to consider whether the Anglo-American alliance against Pakistan from 1974 to 1980 was indeed a 'special relationship'. It is possible to argue that, rather than a special relationship, what can be seen in this period was, at least in part, a marriage of convenience. Out of the United States' major Western European allies of France, West Germany, and the UK, Britain was the only one of the three not to be engaged in major nuclear deals with developing world states. Thus, it could be argued that although there was a significant relationship in play, there was also the simple fact that Britain was the only partner available to the United States when taking action against Pakistan.

When America and Britain engaged in anti-proliferation activity on the sub-continent, the historical record runs counter to popular belief. Contrary to alarmist popular analyses, the Islamic bomb meme was seldom an important factor in American and British non-proliferation policymaking. While the media and peripheral political figures placed great store in the reality of an Islamic bomb, at government level the concept did not affect actual policy directed against Pakistan. Although some individuals feared—and institutions such as the CIA investigated—an Islamic bomb, the majority opinion was that it was a propaganda rather than a policy challenge. Thus, while Washington and London battled popular perceptions of Pakistan's nuclear aspirations as a pan-Islamic project, in private policymaking carried on as normal. What this propaganda problem did do was change public perceptions of Pakistani atomic aspirations geopolitically, from being a regional South Asian problem to a much wider problem that encompassed the Middle East, with all the challenges that implied. For the American and British governments, it was the *publicity* surrounding pan-Islamic nuclear capability and the *perception* this created, not belief in the reality of an Islamic bomb that created problems. In truth, the Islamic bomb exerted little—if any—influence on policymaking regarding Pakistan.

Much more important to non-proliferation policy and action were ideas of credibility and face. America, Britain, France, and Pakistan all— at different times and for different reasons—sought to maintain credibility in one way or another. The British government's initial inaction regarding inverters was, in part, based on a desire not to create a humiliating diplomatic incident. Even more significantly, the entire strategy of mitigation and the requests to Zia not to test a nuclear device were founded in a need to maintain American and British credibility in global non-proliferation affairs. Thus, it was ideas of national standing and the ability to influence on the global stage that exerted the greater pressure on policymakers. This came about not because of fears of Islamic proliferation, but because of the determination and stubbornness of Zia and his government.

However, it is the Islamic bomb paradigm that has become an abiding feature of discussions surrounding the nuclear ambitions of Muslim states. Throughout the 1980s, the 1990s, and into the twenty-first century the threat of an Islamic bomb repeatedly emerged as shorthand for an aggressive, confrontational Islamic quest for power. The phrase crops up in the traditional media, on blogs, in podcasts, and on discussion sites such as Reddit. To cite just one example, in advance of Israeli Prime Minister Binyamin Netanyahu's March 2015 address to the US Congress, an editorial in the conservative *Washington Times* commented:

> Mr Netanyahu has the opportunity to talk in plain speech with no equivocation about the threat that Iran, armed with the Islamic bomb, poses to the survival of the Jewish state and perhaps the United States as well. Perilous times call for strong measures, and these are perilous times.[5]

This book has thrown new light on a key moment in non-proliferation history. By so doing, it illuminates the roots of cultural imaginings, brings Britain into play as a significant force in global non-proliferation policy,

---

[5] 'The World in Peril', March 2, 2015, *The Washington Times*, http://www.washingtontimes.com/news/2015/mar/2/editorial-benjamin-netanyahu-to-expose-obamas-iran (accessed 3 March, 2015).

and moves the timeline of Western acquiescence to Pakistan's nuclear ambitions further back in time. Nuclear non-proliferation remains a contested topic in foreign affairs. The nuclear programmes of Iran and North Korea have also provoked fear, outrage, and no small amount of sabre rattling on all sides. However, if this study has demonstrated anything, it is that framing nuclear proliferation in terms of a Huntingtonian 'clash of civilisations' is both false and diplomatically unhelpful. By eschewing a reliance on this rhetoric it will be possible to comprehend more clearly why certain states seek nuclear capability. Hopefully a more nuanced understanding of the history of proliferation and non-proliferation will contribute to more fruitful, productive dialogue over the issue in the future.

Primary Sources

Unpublished Documentary Sources

United Kingdom

BBC Written Archives Centre, Caversham
  Document Archives
National Archives (Public Record Office), Kew
  Records of the Cabinet Office
    CAB 128: Minutes
    CAB 129: Memoranda
    CAB 130: Miscellaneous Committees: Minutes and Papers
    CAB 148: Cabinet Office: Defence and Oversea Policy Committees
    and Sub-Committees: Minutes and Papers
  Records of the Department of Energy
    EG 8: Atomic Energy Division: Registered Files
  Records of the Foreign & Commonwealth Office
    FCO 21: Far Eastern Department: Registered Files
    FCO 37: South Asia Department: Registered Files
    FCO 66: Disarmament Department and Arms Control and
    Disarmament Department: Registered Files
    FCO 96: Energy Department: Registered Files
  Records of the Ministry of Defence
    DEFE 13: Private Office: Registered Files

© The Author(s) 2017   287
M.M. Craig, *America, Britain and Pakistan's Nuclear Weapons
Programme, 1974–1980*, Security, Conflict and Cooperation
in the Contemporary World, DOI 10.1007/978-3-319-51880-0

DEFE 19: Central Defence Scientific Staff and predecessors:
Registered Files and Papers
Records of the Prime Minister's Office
PREM 16: Correspondence and Papers, 1974–1979
PREM 19: Correspondence and Papers, 1979–1997
Records of the Treasury
T 362: Overseas Finance (Exports) Division
T 390: Domestic Economy Sector, Industrial Policy Files

United States

Gerald R. Ford Presidential Library, Ann Arbor, MI
Philip W. Buchen Files
James M. Cannon Files
National Security Adviser Files
  Memoranda of Conversations
  Presidential Agency File
  Presidential Correspondence File
  Presidential Country File for Middle East and South Asia
  Presidential Briefing Materials for VIP Visits
  Presidential Subject File
  Presidential Transition File
  White House Situation Room, Evening Reports from the NSC Staff
  Study Memoranda and Decision Memoranda
  Trip Briefing Books and Cables of Henry Kissinger
National Security Council Institutional Files
Remote Archives Capture Files
Glenn R. Schleede Files
Jimmy Carter Presidential Library and Museum, Atlanta, GA
  Carter White House Central Files
  National Security Staff Files
  Office of the Chief of Staff Files
  Records of the Domestic Policy Staff
  Records of the Office of Congressional Liaison
  Records of the Office of the National Security Adviser
  Records of the Speechwriter's Office
  Records of the Office of the Staff Secretary
  Remote Archives Capture System

Subject File
Zbigniew Brzezinski Collection
National Archives and Records Administration, Archives II, College Park, MD
  Record Group 59: General Records of the State Department
    Records of Warren Christopher
    Records of Anthony Lake
    Records of Edmund Muskie
National Security Archive, George Washington University, Washington D.C.
  Nuclear Non-proliferation Unpublished Collection

PUBLISHED DOCUMENTARY SOURCES

United Kingdom

Churchill College Cambridge, Churchill Archives
  British Diplomatic Oral History Programme
  (www.chu.cam.ac.uk/archives/collections/BDOHP)

Hansard Online, 1803–2005
(http://hansard.millbanksystems.com)

Margaret Thatcher Foundation Archives
(www.margaretthatcher.org/archive)

The National Archives (Public Record Office) Online Collections
(http://discovery.nationalarchives.gov.uk/SearchUI/Home/OnlineCollections)

United States

Central Intelligence Agency Freedom of Information Electronic Reading Room
(www.foia.cia.gov)
Congressional Record
Declassified Documents Reference System
(http://gdc.gale.com/products/declassified-documents-reference-system)

Digital National Security Archive
(http://nsarchive.chadwyck.com/home.do)

Gerald R Ford Presidential Library online collections
National Security Adviser, Memoranda of Conversations
(www.fordlibrarymuseum.gov/library/guides/findingaid/
Memoranda_of_Conversations.asp)

National Archives and Records Administration, Access to Archival
Databases (AAD) system
Wars/International Relations: Diplomatic Records
(http://aad.archives.gov/aad/series-list.jsp?cat=WR43)

National Security Archive *Nuclear Vault* electronic briefing books
6: *India and Pakistan: On the Nuclear Threshold*
(http://www2.gwu.edu/~nsarchiv/NSAEBB/NSAEBB6)
114: *China, Pakistan, and the Bomb*
(http://www2.gwu.edu/~nsarchiv/NSAEBB/NSAEBB114)
155: *National Intelligence Estimates of the Nuclear Proliferation
Problem, the First Ten Years, 1957–1967*
(http://www2.gwu.edu/~nsarchiv/NSAEBB/NSAEBB155/index.
htm)
240: *In 1974 Estimate, CIA Found that Israel Already Had a Nuclear
Stockpile and that "Many Countries" Would Soon Have Nuclear
Capabilities*
(http://www2.gwu.edu/~nsarchiv/NSAEBB/NSAEBB240/index.
htm)
333: *The United States and Pakistan's Quest for the Bomb*
(http://www2.gwu.edu/~nsarchiv/nukevault/ebb333/index.htm)
352: *Non-papers and Demarches*
(http://www2.gwu.edu/~nsarchiv/nukevault/ebb352/index.htm)
377: *New Documents Spotlight Reagan-era Tensions over Pakistani
Nuclear Program*
(http://www2.gwu.edu/~nsarchiv/nukevault/ebb377)
451: *Proliferation Watch; U.S. Intelligence Assessments of Potential
Nuclear Powers, 1977–2001*
(http://www2.gwu.edu/~nsarchiv/nukevault/ebb451)

485: *Israel Crosses the Threshold II: The Nixon Administration Debates the Emergence of the Israeli Nuclear Programme*
(http://nsarchive.gwu.edu/nukevault/ebb485/)
US Department of State
Foreign Relations of the United States (FRUS) series
   FRUS 1969–1976, Volume E8 'Documents on South Asia, 1973–1976'
   (http://history.state.gov/historicaldocuments/frus1969-76ve08)
   FRUS 1977–1980, Vol.XIII, China
   (http://history.state.gov/historicaldocuments/frus1977-80v13)
Wilson Center History and Public Policy Program Digital Archive
  The Non-proliferation International History Project
  (http://digitalarchive.wilsoncenter.org/theme/nuclear-history)

Newspapers

United Kingdom

The *Daily Express*
The *Guardian*
The *Observer*
The *Times*
The *Sunday Times*

United States

The *New York Times*
The *Washington Post*

India

The *Times of India*

SELECT BIBLIOGRAPHY

AUTOBIOGRAPHIES AND MEMOIRS

Bhutto, Zulfikar Ali, *If I Am Assassinated* (New Delhi: Vikas Press, 1979)

Brzezinski, Zbigniew, *Power and Principle: Memoirs of the National Security Adviser, 1977–1981* (London: Weidenfeld & Nicholson, 1983)

Callaghan, James, *Time and Chance* (London: Collins, 1987)

Carter, Jimmy, *Keeping Faith: Memoirs of a President* (London: Collins, 1982)

———, *White House Diary* (New York: Farrar, Straus, and Giroux, 2010)

Moynihan, Daniel Patrick, *Daniel Patrick Moynihan: A Portrait in Letters of an American Visionary* (New York: Public Affairs, 2010)

Owen, David, *Time to Declare* (London: Penguin, 1992)

Thatcher, Margaret, *The Downing Street Years* (New York: HarperPerennial, 1995)

Vance, Cyrus, *Hard Choices: Critical Years In American Foreign Policy* (New York: Simon & Schuster, 1983)

Wilson, Harold, *Final Term: The Labour Government, 1974–1976* (London: Weidenfeld and Nicholson, 1979)

BOOKS

Aldrich, Richard J., *The Hidden Hand: Britain, America and Cold War Secret Intelligence* (London: John Murray, 2001)

Aldrich, Richard J., and Rory Cormac, *The Black Door: Spies, Secret Intelligence, and British Prime Ministers* (London: William Collins, 2016)

© The Author(s) 2017   293
M.M. Craig, *America, Britain and Pakistan's Nuclear Weapons Programme, 1974–1980*, Security, Conflict and Cooperation in the Contemporary World, DOI 10.1007/978-3-319-51880-0

Anthony, Ian, Christer Ahlstrom, and Vitaly Fedchenko, *SIPRI Research Report No.22: Reforming Nuclear Export Controls: The Future of the Nuclear Suppliers Group* (Oxford: Oxford University Press, 2007)

Armstrong, David, and Joseph Trento, *America and the Islamic Bomb* (Hanover: Steerforth Press, 2007)

Bartlett, C. J., *British Foreign Policy in the Twentieth Century* (Basingstoke: Macmillan, 1989)

Basrur, Rajesh M., *South Asia's Cold War: Nuclear Weapons and Conflict in Comparative Perspective* (London: Routledge, 2008)

Bass, Gary J., *The Blood Telegram: Nixon, Kissinger, and a Forgotten Genocide* (New York: Alfred A. Knopf, 2013)

Baylis, John, *Anglo-American Defence Relations 1939–1980*, 2nd edition (London: Macmillan, 1984)

Brands, Hal, *Making the Unipolar Moment: US Foreign Policy and the Rise of the Post-Cold War Order* (Ithaca: Cornell University Press, 2016)

Braut-Hegghammer, Malfrid, *Unclear Physics: Why Iraq and Libya Failed to Build Nuclear Weapons* (Ithaca: Cornell University Press, 2016)

Brenner, Michael J., *Nuclear Power and Non-proliferation: The Remaking of U.S Policy* (Cambridge: Cambridge University Press, 1981)

Botti, Timothy J., *The Long Wait: The Forging of the Anglo-American Nuclear Alliance, 1945–1958* (New York: Greenwood Press, 1987)

Cain, Frank, *Economic Statecraft During the Cold War* (London: Routledge, 2007)

Callaghan, John, *The Labour Party and Foreign Policy* (London: Routledge, 2007)

Campbell, John L., *Collapse of an Industry: Nuclear Power and the Contradictions of U.S. Policy* (Ithaca: Cornell University Press, 1988)

Campbell, Kurt M., Robert J. Einhorn, and Mitchell B. Reiss, *The Nuclear Tipping Point: Why States Reconsider Their Nuclear Choices* (Washington, D.C.: Brookings Institution Press, 2004)

Chakma, Bhumitra, *Strategic Dynamics and Nuclear Weapons Proliferation in South Asia: A Historical Analysis* (Bern: Peter Lang, 2004)

Clausen, Peter A., *Nonproliferation and the National Interest: America's Response to the Spread of Nuclear Weapons* (New York: HarperCollins, 1993)

Cohen, Avner, *Israel and the Bomb* (New York: Columbia University Press, 1998)

Corera, Gordon, *Shopping for Bombs: Nuclear Proliferation, Global Insecurity, and the Rise and Fall of the A. Q. Khan Network* (London: Hurst & Co., 2006)

Craig, Campbell and Sergei Radchenko, *The Atomic Bomb and the Origins of the Cold War* (New Haven: Yale University Press, 2008)

Dallek, Robert, *Nixon and Kissinger: Partners in Power* (New York: HarperCollins, 2007)

Dobson, Alan P., *The Politics of the Anglo-American Economic Special Relationship 1940–1987* (Brighton: Wheatsheaf Books, 1988)

Dumbrell, John, *A Special Relationship: Anglo-American Relations in the Cold War and After* (Basingstoke: Macmillan, 2001)

Engel, Jeffrey A., *Cold War at 30,000 Feet: The Anglo-American Fight for Aviation Supremacy* (Cambridge: Harvard University Press, 2007)

Ferguson, Niall, Charles S. Maier, Erez Manela, and Daniel J. Sargent (eds.), *The Shock of the Global: The 1970s in Perspective* (Cambridge: Belknap Press of the Harvard University Press, 2011)

Frantz, Douglas, and Catherine Collins, *The Man From Pakistan: The True Story of the World's Most Dangerous Nuclear Smuggler* (New York: Hachette, 2007)

Freedman, Lawrence, *A Choice of Enemies: America Confronts the Middle East* (New York: PublicAffairs, 2008)

Freeman, J. P. G., *Britain's Nuclear Arms Control Policy in the Context of Anglo-American Relations, 1957–68* (Basingstoke: Macmillan, 1986)

Ganguly, Sumit, and S. Paul Kapur, *India, Pakistan, and the Bomb: Debating Nuclear Stability in South Asia* (New York: Columbia University Press, 2010)

Gavin, Francis J., *Nuclear Statecraft: History and Strategy in America's Atomic Age* (Ithaca: Cornell University Press, 2012)

Gerges, Fawaz, *America and Political Islam: Clash of Cultures or Clash of Interests?* (Cambridge: Cambridge University Press, 1999)

GhaneaBassiri, Kambiz, *A History of Islam in America* (New York: Cambridge University Press, 2010)

Glad, Betty, *An Outsider in the White House: Jimmy Carter, His Advisors, and the Making of American Foreign Policy* (Ithaca: Cornell University Press, 2009)

Grasselli, Gabriella, *British and American Responses to the Soviet Invasion of Afghanistan* (Aldershot: Dartmouth, 1996)

Greenwood, Sean, *Britain and the Cold War, 1945–91* (Basingstoke: Macmillan Press, 2000)

Hanhimäki, Jussi, *The Flawed Architect: Henry Kissinger and American Foreign Policy* (Oxford: Oxford University Press, 2004)

Haslam, Jonathan, *Near and Distant Neighbours: A New History of Soviet Intelligence* (Oxford: Oxford University Press, 2015),

Haqqani, Husain, *Pakistan: Between Mosque and Military* (Washington, D.C.: Carnegie Endowment for International Peace, 2005)

Haynes, John Earl, Harvey Klehr, and Alexander Vassiliev, *Spies: The Rise and Fall of the KGB in America* (New Haven: Yale University Press, 2009)

Hecht, Gabrielle, *Being Nuclear: Africans and the Global Uranium Trade* (Cambridge: The MIT Press, 2012)

Hegghammer, Thomas, *Jihad in Saudia Arabia: Violence and Pan-Islamism Since 1979* (Cambridge: Cambridge University Press, 2010)

Hersh, Seymour, *The Sampson Option: Israel's Nuclear Arsenal and American Foreign Policy* (New York: Random House, 1991)

Hunt, Michael H., *Ideology and U.S. Foreign Policy* (New Haven: Yale University Press, 1987)

Hymans, Jaques E. C., *The Psychology of Nuclear Proliferation: Identity, Emotions and Foreign Policy* (Cambridge: Cambridge University Press, 2006)

Jeffreys-Jones, Rhodri, *In Spies We Trust: The Story of Western Intelligence* (Oxford: Oxford University Press, 2013)

———, *The CIA and American Democracy*, 3rd edition (New Haven: Yale University Press, 2003)

Jones, Matthew, *After Hiroshima: The United States, Race, and Nuclear Weapons in Asia, 1945–1965* (Cambridge: Cambridge University Press, 2010)

Jones, Rodney W., *Nuclear Proliferation: Islam, the Bomb, and South Asia* (Beverly Hills: Sage, 1981)

Kapur, Ashok, *Pakistan's Nuclear Development* (Beckenham: Croom Helm, 1987)

Kaufman, Burton I, and Scott Kaufman, *The Presidency of James Earl Carter*, 2nd edition, revised (Lawrence: University Press of Kansas, 2006)

Kaufman, Scott, *Plans Unraveled: The Foreign Policy of the Carter Administration* (DeKalb: Northern Illinois University Press, 2008)

Kepel, Giles, *Jihad: The Trail of Political Islam* (London: I.B. Tauris, 2004)

Khan, Feroz Hassan, *Eating Grass: The Making of the Pakistani Bomb* (Stanford: Stanford University Press, 2012)

Knight, Amy, *How the Cold War Began: The Igor Gouzenko Affair and the Hunt for Soviet Spies* (New York: Carroll and Graf, 2005)

Kux, Dennis, *Estranged Democracies: India and the United States, 1941–1991* (Washington D.C.: National Defense University Press, 1993)

———, *Disenchanted Allies: The United States and Pakistan, 1947–2000* (Washington D.C.: Woodrow Wilson Center Press, 2001)

Levy, Adrian, and Catherine Scott-Clark, *Deception: Pakistan, the United States, and the Secret Trade in Nuclear Weapons* (New York: Walker & Co, 2007)

Little, Douglas, *American Orientalism: The United States and the Middle East Since 1945* (Chapel Hill: University of North Carolina Press, 2008)

Lyons, Jonathan, *Islam Through Western Eyes: From the Crusades to the War on Terrorism* (New York: Columbia University Press, 2012)

Maddock, Shane J., *Nuclear Apartheid: The Quest for American Atomic Supremacy from World War II to the Present* (Chapel Hill: University of North Carolina Press, 2010)

Matinuddin, Kamal, *The Nuclearization of South Asia* (Oxford: Oxford University Press, 2002)

McGarr, Paul M., *The Cold War in South Asia: Britain, the United States, and the Indian Subcontinent, 1945–1965* (Cambridge: Cambridge University Press, 2013)

McMahon, Robert J., *The Cold War on the Periphery: The United States, India and Pakistan* (New York: Columbia University Press, 1994)

Melissen, Jan, *The Struggle for Nuclear Partnership: Britain, the United States, and the Making of an Ambiguous Alliance, 1952–1959* (Groningen: Styx Publications, 1993)

Meyer, Stephen M., *The Dynamics of Nuclear Proliferation* (Chicago: University of Chicago Press, 1984)

Mieczkowski, Yanek, *Gerald Ford and the Challenges of the 1970s* (Lexington: University Press of Kentucky, 2005)

Moran, Christopher, *Company Confessions: Revealing CIA Secrets* (London: Biteback, 2015)

Morgan, Kenneth O., *Callaghan: A Life* (Oxford: Oxford University Press, 1997)

Muehlenbeck, Philip (ed.), *Religion and the Cold War: A Global Perspective* (Nashville: Vanderbilt University Press, 2012)

Mueller, John, *Atomic Obsession: Nuclear Alarmism from Hiroshima to Al-Qaeda* (Oxford: Oxford University Press, 2010)

Nossiter, Bernard, *Soft State: A Newspaperman's Chronicle of India* (New York: Harper & Row, 1970)

Olmsted, Kathryn S., *Real Enemies: Conspiracy Theories and American Democracy, World War I to 9/11* (Oxford: Oxford University Press, 2009)

Ovendale, Ritchie, *Anglo-American Relations in the Twentieth Century* (Basingstoke: Macmillan, 1998)

Palit, D K, and P K S Namboodiri, *Pakistan's Islamic Bomb* (New Delhi: Vikas Press, 1979)

Patrikarakos, David, *Nuclear Iran: The Birth of an Atomic State* (London: I.B. Tauris, 2012)

Perkovich, George, *India's Nuclear Bomb: The Impact On Global Proliferation* (Berkeley: University of California Press, 1999)

Phythian, Mark, *The Politics of British Arms Sales Since 1964* (Manchester: Manchester University Press, 2000)

Pimlott, Ben, *Harold Wilson* (London: HarperCollins, 1993)

Potter, William C., *Nuclear Power and Nonproliferation: An Interdisciplinary Perspective* (Cambridge: Oelgeschlager, Gunn & Hain, Publishers Inc., 1982)

Powaski, Ronald E, *Entangling Alliance: The United States and European Security, 1950–1993* (Westport: Greenwood, 1994)

———, *March to Armageddon: The United States and the Nuclear Arms Race, 1939 to the Present* (Oxford: Oxford University Press, 1987)

Preston, Andrew, *Sword of the Spirit, Shield of Faith: Religion in American War and Diplomacy* (New York: Anchor Books, 2012)

Rabinowitz, Or, *Bargaining on Nuclear Tests: Washington and Its Cold War Deals* (Oxford: Oxford University Press, 2014)

Reiss, Mitchell, *Bridled Ambition: Why Countries Constrain Their Nuclear Capabilities* (Washington, D.C.: Woodrow Wilson Center Press, 1995)

Richelson, Jeffrey, *Spying on the Bomb: American Nuclear Intelligence from Nazi Germany to Iran and North Korea* (New York: W.W. Norton & Co., 2006)

Rossbach, Niklas H., *Heath, Nixon and the Rebirth of the Special Relationship: Britain, the US and the EC, 1969–74* (Basingstoke: Palgrave Macmillan, 2009)

Ruston, Roger, *A Say in The End of the World: Morals and British Nuclear Weapons Policy, 1941–1987* (Oxford: Clarendon Press, 1989)

Said, Edward, *Orientalism* (London: Penguin, 2003)

Salik, Naim, *The Genesis of South Asian Nuclear Deterrence: Pakistan's Perspective* (Oxford: Oxford University Press, 2009)

Sargent, Daniel J., *A Superpower Transformed: The remaking of American foreign relations in the 1970s* (Oxford: Oxford University Press, 2015)

Schrafstetter, Susan, and Stephen Twigge, *Avoiding Armageddon: Europe, the United States, and the Struggle for Nuclear Nonproliferation, 1945–1970* (Westport: Praeger, 2004)

Scott, Andrew, *Allies Apart: Heath, Nixon, and the Anglo-American Relationship* (Basingstoke: Palgrave Macmillan, 2011)

Scott, Len, and Stephen Twigge, *Planning Armageddon: Britain, the United States, and Command of Western Nuclear Forces, 1945–1964* (Amsterdam: Harwood Academic, 2000)

Simpson, John, Jenny Neilsen, Marion Swinerd, and Isabelle Anstey (eds.), *NPT Briefing Book, 2012 Edition* (London: Kings College London Centre for Science & Security Studies, 2012)

Strong, Robert A, *Working in the World: Jimmy Carter and the Making of American Foreign Policy* (Baton Rouge: Louisiana State University Press, 2000)

Suri, Jeremi, *Henry Kissinger and the American Century* (Cambridge: The Belknap Press of Harvard University Press, 2007)

Tannenwald, Nina, *The Nuclear Taboo: The United States and the Non-use of Nuclear Weapons Since 1945* (Cambridge: Cambridge University Press, 2007)

Tate, Simon, *A Special Relationship?: British Foreign Policy in the Era of American Hegemony* (Manchester: Manchester University Press, 2012)

Trachtenberg, Marc, *A Constructed Peace: The Making of European Settlement* (Princeton: Princeton University Press, 1999)

Venter, Al, *Allah's Bomb: The Islamic Quest for Nuclear Weapons* (Guilford: The Lyons Press, 2007)

Vickers, Rhiannon, *The Labour Party and the World, Volume 2: Labour's Foreign Policy 1951–2009* (Manchester: Manchester University Press, 2011)

Walker, William, *Nuclear Entrapment: THORP and the Politics of Commitment* (London: Institute for Public Policy Research, 1999)

Weissman, Steve, and Herbert Krosney *The Islamic Bomb: The Nuclear Threat to Israel and the Middle East* (New York: Times Books, 1981)

Westad, Odd Arne, *The Global Cold War: Third World Interventions and the Making of Our Times* (Cambridge: Cambridge University Press, 2005)

Wittner, Lawrence S., *The Struggle Against the Bomb, Volume 3: Toward Nuclear Abolition* (Stanford: Stanford University Press, 2003)

Yaqub, Salim, *Imperfect Strangers: Americans, Arabs, and U.S.-Middle East Relations in the 1970s* (Ithaca: Cornell University Press, 2016)

Yusuf, Moeed, *Predicting Proliferation: The History of the Future of Nuclear Weapons* (Boston: Brookings Institution, 2009)

JOURNAL ARTICLES AND CHAPTERS FROM EDITED COLLECTIONS

Ahmed, Samina, 'Pakistan's Nuclear Weapons Program: Turning Points and Nuclear Choices', *International Security*, 23:4 (1999), 178–204

Albright, David, and Mark Hibbs, 'Pakistan's Bomb: Out of the Closet,' *Bulletin of the Atomic Scientists*, (July/August 1992), 38–43

Boardman, Robert and Malcolm Grieve, 'The Politics of Fading Dreams: Britain and the Nuclear Export Business', in Robert Boardman and James F. Keeley, *Nuclear Exports and World Politics: Policy and Regime* (London: MacMillan, 1983), 98–119

Brands, Hal, 'Non-Proliferation and the Dynamics of the Middle Cold War: The Superpowers, the MLF, and the NPT', *Cold War History*, 7:3 (2007), 389–423

———, 'Progress Unseen: U.S. Arms Control Policy and the Origins of Detente, 1963–1968', *Diplomatic History*, 30:2 (2006), 253–285

———, 'Rethinking Nonproliferation: LBJ, the Gilpatric Committee, and U.S. National Security Policy', *Journal of Cold War Studies*, 8:2 (2006), 83–113

Braun, Chaim, 'The Nuclear Energy Market and the Nonproliferation Regime', *Nonproliferation Review*, 13:3 (2006), 627–644

Braun, Chaim, and Christopher F. Chyba, 'Proliferation Rings: New Challenges to the Nuclear Nonproliferation Regime', *International Security*, 29:2 (2004), 5–49

Burk, Kathleen, 'The Americans, the Germans and the British: The 1976 IMF Crisis', *Twentieth Century British History*, 5:3 (1994), 351–369

Burr, William, 'A Scheme of "Control": The United States and the Origins of the Nuclear Suppliers' Group, 1974–1976', *International History Review*, 36:2 (2014), 252–276

Chafetz, Glenn, 'The Political Psychology of the Nuclear Nonproliferation Regime', *The Journal of Politics*, 57:3 (1995), 743–775

Chakma, Bhumitra, 'Pakistan's nuclear weapons programme: Past and future', in Olav Njolstad (ed.), *Nuclear Proliferation and International Order: Challenges to the Non-Proliferation Treaty* (Abingdon: Routledge, 2011), 26–38

———, 'Road to Chagai: Pakistan's Nuclear Programme, Its Sources and Motivations', *Modern Asian Studies*, 36:4 (2002), 871–912

Chernus, Ira, 'Operation Candor: Fear, Faith, and Flexibility', *Diplomatic History*, 29:5 (2005), 779–809

Choi, Lyong, 'The First Nuclear Crisis in the Korean Peninsula, 1975–76', *Cold War History*, 14:1 (2014), 71–90

Cordle, Daniel, 'Beyond the apocalypse of closure: nuclear anxiety in the post-modern literature of the United States', in Hammond, Andrew (ed.), *Cold War Literature: Writing the Global Conflict* (Abingdon: Routledge, 2006)

Craig, Malcolm M., '"I Think we Cannot Refuse the Order": Britain, America, Nuclear Non-proliferation, and the Indian Jaguar Deal, 1974–78', *Cold War History*, 16:1 (2016), 61–81

———, '"Nuclear Sword of the Moslem World"?: The United States, Britain, Pakistan, and the "Islamic bomb", 1977–80', *International History Review*, 38:5 (2016), 857–879

Drogan, Mara, 'The Nuclear Imperative: Atoms for Peace and the Development of U.S. Policy on Exporting Nuclear Power, 1953–1955', *Diplomatic History*, 40:5 (2016), 948–974

Emery, Christian, 'The Transatlantic and Cold War Dynamics of Iran Sanctions, 1979–80', *Cold War History*, 10:3 (2010), 371–396

Forden, Geoffrey E., 'How the World's Most Underdeveloped Nations Get the World's Most Dangerous Weapons', *Technology and Culture*, 48:1 (2007), 92–103

Ganguly, Sumit, 'The Indian and Pakistani Nuclear Programmes: A Race to Oblivion?', in Raju G. C. Thomas (ed.), *The Nuclear Non-proliferation Regime: Prospects for the 21st Century* (Basingstoke: Macmillan Press, 1998), 272–283

Gavin, Francis J., 'Blasts from the Past: Proliferation Lessons from the 1960s', *International Security*, 29:3 (2004/5), 100–135

Gerlini, Matteo, 'Waiting for Dimona: The United States and Israel's Development of Nuclear Capability', *Cold War History*, 10:2 (2010), 143–161

Gray, William Glenn, 'Commercial Liberties and Nuclear Anxieties: The US-German Feud Over Brazil, 1975–7', *International History Review*, 34:3 (2012), 449–474

Gummett, Philip, 'From NPT to INFCE: Developments in Thinking About Nuclear Non-Proliferation', *International Affairs*, 57:4 (1981), 549–567

Gusterson, Hugh, 'Nuclear Weapons and the Other in the Western Imagination', *Cultural Anthropology*, 14:1 (1999), 111–143

Hamblin, Jacob Darwin, 'The Nuclearization of Iran in the Seventies', *Diplomatic History* 38:5 (2014), 1114–35

Hastings, Justin V., 'The Geography of Nuclear Proliferation Networks', *Nonproliferation Review*, 19:3 (2012), 429–450

Hecht, Gabrielle, 'The Power of Nuclear Things', *Technology and Culture*, 51:1 (2010), 1–30

Hersman, Rebecca K.C., and Robert Peters, 'Nuclear U-turns', *Nonproliferation Review*, 13:3 (2006), 539–553

Hilfrich, Fabian, 'Roots of Animosity: Bonn's Reaction to US Pressures in Nuclear Proliferation', *International History Review*, 36:2 (2014), 277–301

Hoodbhoy, Pervez, 'Myth Building: The "Islamic Bomb"', *Bulletin of the Atomic Scientists*, (June, 1993), 42–49

———, 'Iran, Saudi Arabia, Pakistan, and the "Islamic Bomb"', in Hoodbhoy, Pervez (ed.), *Confronting the Bomb: Pakistani and Indian Scientists Speak Out* (Oxford: Oxford University Press, 2013), 151–167

Hughes, R. Gerald, and Thomas Robb, 'Kissinger and the Diplomacy of Coercive Linkage in the "Special Relationship" between the United States and Great Britain, 1969–1977', *Diplomatic History*, 37:4 (2013), 861–905

Hull, Christopher, '"Going to War in Buses": The Anglo-American Clash over Leyland Sales to Cuba, 1963–1964', *Diplomatic History*, 34:5 (2010), 793–822

———, 'Our Arms in Havana: British Military Sales to Batista and Castro, 1958–59', *Diplomacy and Statecraft*, 18:3 (2007), 593–616

Jones, Matthew, and John W. Young, 'Polaris, East of Suez: British Plans for a Nuclear Force in the Indo-Pacific, 1964–1968', *Journal of Strategic Studies*, 33:6 (2010), 847–870

Kampani, Gaurav, 'Second Tier Proliferation: The Case of Pakistan and North Korea', *Nonproliferation Review*, 9:3 (2002), 107–116

Kemp, R Scott, 'The End of Manhattan: How the Gas Centrifuge Changed the Quest for Nuclear Weapons', *Technology and Culture*, 53:2 (2012), 272–305

Khalil, Osamah F., 'The Crossroads of the World: U.S. and British Foreign Policy Doctrines and the Construct of the Middle East, 1902-2007', *Diplomatic History*, 38:2 (2014), 299–344

Khan, Feroz Hassan, 'Nuclear Proliferation Motivations', *Nonproliferation Review*, 13:3 (2006), 501–517

Krige, John, 'Atoms for Peace, Scientific Internationalism, and Scientific Intelligence', *Osiris*, 21:1 (2006), 161–181

———, 'US Technological Superiority and the Special Nuclear Relationship: Contrasting British and US Policies for Controlling the Proliferation of Gas Centrifuge Enrichment', *International History Review*, 36:2 (2015), 230–251

Lahey, Daniel James, 'The Thatcher Government's Response to the Soviet Invasion of Afghanistan, 1979–1980', *Cold War History*, 13:1 (2013), 21–42

Lane, Ann, 'Foreign and defence policy', in Seldon, Anthony, and Kevin Hickson (eds.), *New Labour, Old Labour: The Wilson and Callaghan Governments, 1974–79* (London: Routledge, 2004)

Lavoy, Peter, 'Nuclear Myths and the Causes of Nuclear Proliferation', *Security Studies*, 2:3 (1993), 192–212

Ludlam, Steve 'The Gnomes of Washington: Four Myths of the 1976 IMF Crisis', *Political Studies*, 40:4 (1992), 713–727

Luongo, Kenneth N., and Isabelle Williams, 'The Nexus of Globalization and Next-generation Nonproliferation', *Nonproliferation Review*, 14:3 (2007), 459–473

Martinez, J.M., 'The Carter Administration and the Evolution of American Nuclear Nonproliferation Policy, 1977–1981', *Journal of Policy History*, 14:3 (2002), 261–92

Metz, William D, 'Ford-MITRE Study: Nuclear Power Yes, Plutonium No', *Science*, New Series, 196:4285 (April 1, 1977), 41

Montgomery, Alexander H., 'Ringing in Proliferation: How to Dismantle an Atomic Bomb Network', *International Security*, 30:2 (2005), 153–187

Nye, Joseph S, 'Nonproliferation: A Long-Term Strategy', *Foreign Affairs* 56 (1978), 601–623

Oren, Ido, and Ty Solomon, 'WMD, WMD, WMD: Securitisation Through Ritualized Incantation of Ambiguous Phrases', *Review of International Studies*, 41:2 (2015), 313–336

Pelopidas, Benoit, 'The Oracles of Proliferation: How Experts Maintain a Biased Historical Reading that Limits Policy Innovation', *Nonproliferation Review*, 18:1 (2011), 297–314

Popp, Roland, 'Introduction: Global Order, Cooperation between the Superpowers, and Alliance Politics in the Making of the Nuclear Non-Proliferation Regime', *International History Review*, 36:2 (2014), 195–209

Priest, Andrew, 'The President, the "Theologians" and the Europeans: The Johnson Administration and NATO Nuclear Sharing', *International History Review*, 33:2 (2011), 257–275

Ritchie, Nick, 'Relinquishing Nuclear Weapons: Identities, Networks and the British Bomb', *International Affairs*, 86:2 (2010), 465–487

Rosenthal, Alan, 'The Politics of Passion: An Interview with Anthony Thomas', *Journal of Film and Video*, 49:1–2 (1997), 94–102

Rotter, Andrew J., 'Christians, Muslims, and Hindus: Religion and U.S.-South Asian Relations, 1947–1954', *Diplomatic History*, 24:4 (2000), 593–613

———, 'Saidism Without Said: Orientalism and U.S. Diplomatic History', *American Historical Review*, 105:4 (2000), 1205–1217

Roy, Raj, 'Peter Ramsbotham', in Michael F. Hopkins, Saul Kelly, and John W. Young, *The Washington Embassy: British Ambassadors to the United States, 1939–77* (Basingstoke: Palgrave Macmillan, 2009), 209–228

Sagan, Scott, 'Why do States Build Nuclear Weapons? Three Models in Search of a Bomb', *International Security*, 21:3 (1996/97), 54–86

Said, Mohamed Kadry, 'Israel's Nuclear Capability: Implications on Middle East Security', in Olav Njolstad (ed.), *Nuclear Proliferation and International Order: Challenges to the Non-Proliferation Treaty* (Abingdon: Routledge, 2011)

Sarkar, Jayita, 'The Making of a Non-Aligned Nuclear Power: India's Proliferation Drift, 1964–8', *International History Review*, 37:5 (2015), 933–950

Schmidt, Fritz, 'NPT Export Controls and Zangger Committee', *Non-proliferation Review*, 7:3 (2000), 136–145

Schrafstetter, Susanna, 'Preventing the "Smiling Buddha"; British-Indian Nuclear Relations and the Commonwealth Nuclear Force, 1964–68', *Journal of Strategic Studies*, 25:3 (2002), 87–108

Schrafstetter, Susanna, and Stephen R. Twigge, 'Spinning into Europe: Britain, West Germany and the Netherlands: Uranium Enrichment and the Development of the Gas Centrifuge 1964–1970', *Contemporary European History*, 11:2 (2002), 253–272

Shaikh, Farzana, 'Pakistan's Nuclear Bomb: Beyond the Non-proliferation Regime', *International Affairs*, 78:1 (2002), 29–48

Singh, S. Nihal, 'Why India Goes to Moscow for Arms', *Asian Survey*, 24:7 (1984), 707–720

Skjoldebrand, R, 'The International Nuclear Fuel Cycle Evaluation – INFCE', *The IAEA Bulletin*, 22:2, 30–33

Spelling, Alex, 'Ambassadors Richardson, Armstrong, and Brewster, 1975–81', in Holmes, Alison R., and J. Simon Rofe, *The Embassy in Grosvenor Square: American Ambassadors to the United Kingdom, 1938–2008* (Basingstoke: Palgrave Macmillan, 2012), 189–213

Stoddart, Kristan, 'The Labour Government and the Development of Chevaline, 1974–1979', *Cold War History*, 10:3 (2010), 287–314

Suter, Keith D., 'The 1975 Review Conference of the Nuclear Non-proliferation Treaty', *Australian Outlook*, 30:2 (1976), 322–340

Swango, Dane, 'The United States and the Role of Nuclear Co-operation and Assistance in the Design of the Non-Proliferation Treaty', *International History Review*, 36:2 (2014), 210–229

Twigge, Stephen R., 'Operation Hullabaloo: Henry Kissinger, British Diplomacy, and the Agreement on the Prevention of Nuclear War', *Diplomatic History*, 33:4 (2009), 689–701

Vinen, Richard, 'Thatcherism and the Cold War', in Jackson, Ben and Robert Saunders (eds.), *Making Thatcher's Britain* (Cambridge: Cambridge University Press, 2012)

Walker, J. Samuel, 'Nuclear Power and Nonproliferation: The Controversy Over Nuclear Exports, 1974–1980', *Diplomatic History*, 25:2 (2001), 215–249

Walker, William, 'Destination Unknown: Rokkasho and the International Future of Nuclear Reprocessing', *International Affairs*, 82:4 (2006), 743–761

––––––, 'Nuclear Order and Disorder', *International Affairs*, 76:4 (2000), 703–724

Wenger, Andreas, 'Crisis and Opportunity: NATO's transformation and the multi-lateralisation of détente, 1966–1968', *Journal of Cold War Studies*, 6:1 (2004), 22–74

Widén, J.J., and Jonathan Colman, 'Lyndon B. Johnson, Alec Douglas- Home, Europe and the NATO Multilateral Force, 1963–64', *Journal of Transatlantic Studies*, 5:2 (2007), 179–198

Yasmeen, Samina, 'Is Pakistan's Nuclear Bomb an Islamic Bomb?,' *Asian Studies Review*, 25:2 (2001), 201–215

Young, John, 'Killing the MLF? The Wilson Government and Nuclear Sharing in Europe, 1964–66', *Diplomacy & Statecraft*, 14:2 (2003), 295–324

DISSERTATIONS

Pande, Aparna, 'Foreign Policy of an Ideological State: Islam in Pakistan's International Relations' (Ph.D diss., Boston University, 2010)

INDEX

© The Author(s) 2017
M.M. Craig, *America, Britain and Pakistan's Nuclear Weapons Programme, 1974–1980*, Security, Conflict and Cooperation in the Contemporary World, DOI 10.1007/978-3-319-51880-0

Printed by Printforce, the Netherlands